Habitat Conservation

Habitat Conservation
MANAGING THE PHYSICAL ENVIRONMENT

Edited by
A. WARREN AND J. R. FRENCH
Department of Geography
University College London, UK

JOHN WILEY & SONS, LTD

Chichester • New York • Weinheim • Brisbane • Singapore • Toronto

Other Wiley Editorial Offices

John Wiley & Sons, Inc., 605 Third Avenue,
New York, NY 10158-0012, USA

WILEY-VCH Verlag GmbH, Pappelallee 3,
D-69469 Weinheim, Germany

Jacaranda Wiley Ltd, 33 Park Road, Milton,
Queensland 4064, Australia

John Wiley & Sons (Asia) Pte Ltd, 2 Clementi Loop #02-01,
Jin Xing Distripark, Singapore 129809

John Wiley & Sons (Canada) Ltd, 22 Worcester Road,
Rexdale, Ontario M9W 1L1, Canada

British Library Cataloguing in Publication Data

A catalogue record for this book is available from the British Library

ISBN 0 471 98498 1 (cloth)
ISBN 0 471 98499 X (paperback)

Typeset in 10/12pt Times by Dobbie Typesetting Ltd, Tavistock, Devon, UK
Printed and bound in Great Britain by Biddles Ltd, Guildford, Surrey
This book is printed on acid-free paper responsibly manufactured from sustainable forestry,
in which at least two trees are planted for each one used for paper production.

Contents

List of Contributors

C. T. Agnew
Department of Geography, University of Manchester, Oxford Road, Manchester M13 9PL, UK

N. J. Anderson
Department of Geography, University of Copenhagen, Østervoldgade 10, DK-1350, Copenhagen K, Denmark

S. M. Arens
Netherlands Centre for Geo-Ecological Research ICG, University of Amsterdam, Institute for Biodiversity and Ecosystem Dynamics, Physical Geography, Nieuwe Achtergracht 166, 1018 WV Amsterdam, The Netherlands

N. G. Bayfield
Centre for Ecology and Hydrology, Banchory Research Station, Hill of Brathens, Glassel, Banchory, Kincardineshire AB31 4BY, Scotland, UK

L. Carvalho
Environmental Change Research Centre, Department of Geography, University College London, 26 Bedford Way, London WC1H 0AP, UK

N. J. Clifford
School of Geography, University of Nottingham, University Park, Nottingham NG7 2RD, UK

S. Fennessy
Department of Biology, Kenyon College, Gambier, Ohio 43022, USA

M. Finlayson
Environmental Research Institute of the Supervising Scientist, Jabiru, Australia

J. R. French
Coastal & Estuarine Research Unit, Department of Geography, University College London, Chandler House, 2 Wakefield Street, London WC1N 1PF, UK

F. M. R. Hughes
Department of Geography, University of Cambridge, Downing Place, Cambridge CB2 3EN, UK

P. D. Jungerius
Netherlands Centre for Geo-Ecological Research ICG, Landscape and Environmental Research Group, University of Amsterdam, Nieuwe Prinsengracht 130, 1018 VZ Amsterdam, The Netherlands
and
Bureau G & L, Oude Bennekomseweg 31, 6717 LM Ede, The Netherlands

F. van der Meulen
Ministry of Transport, Public Works and Water Management, National Institute for Marine and Coastal Management, Coastal Zone Management Centre, P.O. Box 20907, 2500 EX 's-Gravenhage, The Netherlands

D. J. Reed
Department of Geology and Geophysics, University of New Orleans, New Orleans, LA 70148, USA

S. B. Rood
Department of Biological Sciences, University of Lethbridge, Alberta, Canada T1K 3M4

T. Spencer
Cambridge Coastal Research Unit, Department of Geography, University of Cambridge, Downing Place, Cambridge CB2 3EN, UK

J. R. Thompson
Wetland Research Unit, Department of Geography, University College London, Chandler House, 2 Wakefield Street, London WC1N 1PF, UK

A. Warren
Department of Geography, University College London, Chandler House, 2 Wakefield Street, London WC1N 1PF, UK

Preface

This collection of essays has two purposes. First, it continues a series of books whose aim has been to discuss nature conservation in depth and with reference to up-to-date scientific literature. Earlier books in the series were also published by John Wiley and Sons and include: *Conservation in practice*, (1974), edited by A Warren and FB Goldsmith; *Conservation in perspective*, (1983), also edited by A Warren and FB Goldsmith; and *Conservation in progress* (1993) edited by FB Goldsmith and A Warren. All the volumes arose out of the Masters course in Conservation at University College London (for many years a unique course, at least in Britain), containing contributions from the permanent members of the academic staff at UCL and from the many outside lecturers who taught the course. The present book maintains this tradition, although it draws upon a rather broader authorship.

Second, as geomorphologists, we have been very conscious of the upsurge of interest and research in the relations between the earth surface sciences, ecology and conservation. Geomorphology, hydrology, climatology, limnology and related sciences are finding increasingly wide application in the management of the pressing contemporary problems of nature conservation. This can be seen in contracts from government bodies (such as the Environment Agency and English Nature in Britain, the Rijkwaterstaatt in The Netherlands, the Department of the Environment in the United States and equivalent bodies elsewhere). These organisations clearly need advice from the earth sciences and are buying it in large quantities. Evidence also comes from a profusion of papers, mostly in the earth science literature, but a few in the ecological and biological literature as well. Little of this research, applied or scientific, has reached a general audience in conservation or even (in a widely acceptable form) many earth scientists themselves. Moreover, the new work needs to be consolidated, for, as the following chapters show, there are many common themes, such as new approaches to the quixotic quest for equilibrium, problems of scale and so on. We believe that this book makes a very good start at these tasks.

In a volume of this size, it is not practical to attempt an exhaustive habitat-by-habitat analysis. We have chosen to focus attention primarily on temperate systems, and have thereby excluded a number of important environments (notably tropical reefs and rainforests) that will have to wait for specialist coverage elsewhere. This way of packaging the problem allows for the enforcement of some commonality in presentational style and organisation, such that a collection of individual essays also has value as a fundamental text for masters and undergraduate students studying nature conservation, habitat ecology, geomorphology, hydrology and the related sciences, and environmental management. We believe that the ideas contained in this volume will also be of interest to those working on the margins of professional conservation — from land managers to environmental consultants and engineers.

We must first thank our contributors, many of whom have waited years to see the publication of their chapters. We must also thank Inga Warren for the cover design, and Elanor McBay and Catherine Pyke for the diagrams, many of which are original.

Andrew Warren
Jonathan French
University College London

1 Relations between Nature Conservation and the Physical Environment

ANDREW WARREN AND JON FRENCH
University College London, UK

INTRODUCTION

Nature conservationists have not always been well informed about the dynamics of the physical environment, in either its 'natural' or its constrained state. At the global scale, concern with climate change is often translated into crude, grossly oversimplified assumptions and predictions concerning the likely impacts on wildlife habitats. The treatment of accelerated sea-level rise is a case in point. Deceptively precise 'best estimate' projections have been taken to portend widespread erosion and inundation of coastal and estuarine margins. These scenarios have been used, in turn, as a basis for highly generalised assessments of threats posed to key habitats such as dunes, marshes, mangroves and reefs. Applied locally, many of these simplistic assessments turn out to be both quantitatively flawed and conceptually naïve, and, if taken seriously, would be very poor guides to the successful management of individual sites.

Simplistic assumptions about the variability and dynamics of smaller-scale physical systems can be almost as dangerous. At worst, the physical environment is seen as an unchanging backdrop to the development of the ecosystem, whose patterns are easily predictable. This might have been an acceptable, even necessary, assumption when little was known about these systems, as when McArthur and Wilson developed their classic theory of island biogeography. Against the requirements of modern conservation, however, this approach leads to models that are seriously deficient as management tools.

The relations between conservation and the physical environment have, however, been transformed over recent decades. In particular, there have been major developments in four key areas.

First, there has been an explosion in our understanding and ability to model processes in the physical world at the scale at which ecosystems operate, as we hope this book shows. This is true at a range of scales, from that of global climate to that of ecohydraulics as they relate to micro-organisms, although, as we emphasise below, there are still major gaps in our knowledge.

Second, manipulation of the physical environment, by engineering and other structures, has now reached very threatening proportions. Overconfidence in our ability to manage the physical environment, itself inherited from the nineteenth century, has left, and is still leaving, an alarming legacy of huge dams, straightened rivers, regraded slopes and stabilised shorelines, each with potentially or actually severe consequences for natural habitats. Many habitats have been substantially lost, and many more seriously degraded. New habitats have also been created, as in the freshwater grazing marshes of reclaimed estuarine margins, regulated rivers downstream of dams, and abandoned quarry faces. Many of these have acquired considerable ecological significance in their own right, yet depend for their existence upon past management practices which are now of doubtful economic viability. Managing the physical effects of these interventions requires a much better understanding of how they have altered the environment, and of how it once functioned (and may function again) in their absence.

Third, the 'hard-engineering' mind-set that lay behind such major environmental manipulations is being replaced, in some parts of the world at least, by a new paradigm, in which engineers want to work with rather than against nature. The intention now is that human impacts on the environment are kept within 'sustainable' limits. In some areas, as in the restoration of riverine, saltmarsh and other wetland habitats, this new movement has demonstrated impressive advances. These need to be known more widely so that their underlying philosophy can be translated to other habitats and environments. Concepts involving the functions and 'services' performed by natural ecosystems have been developed most fully for wetlands, but are rapidly being formulated for other major ecosystem types. Conceptual sophistication has been paralleled by the emergence of specialised ecological engineering technologies for habitat remediation and restoration. Ecological engineering has the potential to mitigate the deleterious effects of past practices and the impacts of future urban and industrial expansion. It can also provide a basis for more enlightened approaches to environmental management which incorporate rather than exclude the important physical functions performed by natural ecosystems. The history of the development of engineering thinking about the coastline of the Netherlands and its coastal dunes is an excellent example of the way that the new paradigm has penetrated both engineering and conservation circles.

Despite a potentially beneficial convergence of scientific, engineering, political and public awareness and thinking, the effective incorporation of conservation priorities remains hampered by inadequate knowledge of the natural landform dynamics and the difficulty in devising design criteria which are practical from engineering and economic perspectives, yet which produce desirable geomorphological, hydrological and ecological outcomes. Experiences with river management, and the emergence of ecohydraulics as a coherent subdiscipline, provide good examples of this kind of problem. Thus, engineers charged with the implementation of river-restoration schemes have only recently become aware of the need to mimic the spatial variability inherent in prototype pool–riffle sequences when designing new aquatic habitats. Likewise, ecologists working in tropical rainforests have suddenly become aware that slope dynamics have a vital role to play in the patterning and dynamics of the ecosystem, yet have only just begun to collaborate with geomorphologists. In the lake environment, the association of ecological with physical environmental dynamics has been acknowledged for longer, but lake ecologists still need to know much more than they do about physical processes.

In all these environments there has been a vigorous debate about the distinctions between 'gardening', management for conservation and *laissez-faire* approaches to the perpetuation of natural system functioning, and over the advantages and ethics of each of these contrasting approaches. These debates will stay shallow unless they include some consideration of how the underlying physical environment both influences the functional characteristics of natural ecosystems and imposes constraints on the extent to which these can be managed or engineered. Although many of these process linkages are still very poorly understood, there is also a problem of communication and knowledge uptake. Certainly, as the varied contributions in the present volume show, there is a growing body of research on the physical environment which is relevant to all these debates, and which has not yet been widely accessed or appreciated by ecologists or conservationists.

Fourth and finally, we are at a critical time in the development of both the physical and biological sciences that contribute to ecology. In ecology and conservation the arrival of a new paradigm is widely hailed. Ideas about equilibrium, which dominated the old paradigm, are being replaced by concepts to do with disequilibrium, even chaos, and it is the physical environment from which the stimulus to change is almost always seen to come. Hydrologists, limnologists, pedologists and geomorphologists have seen parallel changes in their conceptual bases, such that much of their disciplinary nomenclature is shared with, and can be appreciated by, ecologists and biologically trained conservationists.

These emerging trends are well illustrated by current approaches to the management of coastal dunes, where a highly protectionist philosophy among both ecologists and engineers is giving way to one that allows for the operation of shifting and dynamic physical processes. Here, an understanding of the latter is as critical to an understanding of the population dynamics of plant and animal species. Both need close scrutiny if dunes are to be managed in a way that conserves their diversity.

For all this, there is still a mismatch between existing research within geomorphology, hydrology and engineering and the demands of conservation managers. As conservation goals become more tightly specified (for example, in response to concerns over endangered species or particular ecosystem functions) the importance of modelling and monitoring the subtleties of landform–habitat interactions grows.

The disciplines of geomorphology and hydrology, in particular, are presently in a sophistication phase. Specialists in these fields can now provide environmental managers with much better answers to questions concerning the impacts of particular human activities, the vulnerability of specific systems to environmental change, and the feasibility of mitigative or remedial strategies. There may well be a financial penalty associated with such knowledge, however. As some of the contributions in this volume show, the construction and monitoring techniques needed to deploy increased scientific understanding in pursuit of more strictly defined conservation goals may carry a high price tag. Such financial pressures may originate in at least two ways. First, the need to cater for the specialised preferences of individual target species militates against *laissez-faire* approaches to habitat 're-establishment' in favour of more explicit engineering of clearly defined habitat characteristics. Second, the pursuit of more tightly defined conservation goals may necessitate more intensive post-project monitoring in order properly to judge the success of restorative works. This is especially important where natural habitats are 'traded' for artificial ones in order to accommodate major urban developments. Both these issues are well exemplified by the experiences with wetland restoration under compensatory mitigation policies in southern California. Here, attempts to replace habitat to support an endangered bird species, the Light-footed Clapper Rail, have highlighted the unpredictable outcome of existing restoration techniques as well as the importance of incorporating scientific understanding of natural wetland dynamics into more precise criteria for post-project appraisal.

THE AIMS OF THIS BOOK

The fundamental aim of this book is to outline an agenda for translational research to bridge the gap between two rapidly evolving disciplinary areas: the narrowly biological and the narrowly physical sides of ecology and conservation. The intention is to chart the emergence of such a dialogue, where one exists, and to stimulate it where it clearly does not. The varied contributions summarise the current state of the art in the understanding of physical process dynamics within selected environments of high conservation interest. Specifically, this distillation of recent scientific research is intended to aid:

(1) In the identification of sites (and patterns of sites) genuinely at risk;
(2) The evaluation of management options and appropriate timescales for planning and intervention; and
(3) The development of improved scientific bases for habitat restoration and/or the design of replacement habitats.

Individual contributions variously emphasise a number of key principles and themes. These include the scales and modes of transient geomorphological and hydrological system behaviour; the magnitude and frequency of significant events and how these vary between habitats; concepts of equilibrium and disequilibrium and how these mesh with similar debates in ecology; and the nature and detection of trends within systems which often exhibit a large degree of short-term variability. Taken as a whole, the book places considerable emphasis upon recent technological advances in physical process understanding across a broad spectrum of scales (from remote sensing to microscopy), and the extent to which these are being applied in specific management and restoration schemes.

2 Mountain Resources and Conservation

NEIL BAYFIELD

Centre for Ecology and Hydrology, Banchory, Scotland

INTRODUCTION

Mountain areas cover about 20% of the earth's land surface (Louis 1975). They are characterised by steeply sloping ground, strong climatic gradients and well-developed altitudinal zonations of vegetation and soils. Many have been designated as of high landscape, geomorphological and nature conservation value. Human population densities are low, but agriculture, transport and extractive industries in the past and skiing and recreation in recent decades have had substantial impacts. The rate of change appears to be accelerating and the threats are considerable.

This chapter uses a pressures–states–responses (PSR) approach to describe the resources and conservation of mountains (MacGillivray and Kayes 1995). This involves the identification of *pressures*, the resultant *states* or conditions of resources, and the management or policy *responses* that can or could be triggered. The resources considered are climate, vegetation, geology and soils, water resources, and landscape. Pressures include forestry and agriculture, water and energy extraction, transport infrastructure, quarrying, construction activities, recreation and nature conservation.

CLIMATE

CHARACTERISTICS

Even modest topographic barriers can influence mountain climates. Relief of 600 m is sufficient to cause substantial vertical differentiation of climate elements and vegetation communities, and can be used as a rule of thumb to distinguish mountains from hills (Thompson 1964). The most significant climatically related features are maximum and minimum timberlines and

Habitat Conservation: Managing the Physical Environment. Edited by A. Warren and J. R. French.
© 2001 John Wiley & Sons Ltd.

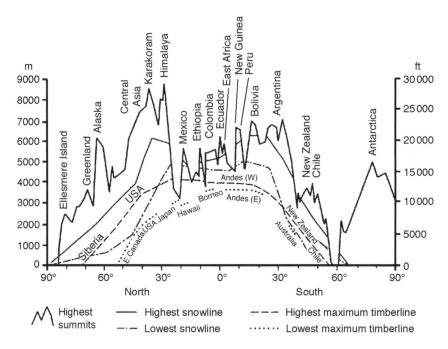

Figure 2.1. Latitudinal cross-section of the highest summits, highest and lowest snowline and upper and lower limits of timberline (Barry and Ives 1974).

snowlines (Troll 1973). On these criteria, the lower limit of mountain landscapes is only a few hundred metres above sea level in Scandinavia, 1600–1700 m in the Alps and 4500 m in the equatorial cordilleras of South America (Figure 2.1). For areas without tree cover or snowlines, the only available criterion for defining mountains is relief.

Temperature generally lapses by 6°C per 1000 m (although rates can be temporarily reversed in temperature inversions). Lapse rates show considerable variability according to climatic zone, and mountains are particularly affected by radiative and turbulent heat exchange which can result in changes in lapse rates at different times of day (Barry 1992), and they can also be affected by the presence of snow-free or snow-covered ground. Both precipitation and windiness tend to increase with altitude, although the patterns of change are much more complex than those for temperature lapse rates, and are strongly influenced by topographic and linked orographic effects. There are pronounced climatic zones on mountains including zones of orographically increased rainfall, especially downwind of mountain peaks, and topoclimates associated with types of terrain and/or plant communities (Figure 2.2).

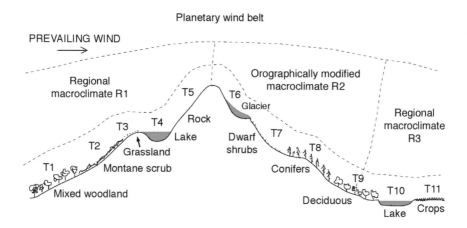

Figure 2.2. Scales of climatic zonation in mountains. R1-3, regional macroclimates; T1-10 topoclimates. Modified after Barry (1992).

PRESSURES

The main pressures on climate are changes in the concentrations of the so-called greenhouse gases, which increase the absorption of infrared radiation. The principal gases, CO_2, CH_4, N_2O and the chlorofluorocarbons (CFCs), are produced in agriculture, manufacturing, transport and other routine activities. The concentrations of these gases, other than that of the CFCs, are expected to continue to increase. Although some of the evidence is contentious, and in spite of uncertainties, the Intergovernmental Panel on Climate Change (IPCC) has concluded that global climate is likely to change, with a projected temperature increase of 1–3.5°C by 2100 AD, or about 0.1–0.2°C per decade (Houghton *et al.* 1996; see also Agnew and Fennessy, 2001). The effects will not be confined to temperature changes, and precipitation increases are expected, particularly in high latitudes and in mid-latitudes during winter months. Snow duration is likely to decline, although, overall, snowfall may increase. Mountain areas are considered to be particularly sensitive to changing climate for several reasons (Guisan *et al.* 1995; Callaghan *et al.* 1998; Price and Barry 1997):

- There have already been substantial changes in snowline and glacier size, consistent with the predictions from changes in greenhouse gases; small changes in temperature could have a substantial impact on snowline and by altering albedo, affect climate over a wider area.
- Many plant and animal species survive at the edge of their physiological range in mountain areas; small changes are therefore likely to have a large impact.

- Mountain areas tend to have higher biodiversity than corresponding lowlands; experimental studies suggest that biodiversity decreases with periods of climate change.

STATES

The global rises in temperature predicted by Global Climate Models (GCMs) indicate that snowmelt will occur earlier because of summer warming, although there may be more snowfall in some western coastal mountains because of increased spring precipitation (Price and Barry 1997). Corroborative evidence is provided by declines in the areas and volumes of glaciers in the last 100 years. It has been estimated that the glaciers of the Alps have lost about 30–40% of their area and 50% in volume since 1850 (Haeberli 1995). The rates of loss appear to be accelerating, with a loss of about 10–20% by volume since 1970, and if warming continues, the ice mass of the Alps could be reduced to 25% of its coverage in 1850 within 30 years. Rates of glacial mass change vary greatly in different parts of the world, and some glaciers are still advancing, but the global trend over the period 1960–1990 shows a net loss of $0.25 \pm 0.10 \, \text{mm yr}^{-1}$ in sea level equivalent (Dyurgerov and Meier 1997).

Snowmelt runoff from mid-latitude mountains is the primary input to river discharge and the major source of water for the populations of adjacent lowlands (Price and Barry 1997). The implications of projected changes in spring precipitation and shorter periods of snowlie are likely to mean increases in peakedness of spring runoff and reduced summer flows (Rango and Van Katwijk 1990). There are also important implications for winter tourism. Snow cover in the 1980s and 1990s has been erratic, and it is not clear what the long-term trends are going to be, but a significant number of ski resorts are currently experiencing financial difficulties because of unreliable snow.

Climate change is likely to affect airflow patterns as well as precipitation and temperature. The changes are likely to vary greatly in different parts of the world, but there have already been increases in the number of winter gales in Sweden since 1970, and there has been a similar trend in Scotland (Harrison 1997).

RESPONSES

Following the 1987 Montreal Protocol, the main responses to climate change have been global reductions since 1994 in production of CFCs, and various national and international targets for reducing emissions of other GHGs. The latter mainly followed the 1992 United Nations Conference on Environment and Development at Rio, and have so far had limited success (these processes are discussed in greater depth in Agnew and Fennesey, 2001).

VEGETATION

CHARACTERISTICS

Most mountains display an altitudinal zonation of vegetation, linked to soil and climatic gradients, though often masked by land use. Aspect can also have a strong influence on the distribution of some mountain species and plant communities (Figure 2.3). Topographic shelter and shade can have marked influence on plant distribution, particularly of moisture-loving cryptogams. These factors influence plants directly by modifying insolation and evapo-transpiration, and indirectly by affecting the pattern and duration of snowlie.

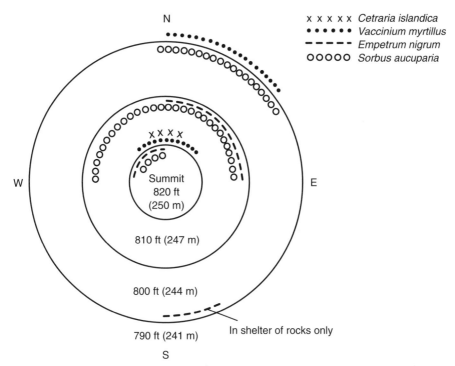

Figure 2.3. Species confined to north-facing slopes on a symmetrical hill at Lochinver, North Scotland (after Gimingham 1964).

PRESSURES

In an EU report on Global Change in Europe's Cold Regions, Heal *et al.* (1998) ranked the most significant pressures on alpine vegetation as (1) land use (grazing), (2) nitrogen deposition and (3) climate change. The effects of the most important of these, land use, can be seen in a study by Cernusca *et al.* (1996). They compared the effects of increasing levels of management and of the converse, abandonment, on mountain plant communities in the Alps, Pyrénées and Scotland (Figure 2.4). In all cases abandonment resulted in an invasion by trees and shrubs or dwarf shrubs. The study pointed out that there was often a close relationship between accessibility and abandonment, with traditional land use persisting only on the most accessible slopes. Increases in scrub cover are perceived to be generally detrimental on the Continent, particularly in countries where traditional land use has been widely abandoned. In Scotland, on the other hand, where overgrazing of mountain vegetation is almost universal, and natural treelines are almost unknown, reducing grazing to a point where tree regeneration is possible is perceived to be an important objective for landscape and nature conservation.

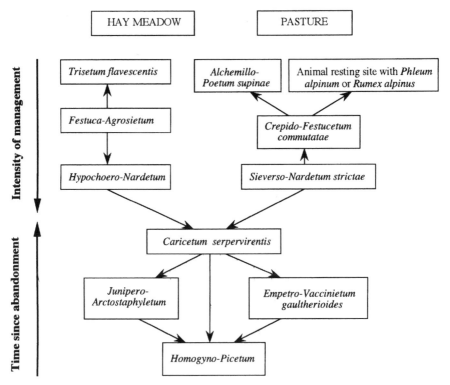

Figure 2.4. Effects of land-use change on vegetation composition on silicate substrates in the Eastern Alps (Cernusca *et al.* 1996).

Other land uses that impact on vegetation include recreational use and management by burning. Many fires in Mediterranean mountain areas are accidental, although some are started to improve grazing and reduce the encroachment of scrub. In Scotland, fire is a widespread tool for managing upland vegetation. Extensive fires are used to improve grazing for sheep and deer. Fine-scale mosaics of small even-aged stands are also created by burning for gamebird habitat. The result is a landscape resembling a patchwork quilt (Figure 2.5). Fires, like grazing, effectively prevent tree regeneration.

Figure 2.5. Patchworks of burns for gamebird habitat management in Scotland on a 5–15-year rotation have a major impact on vegetation cover and structure as well as a striking visual impact on landscape. Heather moorland in Aberdeenshire near Tomintoul.

Recreational use can also have major impacts on vegetation in areas where there are widespread networks of paths. Bayfield (1996) followed the colonisation of bulldozed pistes at Cairn Gorm, in Scotland, over 25 years and found that above the treeline, bryophytes were important colonists within the first few years, but that native vascular species took more than 20 years to replace a cover of sown commercial grass species (Figure 2.6).

Nitrogen deposition in mountains is increasing. It comes from a combination of NO_2 from vehicles and NH_4 from agriculture. Recent European

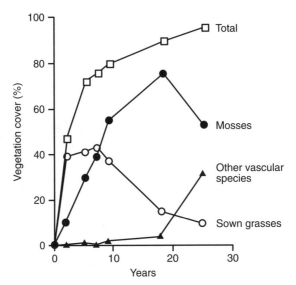

Figure 2.6. Changes from 1968 to 1993 in the cover of mosses, sown grasses and local vascular species on a bulldozed piste at 1000 m in Coire Cas, Cairn Gorm (Bayfield 1996).

research suggests that N deposition, coupled with increasing temperatures, is causing trees to grow 10–30% faster than some decades ago (Speicker, Meilikainen and Skovsgaard 1996). Above the treeline bryophytes are a very effective sink for nitrogen (Heal *et al.* 1998). Accumulating nitrogen will stimulate the decomposition in mires and so drive greater fluxes of C into the atmosphere. At high levels of N input and where bryophytes are absent, vascular plant growth is stimulated. Since the responses of different species to N vary greatly, there may be changes in the species composition of mountain vegetation, probably moving it towards greater dominance by graminoids, with an overall loss of biodiversity (Pitcairn, Fowler and Grace 1991; Cannell, Fowler and Pitcairn 1997).

STATES

Effects of Climate Change

Changes in temperature and precipitation will modify the distribution of montane species, generally resulting in upward shifts of the ecoclimatic zones. Studies by the Institute of Plant Physiology in Vienna, of the nival zone of high peaks in the Alps, re-examined lists of species compiled between the late nineteenth century to the early decades of the twentieth century (Gottfried *et al.*, quoted by Price and Barry 1997). Comparing these with current lists showed that there had been increases in species richness on 90% of sites, with a peak increase of more than 200% on Piz dais Lejs. This supported the hypothesis that temperature changes had resulted in the invasion of the nival zone by species from lower levels. The conclusion was that high mountain species were already reacting to global warming.

Halpen (1994) used Geographic Information Systems (GIS) to predict shifts in mountain vegetation zones in response to GCMs of climate change. The predictions and therefore responses varied between GCMs. He predicted asymmetrical shifts which varied in pattern from zone to zone, because each zone had a different response-profile to changes in temperature and precipitation. Some vegetation zones on mountain peaks would be lost by displacement, but others could be lost by the disappearance of the specific temperature and precipitation regime required by the vegetation zone. The dry temperate mountains in Sierra Nevada, California, would, for example, lose two of their eight vegetation zones under a scenario of $+3.5°C$ and $+10\%$ precipitation (Figure 2.7).

Effects on vegetation have a knock-on effect on animals. French (1996) used GIS to predict the effects of climate warming on the distribution of the small mountain ringlet butterfly *Erebia epiphron* in part of Scotland. The species was confined to a particular wet grassland habitat in specific altitudinal bands, and required predominantly south-facing slopes. He predicted that the species would be dramatically affected by a 1°C rise in temperature, which would bring up to a 27% decline in suitable habitat (Figure 2.8). For a 2°C rise, the loss of habitat area could be 60%.

Effects of Land Use Change

Although vegetation inventories exist for many mountain areas, there have been few attempts to assess the degree of impact on vegetation of changes in land use. Recently the need for this kind of recording has begun to be recognised. An example is the scheme devised by Scottish Natural Heritage to map the impacts of grazing on upland vegetation in quarter-kilometre squares using indicators of condition. The indicators ranged from morphological features such as topiary growth forms, to counts of dung pellets (MacDonald *et al.* 1998). Separate indicators were identified for broad vegetation types and for different altitudinal bands. Another example of a study, this time at a less detailed scale, is the classification of Austrian woodlands by degree of naturalness, again using indicators of impact (BMLF 1997). This study showed that only 3% of the forests were fully natural and 7% were completely artificial, the remainder having various degrees of alteration. This type of scheme needs to be tailored to the resources at hand and probably would not

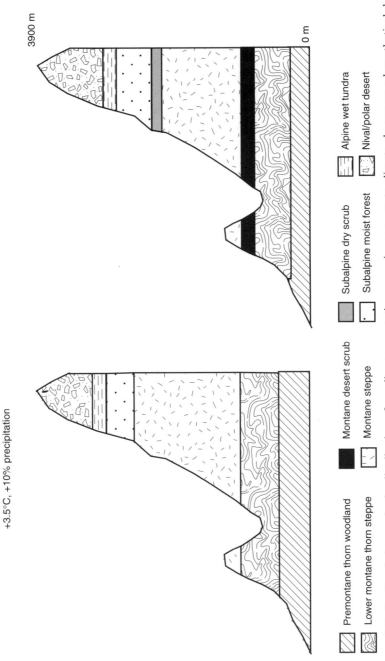

Figure 2.7. Predicted changes in the distribution of ecoclimate vegetation zones in response to climate change on a hypothetical dry temperate mountain site in Sierra Nevada (redrawn after Halpen 1994).

Figure 2.8. Changes in (a) distribution of current potentially suitable habitat for small mountain ringlet butterfly at Ben Lawers, Scotland, and (b) following a 1°C increase in mean temperature. The intensity of shading indicates 'poor', 'fair' or 'good' quality habitat (French 1996).

work elsewhere, but it provides a framework for management and action that is much more relevant than a simple inventory.

RESPONSES

Changes in the traditional agricultural use of mountains are having massive social, economic and ecological repercussions in many countries. Research is helping to identify the levels of grazing required to achieve specific impacts on vegetation, and these levels can be targets for local or national policies to reverse the effects of land abandonment or overgrazing. A number of countries have schemes for subsidising land-use practices that help to maintain traditional landscapes. However, the scale of abandonment is so extensive, for example in parts of the Alps, that it may be too costly to attempt to revert the process except in special circumstances. The smaller-scale impacts of recreation are potentially more tractable. Planning regulations for ski areas now help to control the worst types of impacts on vegetation and soils, and there is a considerable range of techniques available to help restore damaged ground at high altitudes (Urbanska 1995). Unfortunately most are costly, and few have been adopted.

SOILS AND SUBSTRATES

CHARACTERISTICS

Soils respond to the temperature decreases and increased precipitation associated with altitude. The effect is generally to increase the organic matter content, the ratio of carbon to nitrogen and the acidity (Huggett 1995) and to slow the rate of weathering of soil parent material. Slope instability also delays soil development. Alexander, Mallory and Colwell (1993) found that soils on slopes over 30% in northern California were much more immature than those on gentler slopes (this is discussed in greater detail in Warren, this volume).

Parent material is a key factor in determining soil characteristics associated with hydrology and slope stability. In the Alps the juxtaposition of siliceous and carboniferous rocks is often visually very striking. Soils on the limestone are more freely drained, less well vegetated and more jagged than those on the siliceous rocks. Regolith and slope formation on contrasting rocks are well illustrated in a study by Onda (1992) of four experimental drainage basins in the Obara area of Japan (Figure 2.9). The parent rocks were a medium-grained granodiorite and coarse-grained granite. On the granodiorite the high rate of weathering produced a thick regolith with a large water storage capacity and there were few slope failures. On granite, weathering was slower, the regolith thinner and the soils had a smaller water-storage capacity. On these rocks,

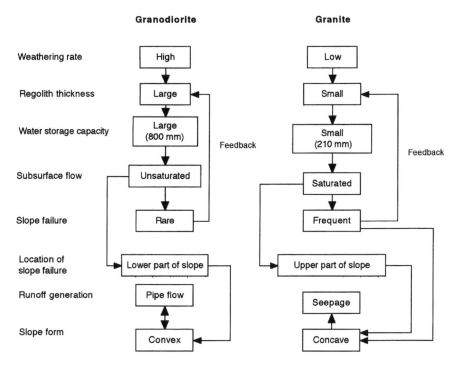

Figure 2.9. Schematic diagram of feedback in regolith-slope systems on granodiorite and granite in experimental drainage basins of the Obara area, central Japan (Onda 1992).

slope failures were common, and this contributed to the maintenance of a thin regolith (see also Warren, this volume).

PRESSURES

Any pressures that disturb vegetation, such as pollution, burning, grazing, timber harvesting or recreation, can affect soils by weakening the protective surface layer of plant material. The result is erosion. Plant cover is vital for the control of erosion by surface wash or by shallow slope failures. The mechanical restraint associated with plants extends only as far as the rooting depth, itself dependent on slope soil and slope type. The stabilizing influence of roots ranges from slight on thick soil mantles over solid bedrock, to very effective on some types of thicker mantles where they penetrate to depths that are close to the bedrock (Figure 2.10). Root protection is, of course, much shallower above the treeline. Destabilisation that extends below the rooting zone is only partially controlled by vegetation cover. Deep-seated circular slope failures are

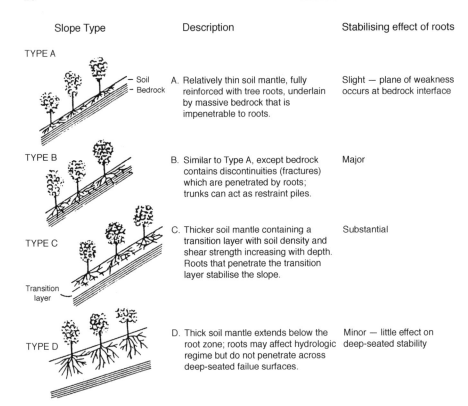

Slope Type	Description	Stabilising effect of roots

TYPE A

— Soil
— Bedrock

A. Relatively thin soil mantle, fully reinforced with tree roots, underlain by massive bedrock that is impenetrable to roots.

Slight — plane of weakness occurs at bedrock interface

TYPE B

B. Similar to Type A, except bedrock contains discontinuities (fractures) which are penetrated by roots; trunks can act as restraint piles.

Major

TYPE C

C. Thicker soil mantle containing a transition layer with soil density and shear strength increasing with depth. Roots that penetrate the transition layer stabilise the slope.

Substantial

Transition layer

TYPE D

D. Thick soil mantle extends below the root zone; roots may affect hydrologic regime but do not penetrate across deep-seated failue surfaces.

Minor — little effect on deep-seated stability

Figure 2.10. Influence of slope stratigraphy on the stabilising effect of roots against slope failure (Gray 1995).

sometimes not prevented at all. If climate change increases precipitation and shortens snowlie, there may be more saturated slope failures. Wasting glaciers can expose very unstable slopes, leading in some cases to massive slope failures (Price and Barry 1997).

In many mountain areas, extensive terracing of lower slopes has been undertaken to increase the area of level ground. This was often for arable agriculture, but sometimes it was to improve grazings. Terracing usually involves very considerable rearrangement of soils: rocks are used to build the retaining walls and finer soil forms the retained slope. Manure might have been added to improve fertility. Terraces represent a major disruption of the original soils and have a massive landscape impact. The tiers of terraces used for rice cultivation that occur in the Far East are particularly spectacular. Many of the terraces built to improve grazing are now of high conservation interest because of the species-rich grasslands they support. However, in the Pyrénées, Alps and elsewhere, the abandonment of the terraces built for arable cultivation and later used for grazing use has resulted in the localised collapse of retaining walls, invasion of terraces by trees and shrubs and the loss of their species-rich grasslands (Figure 2.11).

Figure 2.11. Terraces originally used for cereal production or for grazing, now abandoned and being colonised by trees and shrubs. Fragen, Spanish Pyrénées.

Trampling by tourists is an example of small-scale pressures on mountain soils. Although it affects only a small proportion of most mountain areas, the impacts can be locally severe. Some 470 000 tourists a year are estimated to visit the Aostan Valleys of Gran Paradiso National Park, in North Italy, where in 1995, 41% of the paths were considered in poor or bad condition (Siniscalco 1995). The degree of deterioration is strongly influenced by surface properties such as the vegetation type, soil wetness and ground roughness (Bayfield 1973; Liddle 1998).

More extensive damage can occur at ski areas, where ground surfaces have been smoothed to create ski pistes, and the new surfaces seeded with commercial grasses, or left bare. Disturbances can be visible at a considerable distance (Figure 2.12) and the re-establishment of native vegetation may be very prolonged.

An unusual type of recreational impact on soils and slopes is the construction of a golf course in tropical mountain rainforest at Fraser's Hill, in Malaysia (Figure 2.13). Although carefully constructed, the massive cut-and-

Figure 2.12. Visual impact of slopes engineered for skiing at Mount Yltas, Northwest Finland, in 1995. Photo: Kirsi Myllynen.

fill slopes represent a major disruption of landforms, and in spite of immediate hydroseeding to stabilise the surfaces, some slope failures have occurred.

Large-scale disturbance of vegetation by logging can bring greatly increased risks of erosion, through loss of tree cover, direct damage and exposure of soils and through interception of drainage segments by logging roads and skid trails. The risk of erosion greatly increases in the years immediately after logging (Figure 2.14).

Minerals and rocks have been mined in mountains for millennia. The curious depressions and spoil heaps that litter many mountain ranges often turn out to be long-abandoned mineral workings. Small heaps of shattered rocks may be evidence of gem hunters' activities. Quite small-scale activities can sometimes have a disproportionate impact, as when an outcrop of serpentine was entirely removed by successive visits by geology students (Speight 1973). In the uplands and mountains of Northern Ireland, peat cutting by hand was arguably sustainable, but extraction of shallow blanket peats has now become a commercial activity (Bayfield *et al.* 1991). Machine extraction is in danger of severely depleting or eliminating peats from large tracts of land.

Modern quarrying activities can be on a very large scale. A proposed superquarry on Harris, in the Outer Hebrides, is expected to have a productive

Figure 2.13. Part of a golf course created in tropical rainforest at Fraser's Hill, Malaysia.

life of up to 12 M tonnes per annum for 50 years and involve the removal of most of a mountain.

RESPONSES

Controlling the problems of erosion that result from recreation has been extensively studied in the USA (McEwan, Cole and Simon 1996). Techniques of control include visitor education, closure of worn sites, concentrating use on a few sites and the development of monitoring programmes. Methods of rebuilding footpaths outlined by Bayfield and Aitken (1992) include the use of geotextiles, the machine-building of paths using subsoil materials, and the use of stone pitching and slabs. The selection of suitable routes is seen to be a key to minimising impacts, for example by choosing routes that avoid seepages, gullies and areas of poor vegetation cover.

Statutory or voluntary controls to minimise the risks of erosion due to construction and maintenance work at ski areas are applied widely, although standards vary from country to country. Measures include the avoidance of sensitive slopes, the installation of pylons and buildings by helicopter, and the immediate reinstatement of damaged vegetation (Countryside Commission for Scotland 1988; Florineth 1994).

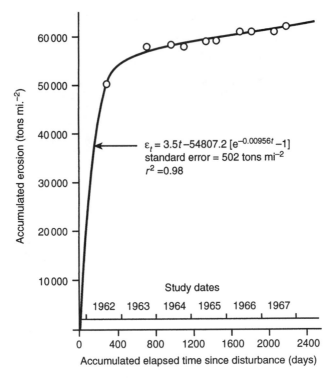

Figure 2.14. Erosion in relation to time since disturbance. Idaho Batholith (Megahan 1974).

Slope stability can be affected by many types of development, including bulldozed ski pistes, road cuttings, paths and cut-and-fill slopes associated with buildings. Engineering solutions such as the use of gabions and crib walls are frequently applied, but they are expensive and visually intrusive. Less intrusive works include the shallow (0.2 m) surface protection of slopes by seeding with grasses, or by transplanting turf. Deeper stability (to about 1.5 m), which is still unintrusive, can be provided by trees and shrubs, planted using so-called bioengineering techniques (Schiechtl and Stern 1996). These include brush layering, live pole planting and fascines (Figure 2.15). The aim is to provide immediate physical strengthening of the slope, followed by increased protection as the plants grow. Deep rooting provides underground stabilisation, and on the surface the leafy growth breaks the impact of precipitation, and traps water-borne soil particles. Tree and shrub cover also have the important role of increasing evapotranspiration from the slope, so reducing the risk of saturation, when slope failure is most likely to occur.

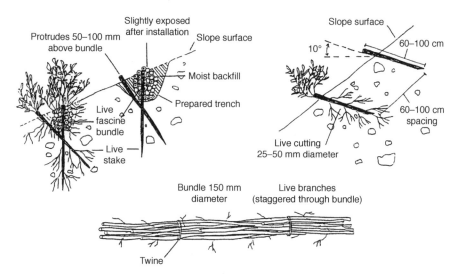

Figure 2.15. Live pole (right) and fascine (left and below) methods of bioengineering for slope stabilisation.

Methods for reducing erosion during logging are also well developed. The United States Department of Agriculture has produced guides for controlling sediment from logging roads (Packer and Christenson undated). A software package (WEPP) has been adapted by the USDA Forest Service for the prediction of the erosion risks of various logging activities. The package takes account of factors such as climate and soil conditions, hydraulic conductivity, types of forest roads and trails, frequency of fires, and the extent of forest regeneration (Elliot and Hall 1997). WEPP has also been used to model the effects of mitigating factors such as retaining surface residues along skid trails.

Statutory controls on quarrying generally require the developer to demonstrate that the impacts of the quarry will be acceptable, and that reinstatement of the site has been taken into account. Large quarries inevitably change the landform, but it is possible to minimise the impacts, for example by landform replication. This involves designing mineral workings so that their final form is consistent with landforms found in their local geomorphological setting (Walton 1995). The proposed superquarry on Harris will remove a large proportion of the Rodel mountain and create a substantial sea loch. The impacts will be mitigated by the progressive creation of corrie landforms around the site and by the reinstatement of much of the natural vegetation (Figure 2.16). The site is important for mosses and lichens and the creation of corries with screes and rock faces should provide extensive habitats for these species as the development proceeds.

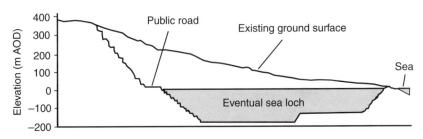

Figure 2.16. Plan of proposed landform replication at a superquarry in Harris (Walton 1995).

WATER RESOURCES

CHARACTERISTICS

Nearly all the world's major rivers originate in mountains, and water from mountain catchments is often the main supply for adjacent lowlands (Figure 2.17). Many mountainous countries supply substantial quantities of water to their neighbours. Switzerland supplies about 67% of its annual precipitation to France, Germany, Austria and Italy by way of the Rhine, Rhône, Danube and Po (Bandyopdhyay et al. 1997). The seasonal pattern of discharge from the mountains is controlled by a complex mix of the seasonal distribution of precipitation and other processes. At high altitudes and latitudes, large quantities of water are stored over winter as snow and ice, and released in the spring and summer. Some may be retained in longer-term storage in glaciers. Switzerland has about 74 km^3 of water stored in glaciers, sufficient to supply its rivers for about five years in the absence of other supplies (Rodda in Bandyopdhyay et al. 1997). Further storage occurs in aquifers and in natural and artificial lakes and reservoirs.

PRESSURES

The geographical distribution and timing of hydrological yields have been widely modified by dams, artificial watercourses and stream and catchment diversion schemes. Such pressures on water resources are continuing in most parts of the world. However, in countries that already generate a significant proportion of their energy needs from hydro sources, such as Norway, Canada and Switzerland, the scope for further water capture is quite limited and in some cases there is considerable opposition to further development.

Climate change is likely to be a major influence on hydrological yield, affecting both the volume and release pattern of water (Price and Barry 1997). Land-use changes are also having a substantial influence on hydrology: the

Figure 2.17. The Spanish Pyrénées are the main source of water for the lowland Ebro valley plain. Aisa Valley, near Jaca.

abandonment of alpine pastures in Europe is resulting in tree and shrub invasion (Cernusca *et al.* 1996), and this tends to reduce yields because of the interception of precipitation, the increase of evapotranspiration, and the shallow surface storage of water (Figure 2.18). Conversely, deforestation has

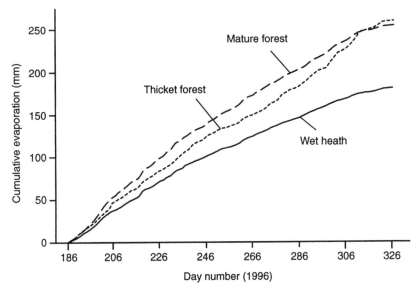

Figure 2.18. Evapotranspiration from mature forest, thicket forest and wet heath in the Cairngorms, Scotland (Price 1999).

been linked to greater peakedness of hydrological flows, erosion and downstream flooding (Hamilton, Gilmour and Cassells 1997).

Road construction can have impacts on slope hydrology and erosion where routes cut across slope-drainage segments. The resulting changed patterns of drainage can result in severe erosion, particularly in areas of high rainfall and low slope stability, as in Nepal (Lawrence 1995). The recreational development of ski areas may also have a small-scale influence on hydrology, by redistributing and compacting snow, and from the use of snow-making equipment. Although there seems to have been little detailed study of these activities, the effects seem likely to result in extended snowlie and the slower release of meltwater.

STATES

In spite of the great importance of mountain catchments, hydrological recording networks are less dense than elsewhere, mainly because of the difficulties of working in remote and difficult terrain (Bandyopadhyay et al. 1997). It is also difficult to record very variable flow rates and to cope with very high sediment loads and low winter temperatures. Snowfall is notoriously difficult to monitor reliably. Alternative approaches to monitoring, which reduce the dependence on ground observations, include the use of remote sensing to assess snow and ice cover, and of weather radar to estimate precipitation (Joss and Lee 1993). Catchment and larger-scale modelling can be used to gauge the consequences of land-use change and of climatic change.

RESPONSES

The use of mountain water by non-mountain peoples and even by people in other countries is the basis for substantial conflicts of interest. Greater public awareness is, however, contributing to changing attitudes. The World Bank (1993) proposed much greater sensitivity to potential social, economic and environmental impacts. They endorsed the need for managing the whole watershed and for identifying land-management practices to minimise degradation of water quality and stream ecosystems (a theme developed at greater length by Thompson, this volume). New proposals are now subject to increasingly detailed investigation and public scrutiny. There is a need for regional, national and even international planning for water resource management, as populations increase and climate changes result in altered availability of water (Commission on Sustainable Development 1995).

LANDSCAPES

CHARACTERISTICS

Mountain landscapes are characterised by high visibility, with prominent areas of sloping ground and many skylines. Mountains are usually considered to be relatively natural and unspoilt, even when extensively terraced or planted. Their landscapes are widely regarded as a resource for recreation and inspiration and they feature extensively in the art of many cultures. Traditional land uses such as transhumance and terracing, villages and vernacular architecture are usually judged to contribute to high landscape value.

PRESSURES

The literature of landscape assessment recognises two main categories of impact: on the landscape fabric and on its visual nature (Landscape Institute 1995).

'Fabric' includes constituent elements such as slopes, skylines, and land cover (forest, pastures, rock etc.). Pressures on the fabric include changes to landforms though quarrying, earth moving on ski areas (Figure 2.19), the construction of roads and tracks with cut-and-fill slopes, building works, dams and the construction of erosion and avalanche control earthworks and fencing. Changes to land cover can include the effects of clear-felling or of tree planting.

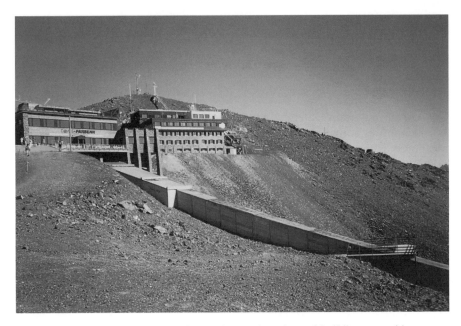

Figure 2.19. Landscape impacts of extensive earthworks and buildings at a ski resort at Davos, Switzerland.

In the UK, post-war planting of geometric blocks of conifers has had a particularly dramatic impact on many mountainous areas of the country (Figure 2.20).

Visual impacts depend on the viewer and the viewpoint. Both the visibility of new features and their aesthetic effects need to be considered. Changes to 'fabric' can have visual impact if they are visible from viewing points such as roads, paths or buildings. The location of features has a crucial effect on their visual impact, and quite small or even 'point' features can be significant. Roads, lines of pylons, wind farms, forest edges and buildings have much more impact if they cut the skyline or conspicuously bisect a field of view. They are also much more obvious if located in the foreground or mid-ground of a field of view.

STATES

There are many schemes for describing landscape characteristics. One of the most straightforward is that of Land Use Consultants (1991), which involves listing the main landform features, land uses, water elements and views. For each element a score is given for dominance (present but inconspicuous, evident, or dominant). There is also a score for whether the visual contribution of a feature is broadly positive ($+$) or negative ($-$). Finally, there is a list of the

Figure 2.20. Impacts of conifer blocks on the landscape fabric of upland Scotland near Tomintoul.

descriptors of aesthetic quality (such as scale, form, and pattern) and of perceptions/impressions (such as rarity, wildness, familiarity and so on) (Table 2.1).

Until recently it has been difficult to monitor landscape impacts except fairly subjectively or at a very local scale, mainly using photomontages. GIS and digital landscape visualisation programs now make visual impact assessment possible at local, regional and even national scales. GIS is particularly suitable for assessing the distribution of the main elements of the landscape fabric, by taking these from existing maps or remotely sensed data, and for the assessment of the 'visual envelope' of individual elements. For example, MLURI (1996) used GIS to assess the visual impact of individual new farm woodland plantings throughout Scotland. Computer-generated 'before' and 'after' views can also be produced for specific locations, to assess the likely impacts of a change such as tree planting (Miller *et al.* 1994). The significance of the changes can then be assessed by panels of observers. Culberston *et al.* (1994) described the use of this approach in Brazil, Japan, the USA and Canada, both as a tool for assessing landscapes and for planning. However, they pointed out that the methodology was still evolving rapidly and that the computer output still needed a great deal of weighting and interpretation. They suggested that there were advantages in combining the training of experts and the experience of lay

Table 2.1. Description of the landscape features of the Eastern Corries of Aonach Mor, Western Scotland following the protocol given by Land Use Consultants (1991)

Landform features			Views		
Peaks	***	(+)	Distant	***	(+)
Valleys	***	(+)	Framed	**	(+)
Precipice/cliffs/scree	***	(+)	Corridor	**	(+)
Plateau	*	(+)			
Ridge/slope/spurs	***	(+)	Recreation		
Terraces	**	(+)			
Coire	***	(+)	Walking/climbing	*	(−)
Morraine	*	(+)	Cross country skiing	*	(−)
Plain/flat	*	(+)			
Rock outcrops	***	(+)			

		Aesthetic qualities		Perceptions/impressions	
Vegetation		Scale:	large	Rarity:	unusual
		Enclosure:	enclosed	Security:	unsettled
Grass moorland	***(+)	Diversity:	diverse	Wildness:	wild
Dwarf shrubs	** (+)	Texture:	rough	Beauty:	attractive
Bog communities	** (+)	Form:	vertical/sloping	Familiarity:	unusual
Bracken	* (−)	Line:	angular	Management:	none
Deciduous trees	* (+)	Colour:	muted	Productivity:	sparse
		Balance:	balanced		
Water		Movement:	calm		
		Pattern:	organised		
Lochans	***(+)				
Streams	***(+)				
Falls/cascades	* (+)				

Notes: Individual features scored as * present but inconspicuous; ** evident; *** dominant. Overall visual contribution positive (+) or negative (−). Principal aesthetic and perceptual qualities are described. After Bayfield, McGowan & Paterson (1993).

observers familiar with the ground. Although these computer techniques can provide broad assessments of large areas, impacts on individual locations such as are required for Environmental Impact Assessments still largely depend on photomontages, albeit sometimes computer-enhanced.

RESPONSES

Measures to minimise visual impacts have been extensively developed by the US Department of Agriculture (1973–7). They involve assessing the desired landscape character (Visual Quality Objectives), and planning the location, design and management of development so as to minimise its impacts both during construction and beyond. Techniques include the siting of roads to avoid skylines and ridges, the following of biological boundaries rather than the bisecting of existing vegetation communities, and making use of topographic shelter to hide routes. In the case of forest harvesting and replanting, impacts can be reduced by taking care to create a natural-looking edge to plantings, and designing cut and planted areas at an appropriate scale and shape to fit the landscape character. Account is taken of the differing visibility of managed areas by having separate criteria for foreground, mid-ground and distant forest blocks.

The visual impacts of ski developments can be minimised by careful siting to avoid skylining, the use of muted colours for pylons and buildings, and by the immediate reinstatement of damaged ground with appropriate turf, transplants or seeding (Countryside Commission for Scotland 1988; Urbanska 1995).

In some mountain areas an uncontrolled proliferation of vehicular tracks is causing concern about erosion, visual intrusion, and penetration of wilderness. In the 1960s and 1970s, hundreds of miles of vehicular tracks were bulldozed in the Cairngorm Mountains in Scotland to improve access for hunting (Watson 1984). Since then planning controls have been imposed on this type of development and in a few instances attempts have been made to reinstate the tracks as footpaths or even to remove them altogether (Figure 2.21). This involves reprofiling the bulldozed surfaces to the original landform, and

Figure 2.21. High-altitude vehicle track above the treeline, with trial section narrowed to footpath width (mid-ground). Beinn a Bhuird, Scotland.

planting the bare surfaces with transplants from previous spoil heaps or from surrounding ground.

DISCUSSION

Rapid changes are taking place in mountain areas. They involve not only climate, land use and population, but also perceptions about their value. The growth of eco-tourism and recreational use of mountains is creating new interest groups with their own concepts of mountain conservation, while there has been a decline in traditional mountain uses in some parts of the world. Infrastructural requirements and the demands of mineral and water exploitation are creating new pressures. Government and local planners have generally responded by increasing controls. Inventories and monitoring of some resources have been undertaken in some countries, but there is little consistency of approach and few attempts to identify quality standards and associated monitoring that would provide a factual basis for planning, or to assess the effectiveness of conservation measures.

The Limits of Acceptable Change (LAC) concept is an approach that has been devised by the US Department of Agriculture Forest Service for wilderness management (Cole and Stankey 1997). First developed in the 1980s, the approach involves setting standards for environmental indicators of change and also identifying the responses to be triggered if standards (the LAC values) are exceeded. An important feature is the involvement of interested parties in identifying key issues and concerns. Another is the assumption that change will continue to occur, so that there has to be regular review of both the key environmental issues and of the LAC values. The concept is thus both flexible and responsive. Typical applications involve setting standards for the numbers of visitors in an area, path widths, damage to campsites, and sediment yields of streams. In each case monitoring is undertaken to check actual values, and the acceptable limits of change are set and reviewed by the panel of interested parties.

It is difficult to adapt the LAC approach to all the issues of change, or to all mountain areas. Some aspects such as landscape quality are not readily adapted to quantitative standards. There are also drawbacks to the approach in terms of the infrastructure and consultative involvement it needs and many countries do not have the social, political or planning control structures that would support it. Nevertheless, there appears to be scope for more widespread gathering of data on environmental variables related to key issues and concerns, and the identification of some kinds of quality standards (local, regional or national) derived from broad consultation, to help plan the conservation of mountain areas for the future (Brunson 1997).

REFERENCES

Agnew C & Fennessy S (2001) Climate change and nature conservation. In: A Warren & JR French (eds) *Habitat conservation: managing the physical environment.* Chichester, John Wiley, 273–304.

Alexander EB, Mallory JI & Colwell WL (1993) Soil–elevation relationships on a volcanic plateau in the Southern Cascade Range, northern California, USA. *Catena* **20**, 113–28.

Bandyopdhyay J, Rodda JC, Kattelmann R, Kunzewicz ZW & Kraemer D (1997) Highland waters — a resource of global significance. In: B Messserli & JD Ives (eds) *Mountains of the world: a global priority.* London, Parthenon, 131–55.

Barry RG (1992) *Mountain weather and climate.* London, Routledge.

Barry RG & Ives JD (1974) Introduction. In: JD Ives & RG Barry (eds) *Arctic and Alpine environments.* London, Methuen, 1–13.

Bayfield NG (1973) Use and deterioration of some Scottish mountain footpaths. *Journal of Applied Ecology* **10**, 635–44.

Bayfield NG (1996) Long-term changes in colonization of bulldozed pistes at Cairn Gorm, Scotland. *Journal of Applied Ecology* **33**, 1359–65.

Bayfield NG & Aitken R (1992) *Managing the impacts of recreation on vegetation and soils: a review of techniques.* Report to the Countryside Commission, Countryside Commission for Scotland, English Nature and Countryside Council for Wales. Banchory, Institute of Terrestrial Ecology.

Bayfield NG, Picozzi N, Staines BW, Crisp TC, Carling P, Robinson M, Gustard A & Shipman P (1991) *Ecological impacts of blanket peat extraction in Northern Ireland.* Report to Countryside and Wildlife Branch, Department of Environment Northern Ireland, Banchory, Institute of Terrestrial Ecology.

Bayfield NG, McGowan GM & Paterson IS (1993) *Aonach Mor: environmental assessment of proposals for ski development in the eastern corries.* Report to Nevis Range Development Company, Banchory, Institute of Terrestrial Ecology.

BMLF (1997) *Bildatlas, Naturnähe österreichischer Wälder.* Wien, Bundesministerium für Land-und Forstwirtschaft.

Brunson R (1997) Beyond wilderness: broadening the applicability of Limits of Acceptable Change. In: SF McCool & DN Cole (eds) *Proceedings — limits of acceptable change and related planning processes: progress and future directions.* USDA Forest Service General Technical Report INT-GTR-371, Ogden, UA, Rocky Mountain Research Station, 44–8.

Callaghan TV, Korner C, Heal OW, Lee SE & Cornelissen JHC (1998) Scenarios for ecosystem responses to global change. In: OW Heal, TV Callaghan, JHC Cornelissen, C Korner C & SE Lee (eds) *Global change in Europe's cold regions.* Ecosystems Research Report No. 27, Directorate-General Science, Research and Development, Brussels, European Union.

Cannell MGR, Fowler D & Pitcairn CER (1997) Climate change and pollutant impacts on Scottish vegetation. *Botanical Journal of Scotland* **49**, 301–3.

Cernusca A, Tappeiner U, Bahn M, Bayfield N, Chemini C, Filat F, Graber W, Rosset M, Seigwolf R & Tenhunan J (1996) ECOMONT. Ecological effects of land use changes on European terrestrial mountain ecosystems. *Pirineos* **147/148**, 145–72.

Cole DN & Stankey, GH (1997) Historical development of Limits of Acceptable Change: conceptual clarifications and possible extensions. In: SF McCool & DN Cole (eds) *Proceedings — limits of acceptable change and related planning processes:*

progress and future directions. USDA Forest Service General Technical Report INT-GTR-371, Ogden, Rocky Mountain Research Station, 5–9.

Commission on Sustainable Development (1995) *Decisions and recommendations adopted by the third session on sustainable development.* New York, United Nations.

Countryside Commission for Scotland (1988) *Environmental design and management of ski areas in Scotland: a practical handbook.* Perth, Countryside Commission for Scotland.

Culbertson K, Hershberger B, Jackson S, Mullen S & Olsen H (1994) Geographic information systems as a tool for regional planning in mountain regions: case studies from Canada, Brazil, Japan and the USA. In: MF Ford & DI Heywood (eds) *Mountain environments & geographic information systems.* London, Taylor & Francis, 99–118.

Dyurgerov MB & Meier MF (1997) Year-to-year fluctuations of global mass balance of small glaciers and their contribution to sea-level changes. *Arctic and Alpine Research* **29**, 392–402.

Elliot WJ & Hall DE (1997) *Water Erosion Prediction Project (WEPP): forest applications.* General Technical Report INT-GTR-365. US Department of Agriculture Forest Service, Ogden, UA, Intermountain Research Station.

Florineth F (1994) Erosion control above the timberline in South Tyrol, Italy. In: DH Barker (ed.) *Vegetation and slope stabilisation: protection and ecology.* London, Thomas Telford, 85–94.

Franzén LG (1991) The changing frequency of gales on the Swedish west coast and its possible relation to increased damage to coniferous forests of southern Sweden. *International Journal of Climatology* **11**, 769–93.

French DD (1996) *Use of GIS to predict the effects of environmental change on the small mountain ringlet at Ben Lawers.* Report to Scottish Natural Heritage. Banchory, Institute of Terrestrial Ecology.

Gimingham CH (1964) Dwarf shrub heaths. In: JH Burnett (ed.) *The vegetation of Scotland.* Edinburgh, Oliver & Boyd, 232–89.

Gray DH (1995) Influence of vegetation on the stability of slopes. In: DH Barker (ed.) *Vegetation and slope stabilisation: protection and ecology.* London, Thomas Telford, 2–25.

Guisan A., Tessier L, Holten JI,. Haeberli W & Baumgartner M (1995) Understanding the impact of climate change on mountain ecosystems: an overview. In: A Guisan, JI Holten, R Spichiger & L Tessier (eds) *Potential ecological impacts of climate change in the Alps and Fennoscandian mountains.* Publication hors-série no. 8 de Conservatoire et Jardin botaniques de la Ville de Genève, 15–37.

Haeberli W (1995) Climate change impacts on glaciers and permafrost. In: A Guisan, JI Holten, R Spichiger & L Tessier (eds) *Potential ecological impacts of climate change in the Alps and Fennoscandian mountains.* Publication hors-série no. 8 des Conservatoire et Jardin botaniques de la Ville de Genève, 97–103.

Halpen PN (1994) GIS analysis of the potential impacts of climate change on mountain ecosystems and protected areas. In: MF Price & ID Heywood (eds) *Mountain environments & geographic information systems.* London, Taylor & Francis, 281–301.

Hamilton LS, Gilmour DA & Cassells DS (1997) Montane forests and forestry. In: B Messserli & JD Ives (eds) *Mountains of the world: a global priority.* London, Parthenon, 281–311.

Harrison J (1997) Changes in the Scottish climate. *Botanical Journal of Scotland* **49**, 287–300.

Heal OW, Callaghan TV, Corneilissen JHC Korner C & Lee SE (1998) *Global change in*

Europe's cold regions. Ecosystems Research Report No. 27, Directorate-General Science, Research and Development, Brussels, European Union.

Houghton JT, Meira Filho LG, Callander BA, Harris N, Kattenberg A & Maskell K (1996) *Climate Change 1995: the science of climate change.* Cambridge, Cambridge University Press.

Huggett RJ (1995) *Geoecology: an evolutionary approach.* London, Routledge.

Joss J & Lee R (1993) Weather radar: operational processing for snowcasting and precipitation estimation. In: Moore R (ed.) *Proceedings of WMO Regional Association IV (Europe) workshop on requirements and applications of weather radar data in hydrology and water resources.* Geneva, WHO.

Land Use Consultants (1991) *Landscape assessment: principles and practice.* Perth, Countryside Commission for Scotland.

Landscape Institute (1995) *Guidelines for landscape and visual impact assessment.* London E & FN Spon.

Lawrence CJ (1995) Low cost engineering and vegetative measures for stabilising roadside slopes in Nepal. In: DH Barker (ed.) *Vegetation and slope stabilisation: protection and ecology.* London, Thomas Telford, 142–51.

Liddle M (1998) *Recreation ecology.* London, Chapman & Hall.

Louis H (1975) Neugefasstes Höhendiagramm der Erde. *Bayerische Akademie der Wissenschaften (naturwissenschaftliche Klasse),* 305–26.

MacDonald A, Stevens P, Armstrong H, Immirzi P & Reynolds P (1998) *A guide to upland habitats: surveying land management impacts* (two volumes). Edinburgh, Scottish Natural Heritage.

McEwan D, Cole DN & Simon M (1996) *Campsite impacts in four wildernesses in the south-central United States.* Research Paper INT-RP-490, Ogden UA, USDA Intermountain Research Station.

MacGillivray A & Kayes R. (1995) *Environmental measures. Indicators for the UK environment.* London, Environmental Challenge Group.

Megahan WF (1974) Erosion over time on severely disturbed granitic soils: a model. USDA Forest Service Research Paper INT-156, Ogden, UA, Intermountain Forest and Range Experimental Station.

Miller DR, Morrice JG, Horne PL & Aspinall RJ (1994) Use of GIS for analysis of scenery in the Cairngorm Mountains of Scotland. In: MF Ford & DI Heywood (eds) *Mountain environments & geographic information systems.* London, Taylor & Francis, 110–32.

MLURI (1996) *Evaluation of the farm woodland premium scheme.* B Crabtree (ed.) Economics and Policy Series No 1, Macaulay Land Use Research Institute, Aberdeen.

Naeberli W (1995) Climate change impacts on glaciers and permafrost. In: A Guisan, JI Holten, R Spichiger & L Tessier (eds) *Potential ecological impacts of climate change in the Alps and Fennoscandian mountains.* Publication hors-série no. 8 des Conservatoire et Jardin botaniques de la Ville de Genève, 15–37.

Onda Y (1992) Influence of water storage capacity in the regolith zone on hydrological characteristics, slope processes, and slope form. *Zeitschrift für Geomorphologie NF* **36**, 165–78.

Packer, PE & Christensen GF (undated) *Guides for controlling sediment from secondary logging roads.* Intermountain Forest and Range Experimental Station, Ogden.

Pitcairn, CER, Fowler D & Grace J (1991) *Changes in species composition of semi-natural vegetation associated with the increase in atmospheric inputs of nitrogen.* CSD Report No. 1246, Peterborough, Nature Conservancy Council.

Price D (1999) Evaporative losses from pine colonization. In: A Cernusca, U Tappeiner & N Bayfield (eds) *Land use changes in European mountain ecosystems: ECOMONT — concepts and results.* Bozen, European Academy, 306–28.

Price MF & Barry RG (1997) Climate change. In: B Messserli & JD Ives (eds) *Mountains of the world: a global priority.* London, Parthenon, 409–46.

Rango A & Van Katwijk K (1990) Climate change effects on the snowmelt hydrology of western North American mountain basins. *IEEE Transactions of Geoscience and Remote Sensing* **GE-38**, 970–4.

Schiechtl HM & Stern R (1996) *Ground bioengineering techniques for slope protection and erosion control.* London, Blackwell Scientific.

Siniscalco C (1995) Impact of tourism on flora and vegetation in the Gran Paradiso National Park (NW Alps, Italy). *Braun-Blanquetia* **14**, 1–59.

Speicker H, Meilikainen K & Skovsgaard JP (eds) (1996) *Growth trends in European forests.* Berlin, Springer-Verlag.

Speight MCD (1973) Outdoor recreation and its ecological effects: a bibliography and review. *Discussion Papers in Conservation* **4**, London, University College London.

Thompson WF (1964) How and why to distinguish between mountains and hills. *Professional Geographer* **16**, 6–8.

Troll C (1973) High mountain belts between the polar caps and the equator: their identification and lower limit. *Arctic and Alpine Research* **5**, 19–27.

Urbanska K (1995) Ecological restoration above the timber line and its demographic assessment. In: KM Urbanska & K Grodzinska (eds) *Restoration ecology in Europe.* Zürich, Geobotanical Institute, SFIT.

US Department of Agriculture, Forest Service (1973–7) *National forest landscape management*, Volumes 1–2, US Department of Agriculture. *Agricultural Handbook* **434**, **462**, **478**, **484**, **483**, Washington, DC, US Government Printing Office.

Walton G (1995) Landform replication in quarry design. *Quarry Management*, March, 47–52.

Watson A (1984) A survey of vehicular tracks in North-east Scotland for land use planning. *Journal of Environmental Management* **18**, 345–53.

World Bank (1993) *Water resources management: a World Bank policy paper.* Washington, DC, International Bank for Construction and Development.

3 Valley-side slopes

ANDREW WARREN
University College London, UK

INTRODUCTION

Valley-side slopes are by far the most extensive terrestrial habitat, and on many, if not most, geomorphology is the primary control on ecological patterns and processes. Yet few conservation textbooks say anything about this. This chapter seeks to show how this kind of disdain for the foundations can threaten the ecological superstructure. The aim is to set out the general geomorphological principles that should underpin any management strategy.

Established textbooks reflect an entrenched neglect of geomorphology in ecological research. Even the classic ecological experiments at Hubbard Brook in New Hampshire paid little attention to spatial variations in slope processes, despite the varied terrain, and despite their focus on processes that had strong geomorphological associations (Borman and Likens 1974). Prominent reviews of ecological succession like those of Glenn-Lewin and van der Maarel (1992) barely mention any geomorphology, let alone that of the valley-side slopes. Textbooks about environments, like tropical rainforests, many of which will be shown in this chapter to have strong patterns across valley-side slopes and very dynamic substrates, virtually ignore both (Kellman and Tackaberry 1997). Even general discussions of patterns of diversity (Huston 1994) and author-itative statements about landscape ecology (Pickett and Cadenasso 1995; Forman 1995), where the main cause of variation is blatantly geomorpho-logical, almost wilfully ignore the pattern and dynamics of the substrate.

One can sympathise with attempts to simplify, and geomorphology undoubtedly introduces complexities, but some well-known ecological theories have had to be shored up because of their simplistic assumptions about valley-side slopes. The theory of Island Biogeography (MacArthur and Wilson 1967), at its simplest, predicted that species numbers were strongly related to the area of a parcel of land. Almost from its publication, the discovery of anomalies challenged the basic proposition, and it was soon evident that their most obvious cause was topographical. Examples include small habitat 'islands' in the prairie–forest ecotone in North America and in the southern Appalachians, which have more species than large neighbouring

Habitat Conservation: Managing the Physical Environment. Edited by A. Warren and J. R. French.
© 2001 John Wiley & Sons Ltd.

areas (Higgs and Usher 1980; Simberloff and Gotelli 1984). In Rhode Island, it has been shown that geomorphological variability within woodlands is a more powerful control on plant species diversity than their size (Nichols, Killingbeck and August 1998). As results like these have accumulated, the debate about the geographical associations of diversity has moved progressively towards recognition of the role of geomorphology (e.g. Burnett *et al.* 1998).

As part of this movement, Kohn and Walsh (1994) made a valuable distinction between two effects of the size of an area on the diversity of the species it contains. The first works through its control on the number and types of habitat, in other words, largely geomorphological variability. The second effect, which they termed the 'truly island effect', works through stability in larger populations. Their data, for dicotyledons on small islands in Shetland, showed that the geomorphological effect was the weaker of the two influences on the species-richness. This is, at first sight, inconsistent with the findings of the studies quoted above that placed much more importance on habitat (chiefly geomorphological) diversity, although none of them made Kohn and Walsh's distinctions. The discrepancy is likely to relate to differences in geomorphological variability of the various study sites, and probably also to the size of the total complement of species. Where both are higher than in Shetland, as in Rhode Island, the prairie or the southern Appalachians, then geomorphological variability is likely to be the stronger control, as has apparently been shown. I will use Kohn and Walsh's distinction below, in several contexts.

The claim that geomorphology should have the attention of ecologists is strengthened when animal populations are considered, for landscape patterns are probably even more important to them than to the plant populations. Invertebrates need different biotopes at different stages in their life cycles, so that there are optimum sizes and patterns of types of habitat (Samways 1994). In California the endangered Bay Checkerspot butterfly needs warm slopes where larvae can develop quickly and contiguous cool slopes where the females can emerge and reach diapause before their host plants senesce as summer proceeds (Murphy, Freas and Weiss 1988). Moreover, in wet years more of these butterflies survive on warmer slopes, while in dry years there are more on cooler slopes. This necessary mix of habitats occurs at the scale of the drainage basin. A similar story can be told about desert isopods in the Negev, many species of which also need habitat variability for survival (Shachak and Brand 1991). The same is true of many bird species. The red grouse needs open ground (found on upper slopes in more natural vegetation than now exists) for feeding, and cover (on lower slopes) for nesting. Where the cutting of the forest has destroyed the varied natural mosaic in northern British moorlands, it has had to be recreated by elaborate systems of moor burning (Lawton 1990; Bayfield, 2001).

Yet more support comes from dry areas. Where water is an ecologically limiting factor, topographic controls on its concentration become the strongest controls on species distribution. The association is so mundane that many ecologists have ignored it, but things are changing (Belsky 1995). One of the perceptions in the 'new paradigm' in ecology, which has now reached dry-land ecology (Ellis and Swift 1988; Warren 1995), is that in these environments, non-biotic controls, specifically water supply, are critical controls on ecological pattern. Where the mean annual rainfall is less than 700–900 mm in the East African savannas, Belsky maintained that topography was more important to ecological patterns than were either fire or the effects of large mammals. In these environments, ungulates, like the butterflies mentioned above, needed different facets of the landscape at different times of year. In the dry season they were on the lower slopes; in the wet season they were on the upper. Belsky believed, therefore, that there was greater animal species diversity and biomass where there was more landscape diversity. This is even truer of yet drier areas, where the slightest hollow can support many times more plant production than a nearby slope (Wiens 1985). An example of a vegetation pattern controlled by water flow in these environments is given later. Some argue that aspects of the ecology, physiology and emotions of *Homo sapiens sapiens* can be traced to this same geomorphologically determined ecological pattern in the semi-arid environments in which our ancestors evolved (Foley 1984; Appleton 1990). If so, the geomorphological pattern may extend its influence, albeit weak, to the way in which I am arguing.

It will be seen later that the case for geomorphology is more about process than pattern, although the two are closely linked. And discussion of process moves the argument from ecology to conservation, for if the argument of Pickett, Parker and Fiedler (1992) is followed, the 'new paradigm' in ecology translates, in conservation, into a greater concern with the integrity of processes than with mere numbers of species. It must then be true that an understanding of geomorphological and pedological processes is vital to the good management of ecosystems. As Scatena and Lugo (1995:211) claim, in relation to a Puerto Rican wet forest: 'Deciphering the interactions of biota and landforms over various temporal and spatial scales is fundamental to understanding and managing these complex systems'.

One final contribution that geomorphologists can make to the conservation of habitats on valley-side slopes (as of other habitats, as shown elsewhere in this book) is that they can help to understand naturalness. There is a strong movement for more naturalness in conservation. One of its claims is to move beyond the syndrome that sees management as always necessary. It is argued that 'natural' reserves would be far less expensive than managed ones (Anderson 1991; Whitbread and Jenman 1995)

The claims for geomorphology in any of these respects should not be overstated. There are at least four reasons why it can only be a part of the explanation of ecological patterns or process and therefore only part of a guide to conservation management. First, the kinds of geomorphology that are discussed here can only explain variations at the landscape scale, not at the global or regional scales or the very small scale. Second, there are landscapes where variation in bedrock lithology almost completely over-rides variation in geomorphic position or process (in the part of Appalachians examined by Stolt, Baker and Simpson 1993). Third, geomorphology is more of an ecological factor in high- than low-energy landscapes, an issue that is developed below. Fourth, land-use history is another major source of variability, one that may over-ride and obscure or confuse geomorphological and pedological variability, at least in low-energy landscapes. But, despite these reservations, geomorphology has a crucial role in ecosystem pattern and process in most systems of valley-side slopes, and especially at the scale at which most species use resources, and its consideration is needed in all management plans for conservation at this vital scale.

A BASIC SET OF SLOPE/SOIL TERMS AND IDEAS

Figure 3.1 gives a basic vocabulary for simple 'fluvial' landscapes (the landscapes of river valleys, which are the framework for valley-side slopes). The various elements (upper slopes, side slopes, coves etc.) are referred to below as 'facets'. The figure adds little to the basic patterns introduced in earlier books in this series (Warren 1974, 1983, 1993), but their interpretation is new.

The patterns of Figure 3.1 are much more evident in some landscapes than others. They occur in temperate and tropical wet areas, which cover a large proportion of the terrestrial earth, but in these same areas there are exceptions to the detail, if not the essentials of the argument below. The most extensive of these partial exceptions are landscapes whose patterns were dominated by processes other than fluvial ones during the cold, dry periods of the Quaternary. These are the areas that were glaciated (large parts of Europe and North America) or invaded by sand dunes (on the margins of the deserts, as in the western third of the state of Nebraska, large parts of the West African Sahel, huge swathes of semi-arid Australia and many others). These landscapes have their own repeating patterns, though many have been strongly overprinted with fluvial patterns, as rivers have re-established their dominance when the world has warmed and wetted again. In this re-establishment, some parts of the pattern of valley-side slopes have reappeared more quickly than others, as explained below.

A second distinction occurs between landscapes with high and those with low levels of fluvial energy (Swanson, Wondzell and Grant 1992). At the

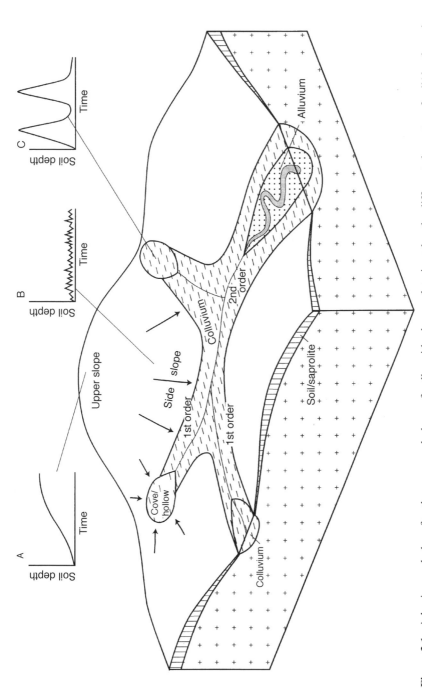

Figure 3.1. A basic vocabulary for the geomorphology of valley-side slopes, showing also the different character of soil/depth against time in different facets.

high-energy extreme are mountainous terrains, as in the Cascades in Oregon, which were used by Swanson and his colleagues as a type-site. In these areas, downslope trends of plant-community boundaries, following the flow lines of streams and landslides, are blatant. Geomorphology is by far the strongest influence on ecological pattern, submerging even human influence. In the Himalaya, it has been asserted, 'the two sets of processes, geophysical and human, are probably several orders of magnitude apart' (Ives and Messerli 1989:121–2). Here Kohn and Walsh's 'truly island effect' is overwhelmed by the geomorphological effect. An ecologically significant distinction within these high-energy landscapes is between weathering-limited and transport-limited slopes. Weathering processes control the form of transport-limited slopes. These slopes are generally rocky or bouldery, provide few good rooting sites, and have somewhat different vegetation patterns to those of transport-limited slopes; they are not discussed further in this chapter. Transport processes, such as concentrated and unconcentrated surface wash or soil mass movement, control the forms of weathering-limited slopes. These slopes are soil-covered, cover much greater areas and are the focus here.

The high-to-moderate energy landscapes stylised in Figure 3.1 are exemplified by a real piece of topography in Figure 3.2. Landscapes like these are termed 'fluvial' for good reason. Their pattern depends on streams carrying away the debris from the bases of slopes. This keeps the facets active and because of this is the dominant process behind ecological patterns. Because stream power has two elements (slope and discharge), two kinds of stream can produce steep slopes. The first is the stream that is itself steep. The second is the large stream (usually in a higher-order basin and with a relatively low slope), which, where it comes in contact with the base of a slope, can move away any debris it produces, and therefore keep it active. River cliffs formed in this way are distinctive, though rare and small facets of fluvial landscapes; they are probably the natural habitat of many of the ruderals in agricultural land today. Except for these occasional steep slopes, larger-order basins share many characteristics with first-order basins, although in different proportions and patterns, and with some distinctive elements, the most important of which, the floodplain (see Hughes and Rood, this volume).

The connecting thread in the soil–slope–stream system in these landscapes is the flow of sediment. Sediment enters the system (input, I) from the breakdown or weathering of the underlying rock (or from added dust or organic matter); it is stored temporarily in the soil or weathered mantle (S), and it is then transferred downslope (output, O), eventually to the stream.

In low-energy geomorphic environments, at the other end of the spectrum, Swanson, Wondzell and Grant (1992) maintained that non-geomorphological processes dominated the formation of ecological pattern. Here, Kohn and Walsh's truly island effect may be a more powerful control on diversity than a geomorphological one. Ecological patterns are more the outcome of fires,

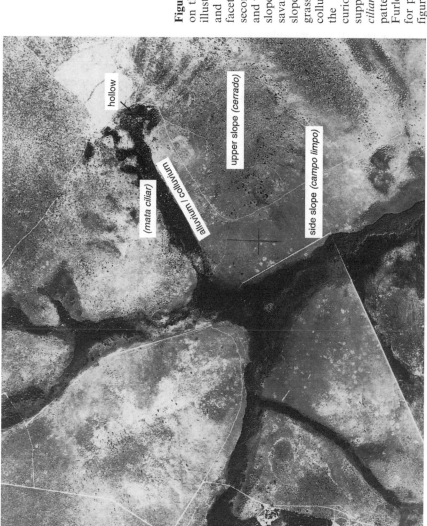

Figure 3.2. A fluvial landscape on the central Brazilian plateau, illustrating typical slope patterns and the strong relation between facet type and vegetation. First-, second- and third-order streams and valleys are shown. The upper slopes are covered in arboreal savanna (*cerrado*). The side slopes support seasonally wet grassland (*campo limpo*). The colluvium and alluvium along the stream-sides and in the curiously 'mop'-shaped hollows support gallery forest (*mata ciliar*). The soils and vegetation patterns are further explained in Furley (1996), whom we thank for permission to reproduce this figure.

winds or biotic processes than of stream incision. One such area is on the gently undulating forest lands on sandy soils in large parts of Michigan and Wisconsin. The sandy soil yields little runoff to initiate geomorphic activity, so that tree-throw mounds, created by occasional strong winds, persist for a thousand years and are much more influential in creating ecological pattern than fluvial erosion (Schaetzl and Follner 1990). In steeplands in the Appalachians, in contrast, tree-throw mounds are eradicated much more quickly by erosion (Hack and Goodlet 1960). Other examples of low-energy landscapes are the extensive lowland rainforests of much of the tropics, as in much of Amazonia, where the weathered mantle is so deep that there is little contact between plants and rock, and where geomorphic energy is too low to create much patterning (Sollins 1998).

However, some low-relief landscapes have impressive geomorphologically determined ecological patterns. One of these exceptions occurs, again, in the drylands. These are the banded vegetation or 'tiger stripe' patterns on low-angle impermeable surfaces in semi-arid lands such as Niger (Figure 3.3). The repeated bands of low trees and bushes, all parallel to the contours, are said to be the result of a water-harvesting process. Plants can only grow where sufficient runoff has been collected from upslope, and they then cut off the supply to the zone immediately downslope of the stripe they form, precluding the growth of the next stripe until sufficient water and nutrient has again accumulated (Thiéry, D'Herbès and Valentin 1995). Similar patterns occur on the low-angle sites in the American SouthWest, which Swanson, Wondzell and Grant (1992) chose to typify landscape with little geomorphologically determined pattern (Ives 1946). Less obvious, but nonetheless patterned vegetation, associated with the same kinds of processes, occurs on many if not most gentle slopes in semi-arid areas, as in Spain (Puigdefábregas and Sánchez 1996). The distinct patterning of the ecology of floodplains is another example of the importance of geomorphology in low-relief landscapes (see Hughes and Rood, 2001).

For all these exceptions, Swanson's distinction between landscapes where there is conspicuous geomorphological control and those where it is more subtle, must be acknowledged. Kennedy (Chorley and Kennedy 1971) captured the gradation between the two kinds of landscape and her study suggests the nature of the relationship between geomorphology and ecological pattern and process. She took two sets of slope in the scarplands of the eastern Paris Basin: one where the slope fed directly into a stream; the other where the stream was distant from the base of the slope. Figure 3.4 shows that in the first set there were strong statistical relations between ecological parameters such as root mass and soil depth and the geomorphic parameters of the slope. In the second set of slopes, the relationships were much weaker, presumably as other factors took over the control of ecological pattern and process.

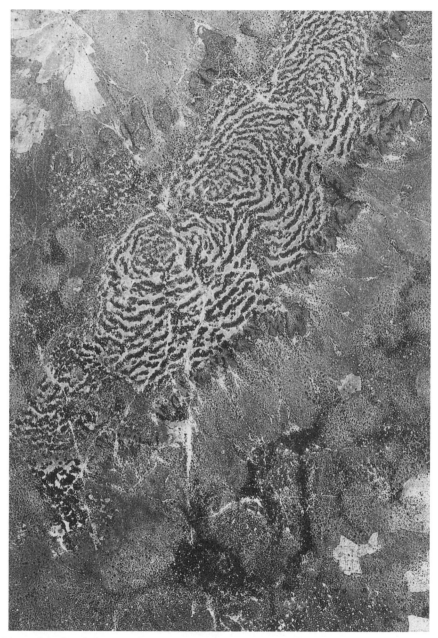

Figure 3.3. Banded vegetation in semi-arid southwestern Niger (mean annual rainfall *c.* 530 mm). Trees and shrubs are arranged in bands parallel to the contours, with almost bare soil between. On these gentle slopes, and impermeable soils, the transfer of water and nutrients is the main control on pattern.

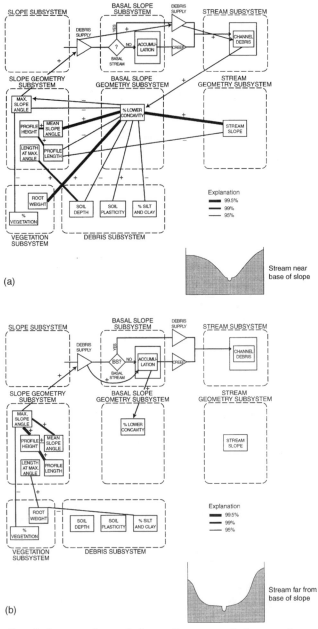

Figure 3.4. Correlation matrices relating soil, geomorphology and some ecological factors in (a) situations where a stream is close to the base of the slope; and (b), where it is far from the base (after Chorley and Kennedy 1971; with permission from the author).

EQUILIBRIUM OR NOT?

This basic vocabulary allows the introduction of concepts that are fundamental to the understanding and management of valley-side slopes. Of these, the most persistent, prominent, even central idea in 'old paradigm' ecology, conservation and geomorphology was the concept of 'equilibrium'. The older interpretation of slope equilibrium in conservation was explained in earlier books in this series (Warren 1974, 1983, 1993). In short, it was argued that ecological equilibrium would depend on slope equilibrium; if one were striving to conserve an equilibrium ecology, one would need equilibrium slopes.

Geomorphology and ecology have kept remarkable theoretical step in their attitudes to equilibrium over the last century (Hack and Goodlet 1960; Drury and Nisbet 1971; Wright 1974; Glenn-Lewin, Peet and Veblen 1992), and the idea is now under serious attack in both. In the early years of the twentieth century W. M. Davis in geomorphology and F. E. Clements in ecology were in remarkable accord that equilibrium was a key concept (Harrison and Warren 1970). Now, nearly a century later, there are both geomorphologists and ecologists who seriously question its utility. The geomorphological argument has been well made by Kennedy (1994), and its application in the valley-side slope context is debated below. The case in ecology is summarised by Glenn-Lewin, Peet and Veblen (1992), Sprugel (1991) and, most accessibly, by Botkin (1990).

Nonetheless, the idea of equilibrium in geomorphology is a starting point for a discussion of the management of valley-side slopes for nature conservation. A simple (though not sufficient) definition of equilibrium would be when input equalled output, or $I = O$ in the terms explained above. This is a form of the 'sediment continuity equation'. Elaborating the equation slightly by adding an element for soil (a much more relevant ecological entity than input or output of sediment), one might say that $I = O + / - \Delta S$, where ΔS is the change in soil depth (or better, as explained below, the depth of the 'weathered mantle' or 'saprolite'). An equilibrium approach to geomorphology and conservation would focus on changes in the depth of the weathered mantle (ΔS), for there might be a critical depth of rooting medium at which the population of some species might be lost. A change in this depth might have been brought about by a change in either the input or the output, one of which might then be managed to maintain or increase the depth of the mantle and so conserve the population.

Some valley-side slope systems do, at first sight, seem to have this kind of balance. The Coweeta watershed, in North Carolina, has been intensively studied by hydrologists from the Forest Service of the United States Department of Agriculture for many decades. Here and elsewhere the weathered mantle is referred to as 'saprolite' (thoroughly altered rock from which material has been dissolved, but is still *in situ*). On the divides between drainage basins at Coweeta, the saprolite is between 6 and 23 m deep. The overlying biologically active soil is about 1 m thick. Velbel (1987) estimated that the rate of production of saprolite (I in the continuity equation) at Coweeta was 4 mm 1000 yr^{-1}, and noted that this appeared to be equal to the rate of removal of sediment (O in the equation), suggesting a system in equilibrium. Most geomorphologists of Velbel's time would have admitted that such a claim was based on flimsy evidence, but it has been surprisingly corroborated by a recent study of rock weathering in a similar, nearby environment. Dating the saprolite with cosmogenic ^{10}Be not only lent support to Velbel's claim of balance but also confirmed his estimate of the rate of saprolite production (Pavich 1989). There also appears to be a balance of input and output in the mountains on the US Pacific coast, where the rate of accumulation of sediment in hollows or coves (see below) seems to equal the rate of output of sediment in the streams (Reneau and Dietrich 1991).

One consequence of the continuity model, and one that would be seen as valuable by those who held to equilibrium thinking in conservation, would be that erosion continually renewed soils on steep slopes, maintaining the supply of nutrients, and Pavich also found corroboration for this notion in the Appalachians. An explanation of the balance, if it existed, would be that there was homeostatic, positive feedback in the slope/soil system. If soil or saprolite were to be removed, solid rock would be brought closer to the surface, and in that way would be subject to faster weathering, replacing the saprolite that had been lost. This is the basis of a well-worn model in geomorphology, now usually attributed to Ahnert (1970), who has developed the idea most thoroughly.

Despite this kind of evidence, most geomorphologists are now very circumspect about claims of equilibrium in slope systems, if they do not reject them entirely (Kennedy 1994). They argue that the idea is useful only if its limitations are realised, and that they are severe. The first and least damaging limitation is the accuracy of the measurements on which the claims are based, and new dating methods (like the ^{10}Be method) do not solve all the problems. They are particularly acute when it comes to estimating the rate of mantle production at the scale of ecosystem change and management intervention, these being far shorter than the enormous stretches of geological time over which Pavich was dating his Appalachian saprolite. The intensive measurements associated with the classic Hubbard Brook experiments were unable to answer the question of whether weathering could keep pace with erosion and so maintain the nutrient supply in a steady-state (equilibrium) forest ecosystem (Borman et al. 1974). The issue of replacement and balance has also stalked American soil conservationists for decades. If they are to advise on acceptable targets for rates of erosion, they must know how fast the soil is being replaced, and yet they are still locked in argument over the issue (Schertz 1983; Johnson 1987). Output (mostly as sediment in streams) is generally easier to measure than input, but the amount of sediment trapped in reservoirs, on which many output estimates depend, may have been overestimated. Sediment produced by wave erosion on the banks of the reservoirs and then added to the total catch of sediment may have been underestimated. It can account for up to 85% of the sediment in some reservoirs, and is commonly of about the same magnitude as the input from streams (Lloyd, Bishop and Reinfelds 1998).

A second, and more important, problem with the concept is the episodicity of sediment-moving events. The output of sediment in a stream is so erratic that it is impossible to know if the results from a few decades of measurement can be taken as a long-term mean. In this light, the balances of input and output found in the studies mentioned above are likely to be no more than fortuitous. There is no denying that, within the period of measurement at Coweeta, most of the sediment was taken in rare high-magnitude/low-frequency events (Velbel 1987). And most geomorphologists would also agree that most of the output over longer periods had probably occurred in still rarer, still larger events, the biggest of which has probably not occurred at Coweeta since measurements began. The occasional rare event can leave an impression on the landscape that lasts for centuries. Even in the gentle Pennine landscapes of northern England, intense cloudbursts, with return periods of anything up to 1000 years, do more damage than countless small showers (Carling 1986).

It is not even that there are differences between gentle and intense events; there are differences along a wide spectrum of intensity. The landslides on the west Dorset coast illustrate this issue. The seaward end of the landslide on Stonebarrow Hill, where the sea undercuts it, experiences almost daily rockfalls. It is fed sediment by mudflows from above, and these are active in most winters. Above them, the upper landslide experiences movement on a much more catastrophic scale, about once in 50 years, the last time being when an observation post slid some 50 m down the hill during World War II, with soldiers in it (all of whom survived). This lifetime-scale landslide is in turn embedded in a set of human-evolution scale landslides that have not moved since the last glaciation. Finally the Quaternary landslides are cutting into a soil that has been essentially unchanged since the mid-Tertiary; a time before humans evolved (Brunsden and Jones 1980). Who is to know if the measuring period is long enough to record the largest or most 'formative' events, or, for that matter, if the frequency of large events is changing or has changed?

This case study leads the argument against equilibrium to a third, and closely related problem: the inheritance of landforms and soils from periods when the climate was different (climatic fluctuation being change at a lower frequency than floods and droughts). The sites of the West Dorset landslides have experienced conditions ranging from near-tropical in parts of the Tertiary, when the upland soils were formed, through the periglacial climate of the first landslides, to the temperate conditions of today. The deep saprolite of the central Appalachians, like the Tertiary soils of western England, was produced in an ancient period when the climate was warmer, and when therefore the rate of weathering was faster than the rate of erosion. Pavich found that the saprolite had remained in place, between production and removal, for about 1 million years, and it is well known that climate has changed hugely during that interval. If the saprolite is an inheritance from a warmer climate, surviving in a cooler one, then it is probably not being formed as quickly as it once was, and would not be replaced if it were to be suddenly stripped away. In Britain, there are many examples of deeply weathered saprolite that is inherited from as far back as the early Tertiary (Fitzpatrick 1963), and many more, if not most, of our soils bear clear signs of inheritance from the Pleistocene (Catt 1991). None would be replaced if it were to be eroded, at least not in its present form or not for many millennia. The other implication of climatic change is that climates have been so variable that equilibrium in large or complex systems never has time to develop (Glenn-Lewin, Peet and Veblen 1992).

An elementary knowledge of earth history suggests that rates of input and output are likely to have varied a great deal, and rarely in concert. Erosion is likely always quicker than replacement. There must have been many short, sharp periods of erosion, and much fewer, but longer periods of replacement. In this kind of rhythm, balance is never likely. Erhart (1955) characterised the alternation as being between periods of '*biostasie*' or landscape stability and periods of '*rhéxtasie*' when erosion was accelerated, and Ibañez, Ballestra and Alvarez (1990) revived these terms in their work in southern Spain. They found much the same general pattern as Follmer (1982) working in Illinois. In both places there were periods in the Late Pleistocene when erosion was much more active than it is now, and these periods had very different patterns of ecological niches in much more irregular landscapes.

In most landscapes, facets in age-old stability are juxtaposed with ones undergoing rapid erosion. In Britain, the ancient, deeply altered soils of the pre- or inter-glacial age on the upper slopes of the chalk Downs contrast with the young, actively eroding slopes and soils just beneath them, as at Ivinghoe Beacon in Bedfordshire and Butser Hill in Sussex (Avery *et al.* 1959). Within short distances, some facets are in near-balance, others distinctly out of balance, some in short-term balance, some in long. In some, catastrophes create landscapes or facets that may take millennia to return to balance and some may be so commonly affected by high-energy events that they never attain balance (Renwick 1992). If 'equilibrium' is applied as a concept to any of these landscapes, it is being used at such a level of generality that it loses any real value.

On top of the problems with the concept of equilibrium in geomorphological theory are the problematic relations between landform stability/instability and the maintenance of species. Examples from sand dune habitats are given by Arens, Jungerius and van der Meulen (2001). There are similar cases on valley-side slopes. In the Howgill Fells of north-western England, periodic changes have important ecological repercussions. Steep gullies, perhaps initiated in some more severe climate of the past, feed sediment into the main stream (Harvey 1992). Although the forces that initiated them may no longer be in operation, some of the gullies and their feeder slopes are active at any one time, others are not. Harvey noted that reactivation occurred only when there was a coincidence between changes on a slope (initiated, for example, by the destruction of vegetation) and changes in the activity of the stream at its base. The gullies support a range of plant and animal communities at various stages in the moorland successional cycle: the most active support higher plants and lichens associated with nearly bare ground; less active gully-basins have been stable for long enough to support a scrubby woodland flora. Geomorphological change therefore renews each community, but to an erratic and chancy rhythm. A change in land use, say removal of sheep grazing, or a change in climate, producing, say, more runoff, faster growth in vegetation or more stream activity, might alter the balance between activity and inactivity, either on the slopes or in the streams. Habitats might be lost either by a deceleration of geomorphological activity or by an acceleration. A management policy based on some theoretical notion of 'equilibrium' might be quite as dangerous if were to promote the wrong kind of equilibrium.

A final major problem with the concept of equilibrium is that few landscapes are undisturbed. Conservationists have accepted that they seldom, if ever, deal with pristine ecosystems; they may not realise that they also seldom deal with a pristine geomorphology or soil pattern on which their ecosystems depend. In the north-western European wildwood, erosion rates were so low that late-Pleistocene palaeosols can be preserved on slopes of up to 5° (Imeson and Jungerius 1974). Clearance, ploughing, burning and grazing all create very considerably accelerated erosion. Sediment discharge is five times greater from agricultural land than from forest in the Luxembourg Ardennes (Imeson 1985). Slippage scars, with their ruderal plant communities, became more common on cleared land, whether in the English Pennines (Evans 1993), the hills of West Virginia (Jacobson *et al.* 1993) or the North Island of New Zealand (Selby 1967). The peat, which covered many upland slopes after they were cleared of trees, is much less stable than the older surfaces it blanketed, and peat slides and their habitats are now quite a common feature of the uplands (Stevenson, Jones and Batarbee 1990). On many sites, the faster erosion has left thinner soils and gullies on steeper slopes, as in the New Forest (Tuckfield 1986) or the North Downs in Kent (Kerney, Brown and Chandler 1964). Downslope, the pulse of erosion released by all this induced erosion has built up deep colluvium (defined below) at the base of slopes, again as in the New Forest and Kent. The period of high erosion rates may not have peaked until the early twentieth century, let alone have established a new 'equilibrium', if we can generalise from the sedimentary record of Slapton Ley in Devon (Heathwaite 1990).

Thus, if we still clung to equilibrium as a concept, we would be hard put to establish what it meant in practice.

It is for reasons like these that geomorphologists and ecologists have put the equilibrium concept to one side. It also provides too vague and distant a target for conservationists. It is now believed to be more important, as Pickett and his colleagues maintained, to conserve process.

GEOMORPHOLOGICAL AND ECOLOGICAL PROCESSES ON VALLEY-SIDE SLOPES

Because of the great variety in landscapes, this section can only be a generalised description of the ecology of valley-side facets. The implications for conservation are discussed in the final section. The pattern of valley-side slopes is important to ecology and conservation at two scales. At the wider scale is the network of higher-order valleys, say up to fourth or fifth order (valleys/streams with more tributaries than those in Figure 3.1), which has two kinds of value. First, large river systems create habitats that do not occur in first-order valleys. Second, larger animals, as in the systems discussed by Belsky in the savannas, and described briefly above, need a range of habitats that is only provided in landscapes at the wider scale. However, this scale of pattern will get little attention here, mainly because most of these habitats are discussed in other chapters in this book, such as those on floodplains (Hughes and Rood, 2001), river channels (Clifford, 2001) and wetlands (Thompson and Finlayson, 2001).

First-order valleys, to which most of the rest of the discussion is devoted, are still a hugely important habitat. They cover at least 60% of most landscapes and, moreover, have facets that also occur over a high proportion of the remainder of higher-order valley systems. Within first-order valleys, there is a variable mix of three groups of vital ecological factors: stability, moisture and nutrient supply. A simplification, but not one that loses much information, is to see these factors in relation to the types of slope shown on Figure 3.1. Figure 3.5(a) shows the distribution of these slope types in a Puerto Rican wet forest.

UPPER SLOPES

Upper slopes are what Hack and Goodlet (1960) called 'noses' and what have been termed by others 'interfluves', 'spurs' or 'ridges'. They are almost always the most stable sites in a fluvial landscape, as in the Puerto Rican forest of Figure 3.5(a) where biotic processes are much more important than physical ones in controlling fertility (Silver *et al.* 1994). Within these facets Kohn and Walsh's 'truly island effect' is likely to be the main determinant of species diversity. In the Puerto Rican forest of Figure 3.5(a) the upper slopes occupy

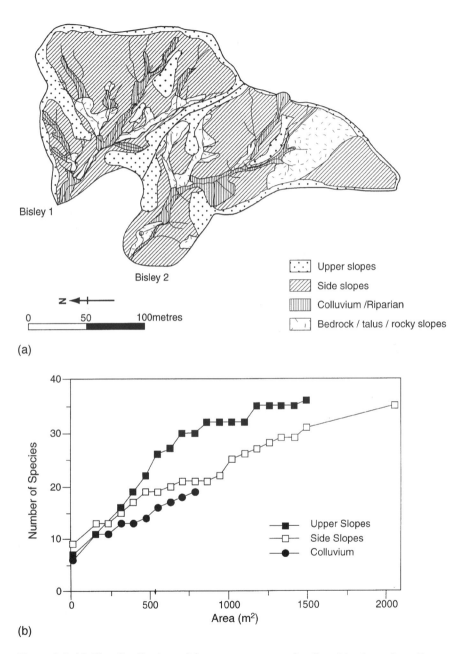

(a)

(b)

Figure 3.5 (a) The distribution of facets on systems of valley-side slopes in a Puerto Rican rainforest; (b) plant species-area curves for the different facets in (a) (both after Scatena and Lugo, 1995; with permission from Elsevier Science).

about 17% of the landscape. In gentler terrain they may occupy much more, up to 90%. Runoff is rare here, and when it does occur, it is geomorphologically ineffectual, because there is little area to feed it and because it is dispersed over convex slopes. Slope angles are not steep enough for landslides, nor is there usually enough moisture to lubricate them. Slow downslope movement of the soil mantle allows deep soils or saprolites to build up (the 6–10 m deep saprolite of central Appalachian divides, mentioned above, is similar to depths found in Puerto Rico (Simon, Larson and Hupp 1990)). The stable soils, with low turnover rates, are generally thoroughly leached of nutrients. In the Puerto Rican forest of Figure 3.5(a), the soils on the upper slopes have the most organic matter, but the least phosphorus and potassium and the lowest pH of soils on all the facets.

Ecosystems respond to all these characteristics. In Puerto Rico, the highest biomass and stem density and the most diverse stands of trees occur on the upper slopes (Figure 3.5(b); Scatena and Lugo 1995), although total plant diversity appears to occur lower down the system (see below). In the Appalachians upper slopes support tree-species mixtures that differ from those on the lower slopes (Hack and Goodlet 1960; Olson and Hupp 1986; Dacy, Phillips and Monk 1987). These are what I called 'K landscapes' in an earlier book in this series: areas that support long-lived plants (Warren 1993). 'K facets' is a better term. In the Puerto Rican forest, the tree vegetation of the upper slopes is dominated by the *tabonuco* (*Dacryodes excelsa*), with a strongly skewed age distribution and some very old trees: 4% of stems are older than 125 years. There is none of this age on lower facets. The old trees are the few that survive the hurricanes, which return every 50 to 60 years (Scatena and Lugo 1995), and reach their highest windspeeds and turbulences on these facets (Boose, Foster and Fluet 1994). In dry, open environments, these windy facets suffer the most extensive and damaging fires (Swanson, Wondzell and Grant 1992). Lower sites have fewer fires in many environments (Runkle 1985).

SIDE SLOPES

Geomorphological processes accelerate as surface angles steepen towards what Hack and Goodlet (1960) called the 'side' slopes. These occupy about 65% of the basins of Figure 3.5(a); in gentler landscapes they occupy much less, perhaps as little as 10%. As the drainage area upslope of a point increases and yields more runoff, and as the slopes steepen, the power of water running over the surface is also increased. Small gullies or rills may appear in intense showers. Soils are thinner than on the ridges, and therefore offer plants only shallow rooting. The shallow soils also hold less water, but because they receive water enriched by having been leached through the soils upslope (Burt, Crabtree and Fielder 1984), and because of the faster input of weathering and output of erosion, they have more available nutrients.

Side-slopes are 'r' facets: ones that favour short-lived species. Physical processes become more important in controlling fertility (Silver *et al.* 1994). Hurricanes may blow more strongly on the upper facets, but the poorly anchored trees on the unstable side slopes in Puerto Rico suffer more, but also recover more quickly (Scatena and Lugo 1995). Many species only germinate in 'autocultivated' soils like these. On the upper slopes autocultivation can only be provided by tree-throw (Armson and Fessenden 1973), but on the side slopes and in the coves it is provided by geomorphological processes as well.

A distinction can be made between two types of side slope: those that have large landslides and those that have only small 'scars'. Both kinds of 'mass movement' are more common on side than upper slopes, because not only is there greater relief for the operation of gravity, but also greater drainage area to provide lubrication. In the forest in Figure 3.5(a), which is developed over hard rock, scars are unknown on the ridges, but occur at $0.05\,\text{ha}^{-1}\,\text{yr}^{-1}$ on the side slopes and on the colluvium in the small valleys (discussed below). This kind of side slope supports younger, less diverse stands of tree than do the upper slopes, their youth being a consequence of instability (Figure 3.5(b); Scatena and Lugo 1995). Where the rock is soft, as in other parts of Puerto Rico and in the central Appalachians, slope failure by landsliding may be extensive. Some of the largest disturbances in the Borneo rainforests are associated with these large landslides on side slopes (Spencer *et al.* 1990). The distinctive habitats of landslides are described below.

'COVES' OR 'HOLLOWS' AND OTHER COLLUVIAL SLOPES

In most landscapes coves hold a much greater share of the total species diversity than their small area suggests. This is true in both the Appalachians (Denslow 1985; Dacy, Phillips and Monk 1987) and in Puerto Rico (Simon, Larson and Hupp 1990). Coves are the concave bowls at the heads of valleys, which collect sediment washed or slid down from above. Although there are differences, they can be grouped with other deposits of 'colluvium' at the base of the slope, some of which parallel the axis of the valley. Colluvium is debris from upslope in storage at the base. In general it has a sharp boundary with the side slopes above and the stream below, as in the Luxembourg Ardennes, the English chalklands, the Brazilian savanna and the central Appalachians (Hack and Goodlet 1960; Imeson and Jungerius 1974; Anderson and Furley 1975; Furley 1997). Colluvium occupies only about 10% of the Puerto Rican basins of Figure 3.5(a). It is said to occupy higher proportions of the landscape in steeper terrain (Dietrich, Wilson and Renau 1986), but is never a major element in the landscape. Colluvial soils are deep, and being fed from upslope, are also the most nutrient-rich of the sequence. Colluvium, especially in coves, is usually wet, and it may be riddled with subsurface 'pipes' carrying running water. Shallow groundwater may come to the surface of the colluvium, especially after rainstorms.

Colluvium is a consequence of the episodicity of the climatic and tectonic environment, accumulating in periods of relative inactivity and being cleared out when activity increases, usually suddenly. Coves, which concentrate water and sediment at the heads of valleys, experience the most frequent movements (Simon, Larson and Hupp 1990) and mobility is one of their most important ecological characteristics. The range of frequencies of the 'flushing' cycles in coves, though everywhere shorter than on other slopes, varies enormously. In the Pacific North-West of the USA, coves are cleared out about once in 60–70 years (Orme 1990), and in small coves in Marin County, just north of San Francisco, the cycle varies from 33 and 1950 years (Ellen, Cannon and Renau 1988). Some cove sediments are much older, dating from periods when the climate was different. Some of the deposits at the base of coves in the Californian Coastal Ranges are as old as 29 k yr. Many of these coves seem to have been completely cleared 9–14 k yr BP in a period of greater activity (Reneau *et al.* 1990).

The flushing of colluvium can only occur when two geomorphological processes reach critical points and when there is also a severe enough storm (as in Harvey's example given above). The two geomorphological processes are the slow build-up of debris to some critical mass, and its 'maturation' or weathering to a point where it loses enough cohesion. Both of these processes are probably faster in hotter, wetter climates and in steeper terrains. A storm must then raise soil pore-water pressure to a critical level (Crozier, Vaughan and Tippett 1990). The three processes may coincide during hurricanes in Puerto Rico and the Appalachians (Williams and Guy 1973), or the El Niño events in California, but only if these external events are far enough apart to allow for enough build-up and maturation of the sediment. Because coves concentrate debris and water, they are much more likely to fail than other colluvial deposits. In the Puerto Rican mountainous rainforest, landslides, most of them in the coves, account for a very significant proportion of disturbances to the forest canopy. They disturb between 0.3% and 0.08% of the landscape per century, depending on lithology (Guariguata 1990).

Thus, in contrast to the direct effects of hurricanes on the vegetation of upper slopes and their damaging effects on the shallow-rooted trees of side slopes, their effect in hollows is often to remove the substrate itself *en masse*. These catastrophic disturbances have complex rhythms that are not in synchrony across a landscape, because of the complexity of their controls. Only a few hollows are flushed in each storm, and most landscapes have a range of coves at varying stages of vegetational succession. Coves are much the most vulnerable parts of the landscape when forests are cleared, as aerial photographs of cleared and devastated landscapes show very clearly, whether they are in New Zealand (Crozier, Vaughan and Tippett 1990), Tanzania (Temple 1972) or the Polish Carpathian foothills (Starkel 1976).

This high level of disturbance, in coves more than on other colluvial deposits, selects for fast-growing, short-lived species, which reproduce early in their life cycles. These are therefore another example of 'r' facets. The most disturbed sites support only shrubs and young trees in the Appalachians (Runkle 1985). In Puerto Rico, mountain palms and tree ferns are distinctive, pioneer occupants. The palm is so characteristic that it can be used as an indicator of landsliding (Simon, Larson and Hupp 1990). The rate of disturbance, perhaps near the optimum, 'intermediate' level for the highest species diversity (Kolasa and Rollo 1991) is the probable reason for the species richness of coves. Another reason is the extremely variable spatial pattern within these sites (Silver et al. 1994). The lower slides, where the thoroughly mixed debris accumulates, are very rich in nutrients and organic matter and re-colonise quickly. The bare upper parts of slides, where the bedrock may be exposed, are re-colonised more slowly, and by different species (Guarigata 1990). Between these two types of site there may be very wet sites, even ponds, where drainage is impeded. All these habitats are temporary, being re-created or rearranged when the landslide moves. Their lifetimes depend on both the frequency of the external disturbances (such as hurricanes) and on the build-up and maturation of the sediment (Simon, Larson and Hupp 1990; Jacobson et al. 1993). A third reason for species diversity is the wetness of these sites compared to others in the landscape. This probably explains the richer flora of the wet colluvial sites in the low-relief New Forest heaths in southern England (Gurnell 1981). A final reason for biological diversity may be the concentration of nutrients.

In landscapes in which fluvial conditions have been re-established after glaciation (for example, in East Anglia or North-East England) or the invasion of sand dunes (as in the High Plains of the USA), rivers quickly re-cut narrow valleys. Coves rapidly re-achieve the proportion they had in unmodified landscapes. Side slopes evolve more slowly, and upper slopes so slowly that non-fluvial forms are preserved there for many thousands of years (Clayton 1997).

CONSERVING VALLEY-SIDE HABITATS

If the slogan for the new conservation is "concern of process" (Pickett, Parker and Fiedler 1992), and if geomorphological processes are as vital to ecological processes, as they have been shown here to be, then conservation must include concern for geomorphological processes. Including geomorphology in management, takes conservation beyond the simple manipulation of succession, which is seen to be its main focus by many conservationists (for example,

Glenn-Lewin and van der Maarel 1992). Succession may well have to be manipulated to conserve diversity, but a recognition that it occurs in nature at rhythms dictated by geomorphological processes (among others) should minimise the need for manipulation or at least allow it to be more effective when it is necessary. This, then, is another argument in the armoury of those who favour naturalness in conservation (see above).

Knowledge of geomorphological patterns can first be used in the selection of management units. It has been claimed that diversity can be more efficiently maximised by selection on the basis of geomorphological units than through surveys of species (Nichols, Killingbeck and August 1998). The claim is corroborated by research in eastern Canadian woodlands (Meilleur, Bouchard and Bergeron 1994), in woodlands in New England (Strahler 1978) and Virgina (Olson and Hupp 1986). Beyond pattern, selection needs to be concerned with the integrity of the processes that maintain diversity. Plant and animal species have adapted to and therefore now need the spatial variety and scale that geomorphological processes provide. They also need temporal variety, and geomorphology plays a part here as well. The most important driving forces for change may be climatic in origin, but those that concern the substrate are mediated or buffered by geomorphological processes. The periodic emptying of coves, described above, is the best example in the valley-side slope context. The maintenance of the essential spatio-temporal contexts of ecological processes, the prime concern in the selection of management units, therefore requires attention to geomorphology.

Essential or vulnerable processes need the closest attention. There are two sets of these: the flow of sediment and water, which carry nutrients from upper slopes, through side slopes to colluvial slopes and coves; and the portfolio of disturbance regimes. It is much easier to maintain the flows in whole basins than in their separate parts. Hence the familiar call for 'basin management', a better-known strategy in relation to wetland management (Thompson and Finlayson, this volume), but no less important for conservation on valley-side slopes. If an upper slope is excluded from control, the ecology of a lower facet may not function as before; flows may be interrupted or contaminated.

Drainage basins, or better a series of basins of different size and lithology, contain a range of disturbance regimes and this range allows for the maintenance of a series of different plant and animal communities, at different stages in the successional cycle. If only small basins are managed, no rare river cliff habitats may be conserved; if upper slopes are excluded there may be no very old trees (like the ancient *tabonucos* in Puerto Rico), or distinctive upper-slope habitats with their deep, leached soils. In moderate-energy environments, upper-slope, stable sites are the most likely to be excluded from conservation management, because their gentle slopes are the most attractive for development. In both the Lathkildale NNR in the Peak District of Derbyshire and the Badlands National Monument in South Dakota, agricultural fields cover the upper slopes, leaving only the steep slopes for conservation. In the Puerto Rican forest of Figure 3.5(a), the upper slopes were the most disturbed for timber harvesting before the area was declared a reserve (Scatena and Lugo 1995). The upper slopes may not have the greatest diversity, but they may be needed in a reserve because they may take hundreds of years to re-create. If one or more facet of the 'natural' landscape, like an upper slope, is excluded from a conservation unit, there will almost certainly be pressure to re-create its communities on landscape facets that are unsuitable. Finally, these processes, the geomorphological and the ecological together, create a total mosaic in the basin context. The Bay Checkerspot butterfly and East African savanna ungulates (and probably even *Homo sapiens sapiens*), mentioned above, all need the mosaic of open and closed sites, cool and dry slopes at the scale that valley-side ecosystems provide.

Once selected, the management of the conservation unit must also account for the flows, the regimes and the patterning that are the geomorphological birthrights of the ecosystem. Anything that interrupts or diverts a flow of water, sediment or nutrients needs to be carefully considered. Dams, roads, deep foundations, diversions of drainage or wells might all do damage. An otherwise dry upper slope may become waterlogged, or a valuable cove desiccated. The most sensitive facets are the upper slopes, for the soil on many has taken millennia to develop, and would take more millennia to re-create. This idea was developed in relation to the Abernethy reserve in Inverness-shire, in an earlier book in this series (Warren 1993). In coves and on side slopes, an occasional landslide should be considered as an integral part of the disturbance regime. It is likely to be vital to the maintenance of the full range of species, by refreshing rare habitats, such as the flooded reverse slopes of landslides, the bare ground on their upslope sides, or wet, rich, periodically renewed habitats. The mosaic of open and closed habitats is probably at its most ecologically effective when it is at the scale of the slope units described above. If artificial disturbance is necessary, it is best at the scale of the valley framework and is best located where disturbance has most commonly occurred previously, as on side slopes and in coves. In short: conservation is most effective when it is fitted into the landscape.

REFERENCES

Ahnert F (1970) Functional relationships between denudation, relief and uplift in large mid-latitude drainage basins. *American Journal of Science* **268**, 243–63.

Anderson JE (1991) A conceptual framework for evaluating and quantifying naturalness. *Conservation Biology* **5**, 347–52.

Anderson KE & Furley PA (1975) An assessment of the relationships between the surface properties of chalk soils and slope form using principal components analysis. *Journal of Soil Science* **26**,130–43.

Appleton J (1990) *The symbolism of habitat: an interpretation of landscape in the arts.* Seattle, University of Washington Press.

Arens SM, Jungerius PD & van der Meulen F (2001) Coastal dunes. In: A Warren & JR French (eds) *Habitat conservation: managing the physical environment.* Chichester, John Wiley, 229–72.

Armson KA & Fessenden RJ (1973) Forest windthrows and their influence on soil morphology. *Proceedings of the Soil Science Society of America* **37**, 781–3.

Avery BW, Stephen I, Brown G & Yaalon DH (1959) The origin and development of brown earths on clay-with-flints and Coombe deposits. *Journal of Soil Science* **10**, 177–95.

Bayfield N (2001) Mountain resources and conservation. In: A Warren & JR French (eds) *Habitat conservation: managing the physical environment.* Chichester, John Wiley, 7–38.

Belsky AJ (1995) Spatial and temporal landscape patterns in arid and semi-arid savannas. In: L Hansson, L Fahrig & G Merriam (eds) *Mosaic landscapes and ecological processes.* London, Chapman & Hall, 31–56.

Boose ER, Foster DR & Fluet M (1994) Hurricane impacts to tropical and temperate forest landscapes. *Ecological Monographs* **64**, 396–400.

Bormann FH & Likens GE (1979) *Pattern and process in a forested ecosystem.* New York, Springer-Verlag.

Bormann FH, Likens GE, Siccama JG, Pierce RS & Eaton JS (1974) The export of nutrients and recovery of stable conditions following deforestation at Hubbard Brook. *Ecological Monographs* **44**, 255–77.

Botkin DB (1990) *Discordant harmonies: a new ecology for the 20th century.* New York, Oxford University Press.

Brunsden D & Jones DKC (1980) Relative time scales and formative events in coastal landslide systems. *Zeitschrift für Geomorphologie, Supplement Band* **34**, 1–9.

Burnett MR, August PV, Brown JH & Killingbeck KT (1998) The influence of geomorphological heterogeneity on biodiversity. I: A patch-scale perspective. *Conservation Biology* **12**, 363–70.

Burt TP, Crabtree RW & Fielder NA (1984) Patterns of hillslope solutional denudation in relation to the spatial distribution of soil moisture and soil chemistry over a hillslope hollow and spur. In: TP Burt & DE Walling (eds) *Catchment experiments in fluvial geomorphology.* Norwich, GeoBooks, 431–45.

Carling PA (1986) The Noon Hill flash floods, July 17th 1983. Hydrological and geomorphological aspects of a major formative event in an upland landscape. *Transactions of the Institute of British Geographers* NS **11**, 105–18.

Catt JA (1991) Soils as indicators of Quaternary climatic change in mid-latitude regions. *Geoderma* **51**, 167–87.

Chorley RJ & Kennedy BA (1971) *Physical geography: a systems approach.* London, Prentice Hall.

Clayton KM (1997) The rate of denudation of some British lowland landscapes. *Earth Surface Processes & Landforms* **22**, 721–32.

Clifford NJ (2001) Conservation and the river channel enviornment. In: A Warren & JR French (eds) *Habitat conservation: managing the physical environment.* Chichester, John Wiley, 67–104.

Crozier MJ, Vaughn EE & Tippett JM (1990) Relative instability of colluvium-filled bedrock depressions. *Earth Surface Processes & Landforms* **15**, 329–39.

Dacy FP Jr, Phillips DL & Monk CD (1987) Forest communities and patterns. In: WT Swank & DA Crossley Jr (eds) *Forest hydrology and ecology at Coweeta, Ecological Studies 66.* New York, Springer-Verlag, 141–50.

Denslow JS (1985) Disturbance-mediated coexistence of species. In: STA Pickett & PS White (eds) *The ecology of natural disturbance and patch dynamics*. London, Academic Press, 307–23.

Dietrich WE, Wilson CJ & Renau SL (1986) Hollows, colluvium and landslides in soil-mantled landscapes. In: AD Abrahams (ed.) *Hillslope processes*. London, Allen & Unwin, 361–88.

Drury WH & Nisbet ICT (1971) Interrelationships between developmental models in geomorphology, plant ecology and animal ecology. *General Systems* 16, 57–68.

Ellen SD, Cannon SH & Reneau SL (1988) Distribution of debris flows in Marin County. In: SD Ellen & GF Wieczorek (eds) *Landslides, floods and marine effects of the storm of January 3–5 1982, in the San Francisco Bay region, California, United States Geological Survey, Professional Paper* **1434**, 63–112.

Ellis JE & Swift DM (1988) Stability of African pastoral ecosystems: alternate paradigms and implications for development. *Journal of Range Management* 41, 450–9.

Erhart H (1955) 'Biostasie' et 'rhéxistasie', esquisse d'une théorie sur le rôle de la pédogenèse en tant que phénomène géologique. *Comptes rendus de l'Académie des Sciences, Paris* D **241**, 1218–20.

Evans R (1993) Sensitivity of the British landscape to erosion. In: DSG Thomas & RJ Allison (eds) *Landscape sensitivity*. Chichester, John Wiley, 189–210.

Fitzpatrick EA (1963) Deeply weathered rock in Scotland: its occurrence, age and contribution to soils. *Journal of Soil Science* 14, 33–3.

Foley R (1984) Putting people into perspective: an introduction to community evolution and ecology. In: R Foley (ed.) *Hominid evolution and community ecology*. London, Academic Press, 1–24.

Follmer LR (1982) The geomorphology of the Sangamon surface: its spatial and temporal attributes. In: CE Thorn (ed.) *Space and time in geomorphology*. Hemel Hempstead, Allen & Unwin, 117–46.

Forman RTT (1995) *Land mosaics: the ecology of landscapes and regions*. Cambridge, Cambridge University Press.

Furley PA (1996) The influence of slope on the nature and distribution of soil and plant communities in the Central Brazilian cerrado. In: UG Anderson & S. M. Brooks (eds) *Advances in hillslope processes*. Chichester, John Wiley, 327–46.

Furley PA (1997) Plant ecology, soil environment and dynamic change in tropical savannas. *Progress in Physical Geography* 21, 257–84.

Glenn-Lewin DC & van der Maarel E (1992) Patterns and processes of vegetation dynamics. In: DC Glenn-Lewin, RK Peet & TT Veblen (eds) *Plant succession: theory and prediction*. London, Chapman & Hall, 11–59.

Glenn-Lewin DC, Peet RK & Veblen TT (1992) Prologue. In: DC Glenn-Lewin, RK Peet & TT Veblen (eds) *Plant succession: theory and prediction*. London, Chapman & Hall, 1–10.

Guariguata MR (1990) Landscape disturbances and forest regeneration in the upper Luquillo Mountains of Puerto Rico. *Journal of Ecology* 78, 814–32.

Gurnell AM (1981) Heathland vegetation, soil moisture and dynamic contributing area. *Earth Surface Processes & Landforms* 6, 553–7.

Hack JT & Goodlet JC (1960) *Geomorphology and forest ecology of a mountain region in the central Appalachians, United States Geological Survey, Professional Paper* **347**.

Harrison CM and Warren A (1970) Conservation, stability and management. *Area* 2, 26–32.

Harvey AM (1992) Process interactions, temporal scales and the development of hillslope gully systems: Howgill Fells, northwest England. *Geomorphology* 5, 323–44.

Heathwaite AL (1990) Catchment controls on the recent sediment history of Slapton

Ley, south-west England. In: JB Thornes (ed.) *Vegetation and erosion*. London, Academic Press, 241–60.

Higgs AJ & Usher M.B (1980) Should nature reserves be large or small? *Nature* **285**, 568–9.

Hughes FMR & Rood SB (2001) Floodplains. In: A Warren & JR French (eds) *Habitat conservation: managing the physical environment*. Chichester, John Wiley, 105–121.

Huston M (1994) *Biological diversity: the co-existence of species in changing landscapes*. Cambridge, Cambridge University Press.

Ibañez JJ, Ballestra RJ & Alvarez AG (1990) Soil landscapes and drainage basins in Mediterranean mountain areas. *Catena* **17**, 573–83.

Imeson AC (1985) Geomorphological processes, soil structure and ecology. In: A Pitty (ed.) *Themes in geomorphology*. Beckenham, Croom Helm, 72–84.

Imeson AC & Jungerius PD (1974) Landscape stability in the Luxembourg Ardennes as exemplified by hydrological and micropedological investigations of a catena in an experimental watershed. *Catena* **1**, 273–95.

Ives JD & Messerli B (1989) *The Himalayan dilemma: reconciling development and conservation*. London, Routledge.

Ives RL (1946) Desert ripples. *American Journal of Science* **244**, 492–501.

Jacobson RB, McGeehin JP, Cron ED, Carr CE, Harper JM & Howard AD (1993) Landslides triggered by the storm of November 3–5 1985, Wills Mountain Anticline, West Virginia. In: RB Jacobson (ed.) *Geomorphic studies of the storm and flood of November 3–5 1985, in the upper Potomac and Cheat River basins in West Virginia and Virginia, Bulletin* **1981**, United States Geological Survey, C1–C33.

Johnson LC (1987) Soil loss tolerance: fact or myth. *Journal of Soil & Water Conservation* **42** 155–60.

Kellman M & Tackaberry R (1997) *Tropical environments: the functioning and management of tropical ecosystems*. London, Routledge.

Kerney MP, Brown EH & Chandler RJ (1964) The late-glacial and post-glacial history of the Chalk escarpment near Brook, Kent. *Philosophical Transactions of the Royal Society of London* **B 248**, 135–204.

Kennedy BA (1994) Requiem for a dead concept. *Annals of the American Association of Geographers* **84**, 702–5.

Kohn DD & Walsh DM (1994) Plant species richness — the effect of island size and habitat diversity. *Journal of Ecology* **82**, 367–77.

Kolasa J & Rollo CD (1991) Introduction. The heterogeneity of heterogeneity: a glossary. In: J Kolasa & STA Pickett (eds) *Ecological Heterogeneity, Ecological Studies* **86**. New York, Springer-Verlag, 1–23.

Lawton JH (ed.) (1990) *Red grouse populations and moorland management, Ecological Issues* **2**. British Ecological Society/Field Studies Council, Shrewsbury.

Lloyd SD, Bishop P & Reinfelds I (1998) Shoreline erosion: a cautionary note in using small farm dams to determine catchment erosion rates. *Earth Surface Processes & Landforms* **23**, 905–12.

MacArthur RH & Wilson EO (1967) *The theory of island biogeography*. Princeton, Princeton University Press.

Meilleur A, Bouchard A & Bergeron Y (1994) The relation between geomorphology and forest community types of the Haut-Saint-Laurent, Quebec. *Vegetatio* **111**, 173–92.

Murphy DD, Freas KE & Weiss SB (1988) An environment-metapopulation approach to population viability analysis for a threatened invertebrate. *Conservation Biology* **4**, 41–51.

Nichols WF, Killingbeck KT & August PV (1998) The influence of geomorphological heterogeneity on biodiversity. II. A landscape perspective. *Conservation Biology* **12**, 371–9.

Olson CG & Hupp CR (1986) Coincidence and spatial variability of geology, soils and vegetation, Mill Run watershed, Virginia. *Earth Surface Processes & Landforms* **11**, 619–30.

Orme AR (1990) Geomorphic effects of vegetation cover and management: some time and space considerations in prediction of erosion and sediment yield. In: JB Thornes (ed.) *Vegetation and erosion*. Chichester, John Wiley, 67–85.

Pavich MJ (1989) Regolith residence time and the concept of surface age of the Piedmont. *Geomorphology* **2**, 181–96.

Pickett STA & Cadenasso ML (1995) Landscape ecology: spatial heterogeneity in ecological systems. *Science* **269**, 331–4.

Pickett STA, Parker TV & Fiedler LP (1992) The new paradigm in ecology: implications for conservation biology above the species level. In: PL Fielder & KS Jain (eds) *Conservation biology*. New York, Chapman & Hall, 65–89.

Puigdefábregas J & Sánchez G (1996) Geomorphological implications of vegetation patchiness on semi-arid slopes. In: MG Anderson & SM Brooks (eds) *Advances in hillslope processes*. Chichester, John Wiley, 1027–60.

Reneau SL & Dietrich WE (1991) Erosion rates in the southern Oregon Coast Range: evidence for an equilibrium between hillslope erosion and sediment yield. *Earth Surface Processes & Landforms* **16**, 307–22.

Reneau SL, Dietrich WE, Donanhue DJ, Jull AJT & Rubin M (1990) Late Quaternary history of colluvial deposition and erosion in hollows, central California Coastal Ranges. *Bulletin of the Geological Society of America* **102**, 969–82.

Renwick WH (1992) Equilibrium, disequilibrium, and nonequilibrium landforms in the landscape. *Geomorphology* **5**, 265–76.

Runkle J (1985) Disturbance regimes in temperate forests. In: STA Pickett & P. White (eds) *Natural disturbance and patch dynamics*. London, Academic Press, 17–33.

Samways MJ (1994) *Insect conservation biology*. London, Chapman & Hall.

Scatena FN & Lugo AE (1995) Geomorphology, disturbance, and the soil and vegetation of two subtropical wet steepland watersheds in Puerto Rico. *Geomorphology* **13**, 199–213.

Schaetzl RJ & Follner LR (1990) Longevity of tree-throw micro-topography: implications for mass-wasting. *Geomorphology* **3**, 113–23.

Schertz DL (1983) The basis of soil loss tolerance. *Journal of Soil & Water Conservation* **38**, 10–14.

Selby MJ (1967) Erosion by high-intensity rainstorms in the lower Waikato Basin. *Earth Science Journal* **1**, 153–6.

Shachak M & Brand S (1991) Relations among spatiotemporal heterogeneity, population abundance, and variability in a desert. In: J Kolasa & STA Pickett (eds) *Ecological Heterogeneity, Ecological Studies* **86**. New York, Springer-Verlag, 202–23.

Silver WL, Scatena FN, Johnson AH, Siccama TG & Sànchez MJ (1994) Nutrient availability in a montane wet tropical forest in Puerto Rico: spatial patterns and morphological considerations. *Plant & Soil* **164**, 129–45.

Simberloff D & Gotelli N (1984) Effects of insularization on plant species richness in the prairie-forest ecotone. *Biological Conservation* **29**, 27–46.

Simon A, Larson MC & Hupp CR (1990) The role of soil processes in determining mechanisms of slope failure and hillslope development in a humid tropical forest, eastern Puerto Rico. *Geomorphology* **3**, 263–86.

Sollins P (1998) Factors influencing species composition in tropical lowland rain forest: does soil matter? *Ecology* **79**, 23–30.

Spencer T, Douglas I, Greer T & Sinun W (1990) Vegetation and fluvial geomorphic processes in south-east Asian tropical rainforests. In: J Thornes (ed.) *Vegetation and erosion*. Chichester, John Wiley, 451–70.

Sprugel DG (1991) Disturbance equilibrium and biological diversity: what is 'natural' vegetation in a changing environment? *Biological Conservation* **58**, 1–18.

Starkel L (1976) The role of extreme (catastrophic) meteorological events in contemporary evolution of slopes. In: E Derbyshire (ed.) *Geomorphology and climate*. Chichester, John Wiley, 203–46.

Stevenson AC, Jones VJ & Batarbee RW (1990) The cause of peat erosion: a palaeolimnological approach. *New Phytologist* **114**, 727–35.

Stolt MH, Baker JC & Simpson TW (1993) Soil–landscape relationships in Virginia. I. Soil variability and parent material uniformity. *Journal of the Soil Science Society of America* **57**, 414–21.

Strahler AH (1978) Response of woody species to site factors of slope angle, rock type and topographic position in Maryland as evaluated by binary discriminant analysis. *Journal of Biogeography* **5**, 403–23.

Swanson FJ, Wondzell SM & Grant GE (1992) Landforms, disturbance and ecotones. In: AJ Hansen & F di Castri (eds) *Landscape boundaries: consequences for biotic diversity and ecological flows, Ecological Studies* **92**. Berlin, Springer-Verlag, 304–23.

Temple PH (1972) Soil and water conservation policies in the Uluguru Mountains, Tanzania. In: A Rapp, L Berry & PH Temple (eds) *Studies of soil erosion and sedimentation in Tanzania*. Geografiska Annaler **A 54**, 110–23.

Thiéry JM, D'Herbès J-M & Valentin C (1995) A model simulating the genesis of banded vegetation patterns in Niger. *Journal of Ecology* **83**, 497–507.

Thompson JR & Finlayson CM (2001) Freshwater wetlands. In: A Warren & JR French (eds) *Habitat conservation: managing the physical environment*. Chichester, John Wiley, 147–78.

Tuckfield CG (1986) A study of dells in the New Forest, Hampshire, England. *Earth Surface Processes & Landforms* **11**, 23–40.

Velbel MA (1987) Weathering and soil forming processes. In: WT Swank & DA Crossley, Jr (eds) *Forest hydrology and ecology at Coweeta, Ecological Studies* **66**. New York, Springer-Verlag, 93–102.

Warren A (1974) Managing the land. In: A Warren and FB Goldsmith (eds) *Conservation in practice*. Chichester, John Wiley, 37–56.

Warren A (1983) Conservation and the land. In: A Warren and FB Goldsmith (eds) *Conservation in perspective*. Chichester, John Wiley, 19–39.

Warren A (1993) Naturalness: a geomorphological approach. In: FB Goldsmith & A Warren (eds) *Conservation in progress*. Chichester, John Wiley, 15–24.

Warren A (1995) Changing understandings of African pastoralism and environmental paradigms. *Transactions of the Institute of British Geographers* NS **20**, 193–203.

Whitbread A & Jenman W (1995) A natural method of conserving biodiversity in Britain. *British Wildlife* **7**, 84–93.

Wiens JA (1985) Vertebrate responses to environmental patchiness in arid and semiarid ecosystems. In: STA Pickett & PS White (eds) *The ecology of natural disturbance and patch dynamics*. London, Academic Press, 169–93.

Williams GP & Guy HP (1973) *Erosional and depositional aspects of Hurricane Camille in Virginia 1969. United States Geological Survey, Professional Paper* **804**.

Wright HE Jr (1974) Landscape development, forest fires and wilderness management. *Science* **186**, 487–95.

Yair A & Shachak M (1987) Studies in watershed ecology of an arid area. In: L Berkofsky & MG Wurtele (eds) *Progress in desert research*. Totowa, Rowan & Allenheld, 145–93.

4 Conservation and the River Channel Environment

N. J. CLIFFORD
University of Nottingham, UK

INTRODUCTION

The need for a conservation perspective in river management has increased with industrialisation and economic maturity. Both in Europe and in the USA, impacts on river environments have followed a similar historical pattern, and the same pattern can be expected in other parts of the world where rapid economic development is taking place. Primary conservation goals include the maintenance of essential ecological processes and life-supporting systems; the preservation of ecological diversity; and the sustainable utilisation of species and ecosystems (IUCN 1980). Management is now required to compromise between or balance the interests of many different users, and, more basically, to ensure enough appropriate information to sustain river environments. Conservation and management are, therefore, difficult to separate, and perspectives on both of these forms of intervention in river environments are broadening in scope and complexity (for a review, see Boon 1992). Impacts may be intentional (planned) or accidental (unplanned). They may have effects which are short-term or long-term, and which are felt at all scales from a channel cross-section, through the floodplain, to the entire river catchment.

In the river environment, successful conservation and management demands the integration of viewpoints and expertise of many types of physical and social scientist. Because activities which affect the river at one scale may have effects at others, there is also a need for an integrated and multi-purpose policy at international, national, catchment and channel levels. It is vital that intervention is broadly based, and that the physical dynamics of channels are considered together with other imperatives.

Very broadly, the history of river use and associated intervention/control strategy has been characterised by the following principal themes:

- Engineering control and regulation
- Preservation or conservation of the natural system

Habitat Conservation: Managing the Physical Environment. Edited by A. Warren and J. R. French.
© 2001 John Wiley & Sons Ltd.

- Sustainability of function
- Restoration of degraded environment.

These themes have now converged, and the weighting given to any one, or any combination, is time- and place-specific. In the developed world, few environments are truly natural, and the anthropogenetic impact has rapidly increased in both scale and complexity. Similarly, true conservation is rare. In addition, with growing legislative regulation of the environment, and public concern and participant involvement, 'successful' conservation management is often defined with more regard for non-technical inputs to the management process. In this chapter, therefore, the term 'conservation management' is adopted as a better representation of the role which conservation concerns play in contemporary river-intervention works and management strategies.

This chapter aims to:

- Characterise the changing nature of the human impact on channels in time and space, and to draw attention to the importance, both scientifically and politically, of recognising geographical and historical contingency.
- Introduce those physical aspects of the channel environment that are most relevant to conservation strategies, principally through the emergent discipline of ecohydraulics and an explanation of the geomorphological characteristics and functioning of the channel system.
- Describe the current status of conservation within river rehabilitation and restoration works, and to draw attention to the wider, non-technical context within which conservation involvement in river works must now be set.

THE NATURE OF THE HUMAN IMPACT ON RIVERS AND THE CONSERVATION MANAGEMENT RESPONSE

Figure 4.1 shows three important aspects of the human impact upon rivers and their catchments: (a) the impacts have changed in kind and scale through time; (b) at any given time or place, the impact may be primarily local (within a specific reach of the channel); on the adjacent floodplain; distributed throughout one or more areas of the catchment or originate outside the catchment; and (c) taking a global perspective, the sequence of changes in human impact may be expected to repeat itself as countries develop. Impacts may vary in character, and the way in which they are mitigated or resolved depends on whether they are 'planned' or 'unplanned'.

'Planned impacts' are usually direct, quickly felt effects, such as land-use change or in-channel construction. Clear-felling of forest cover on slopes close to channel margins may, for example, increase the amount and speed of overland flow, and with this, the potential erosion and transfer of sediments

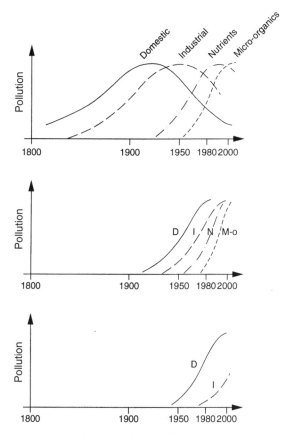

Figure 4.1. Problems of pollution associated with the changing nature of human use of rivers. Source: Maybeck, Chapman and Helmer (1980:306).

into the channel (e.g. Loughran, Campbell and Elliott 1986). Placement of bridge piers frequently modifies local flow behaviour in channels, causing scour of adjacent bed and bank sediments. Wholesale impoundment of flows by dams is the most radical form of this kind of impact (e.g. Elliott and Parker 1997; Hadley and Emmett 1998). However, although this form of intervention can exert long-term effects, most of them are, in principle, reversible, or at least amenable to mitigation. Channel works can be removed or redesigned; land-use change can be reversed. Heavy metal (e.g. Balough, Meyer and Johnson 1997; Miller, Lechler and Desilets 1998) or other mining debris pollution (Richards 1979) may pose more intractable problems, since the pollutants can be incorporated into natural valley fills and then be subject to episodic reworking and storage. This may operate over timescales of hundreds or even thousands of years (Graf 1987). Other land-use changes have more subtle effects. Such changes need to be disentangled from climatic-environmental controls on river-catchment dynamics, and this necessitates a clear conceptual model of sediment sources, storages and delivery within catchments, and an effective monitoring and modelling programme to disclose them (Walling 1983; Ongley 1992).

'Unplanned' effects are usually delayed in their onset, are often more difficult to identify, and may be cumulative. Examples are pollution from diffuse sources, as when herbicides and pesticides in run-off into rivers affect the aquatic food chain; or where wetland habitat is lost as a result of water-level alteration in the channel. Management of these problems must be sensitive to these contrasts and changes. The feasibility of an option will often depend as much on social or political/economic concerns, as on well-informed scientific advice. Figure 4.2 shows that the planned/unplanned typology of human impacts is geographically and historically contingent.

Geographical and historical contingency is not simply a problem for the development of expertise and scientific awareness; there is also a need to develop awareness with respect to political (legislative) agendas covering environmental standards. Geographically, rivers and their catchments frequently transcend national political boundaries, and within countries, the administrative units relevant to environmental legislation. A particular feature of conservation and restoration approaches to rivers relates to emerging legislation at a transnational scale and its explicit consideration of ecological functioning. The European Commission Water Framework Directive, for example, introduces catchment management throughout Europe, and emphasises the biological rather than simply the chemical quality of water, together with consideration of ecological function and structure. However, how policy is to be integrated at different levels in different member states, and the role of public involvement in ecosystem management, is less clear than the actual directives (Pollard and Huxam 1998). In reviewing US experience in the management of water resources, Howe (1990) concluded that, despite agreement on objectives, institutional decision making had been at least a decade behind enlightened opinion and the possibility of efficient implementation. However, the right goals were increasingly being pursued, if inefficiently, rather than pursuing the wrong goals efficiently!

Recognition of these geographical and historical contingencies should result in a reflexive, rather than prescriptive approach to channel management, because management always needs also to be context-dependent. In water resource conflicts, for example, the complexity and deep-seated nature of the problems are often forgotten in the rhetoric of the exchanges between environmental and other interest groups. Widening the 'ownership' of conservation problems, therefore, does not necessarily bring better solutions (Rigg 1995). Nevertheless, conservation involvement in the management of river systems

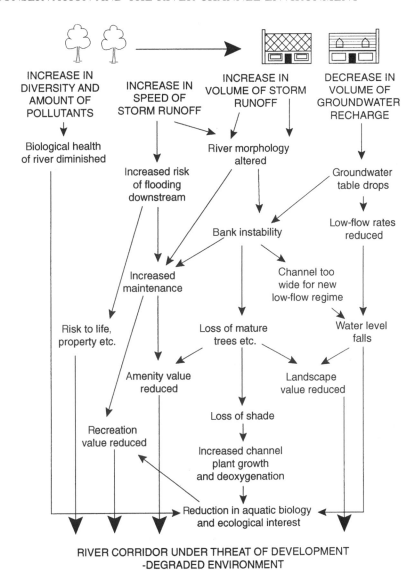

Figure 4.2. Planned and unplanned consequences of human modification of the catchment and river channel environment. Source: Gardiner (1991:51) after an original by Woolhouse (1989).

has achieved notable successes, and the complexity of the task of application should not prevent attempts at outlining coherent conservation management principles which are soundly underpinned technically. In the following sections of this chapter, some essentials of these issues are outlined.

TRADITIONAL ENGINEERING IMPACTS

Engineering intervention in rivers has traditionally been directed towards:

- Flood prevention and mitigation
- Channel stabilisation
- Flow regulation for navigation
- Water supply and quality
- Effluent disposal.

The aim of conventional river engineering is to create a channel in which a 'design discharge' can be conveyed, while maintaining the stability of the banks and controlling erosion of the bed. In practice, this involves a combination of enlarging the channel cross-section and/or making it more efficient at conveying flood water by altering its shape or boundary material. Favoured techniques include: channel deepening and/or straightening (resectioning and realignment); the provision of diversion or relief channels to pass large flows, or the creation of two-stage channels; lining of the channel to reduce channel roughness and hence, speed the passage of flows; controlling sediment transport; or promoting channel stability. Collectively, these activities are known as channelisation. Common 'conveyance' channel shapes and materials are: the trapezoidal cross-section (normally excavated with earth banks and bed and left unlined); or rectangular cross-sections built of stable materials such as concrete or brick. Lined channels are most frequently seen in urban areas, where the greatest degree of stability and maximum predictability is required. The smallest channels may simply be excavated as triangular cross-sections. The water table in adjacent floodplains may be lowered when the channels are deepened, and where drainage into channels is made more efficient through pipe networks. Flood embankments, or levees, may be created and entire rivers may be rerouted underground in culverts, although maintenance costs for these can be high.

The history and consequences of channelisation in Europe have been comprehensively reviewed by Brookes (1985). From a conservation standpoint, traditional engineering approaches have been almost universally condemned. Channelisation degrades habitat and habitat potential directly, by removing bed and bank sediments and vegetation and by altering flow dynamics. Once a habitat is destroyed, altered flow dynamics (with the engineering goal of transferring sediment and water more rapidly) are unlikely to allow gradual recolonisation and rebuilding of pre-existing flora and fauna. Indirectly, channelisation may also have more widespread negative consequences, by altering the sediment and flow dynamics upstream or downstream of the engineering works, and by altering or removing the interaction of the channel and its adjacent floodplain (see Hughes and Rood, 2001).

ENGINEERING OPTIONS FOR CONSERVATION MANAGEMENT

Alternatives to channelisation, or its mitigation are now incorporated into many manuals aimed at professionals (e.g. Gardiner 1991; Rosgen 1996; RRC 1999) or at those with a broader conservation interest (Boon, Callow and Petts 1992; RSPB 1994; Brookes and Shields, 1996; Petts and Calow 1996). Conservation management options may be directed towards structural and non-structural works.

Structural work involves the 'softening' of 'hard' bank protection and river stabilisation, and the accommodation of floods by enlarging the channel using a two-stage design. An excellent source of reference for the design, implementation and appraisal of this kind of measure has recently been produced by the River Restoration Centre (RRC 1999). 'Soft' engineering seeks to enhance the aesthetic value of the river margin, and to accompany this by an increased ecological potential. Uniform concrete revetments might be partially replaced with pockets which allow for some vegetation regrowth; steel sheet pilings might be replaced with wooden boards backfilled with natural sediments and planted with reeds; and, where the risk of erosion is low, channels might be lined with a variety of geotextiles (including, again, pocket areas for vegetation regrowth) or willow-spilings, which might also allow some hydrological connection between the channel and its adjacent floodplain. In all these cases, however, applicability depends upon the combination of local stream power and the frequency of occurrence of significant channel-modifying flow events. None of the softer approaches matches traditional measures in terms of bank protection, and there are few data which may be used definitively to identify which form of protection is more or less costly (or indeed, on how longer-term cost is to be estimated). In circumstances where risk is relatively low, traditional measures may be seen as 'over-designs', and the softer measures may well be successful. But in urban environments, where land costs are high, risks to property and life are greater, and where rivers are already severely modified in box-sections and culverts, soft measures find much less application. Nevertheless, some potential for redesign of cross-sections is possible (see below), and, at least in small, isolated parts of these river systems, wholesale restoration of river planforms, including reinstatement of wetlands, may be possible.

Two-stage channels are designed to enlarge the cross-section during high flows, but to provide a low-flow cross-section which is much closer to a natural river section in terms of its flow behaviour and ecological potential. The redesign of the low-flow channel to accommodate ecological interests must usually still satisfy flood and drainage requirements, but attempts to model actual and prospective performance from both perspectives are now being made (Downs and Thorne 1998). In the low-flow channel, flow may be diverted around rock obstructions, artificial bedforms and channel constrictions (Newbury 1995). In this manner, velocities and depths are varied as a 'first-order' response, and sediment calibre and texture may follow as 'second-order' responses. Cross-section redesign may be accompanied by variation of channel planform, particularly the creation of meandering, which also promotes differentiation of flow and sediment.

Heterogeneity of flow and sediment is essential and beneficial with respect to fish and invertebrate abundance and diversity. Increased heterogeneity may be created by inserting artificial bedforms and constrictions. These may also be used for planting riparian vegetation. Investigations into the ecological response to such measures are becoming more common, and they generally show positive results. An unusually long data set relating to enhancement of fish habitat in Wisconsin was assessed by Trimble (1997). This demonstrated that fish structures enhanced sediment deposition along the stream and may have retarded the lateral migration of the channels, although their utility did depend on the very local erosional/depositional controls. A well-controlled test of the hydraulic and ecological effects of artificial riffle emplacement in a channel was reported by Gore, Crawford and Addison (1998) who found a threefold increase in benthic habitat at low flows, and a 40% increase in the channel area supporting high benthic community diversity.

Non-structural approaches focus on raising awareness of the ecological and environmental consequences of channel management. Emphasis is placed upon post-project appraisal, which allows successive river works to build on positive and negative experiences. The goal is to contribute to the improvement and popularity of ecologically aware river project intervention on the part of both engineers and other policy-making professionals, and wider interest groups. Non-structural approaches to conservation management are thoroughly reviewed in Gardiner (1991) and may include:

- Best-practice guidelines — available for anyone involved in river engineering and river use
- Recommendations of widening consultation to all potentially interested parties
- Classification schemes for river sensitivity and potential for restoration/ rehabilitation and the use of ecologically aware soft engineering techniques
- Demonstration and implementation of alternative engineering works so that construction and maintenance procedures, as well as the actual type of work, may be improved
- Codification and enhancement of the entire project planning, implementation and post-implementation appraisal process.

Both structural and non-structural responses are enhancing the experience and advocacy of river restoration. Because pristine states are now so rare, they are increasingly seen as reference points, rather than goals for stream-restoration efforts. Rather, river restoration is seen as the most feasible way of sustaining river ecology. Conservation is moving beyond the mere preservation of refugia to the creation and maintenance of self-sustaining and functionally diverse community assemblages (Osbourne *et al.* 1993). While river restoration is not only a conservation practice, it can become increasingly so in the longer term, and the criteria for restoration design rely heavily on the same principles which inform conservation management in general. The increasing number and scope of restoration works also means that they are the most likely places to see conservation practice at work outside traditional reserve/refugia environments. An awareness of current debates, and of the principles underlying emerging best practice is, therefore, vital from the conservation perspective.

EMERGING PRINCIPLES FOR RIVER RESTORATION

Over the last two decades, approaches to the restoration of rivers and to the conservation of remaining areas of natural rivers and wetlands have increased enormously in scope and sophistication. The value of multi-disciplinary and multipurpose intervention in river works is now apparent (Whelan 1991). Some rivers have become the focus of teaching and research programmes directed at specialist scientists, riparian communities and interest groups. River restoration and conservation has both responded to and shaped a changing agenda which is an amalgamation of the scientific/technical, political/legislative and socio-economic. In practice, the scientific/technical has rarely been at the forefront (Banks 1991), but with time, the need for accountability and post-project appraisal, together with the momentum of the sustainability debate, have required a better technical/scientific base.

Boon (1998) has summarised the emerging best practice in river restoration using a 'five dimensions' conceptualisation of the river system. They are:

- *Conceptual/motivational*: should river restoration be undertaken, and if so, why? If restoration satisfies a particular sectoral interest, then the wider environmental framework should be identified.
- *Spatial*: recognising the lateral, longitudinal and vertical connectivity of the river system, and the need to fit restoration into this spatially functioning system. Structure and heterogeneity in the channel regulate biodiversity, productivity and nutrient retention (e.g. Zalewski *et al.* 1998), but the corridor characteristics may rely on channel–floodplain interactions, which themselves are characteristic of location within a given river catchment and catchment system (see Hughes and Rood, 2001).
- *Temporal*: the history of exploitation together with the outcome of natural river dynamics must be taken into account for sustainable restoration. Prospectively, there is a need for post-project appraisal of intervention, over short and long timescales.

- *Technological*: the need for sound scientific data to inform all the above, together with growing integration of field, laboratory, GIS-based and numerical simulations; and appreciation and testing of the range of modified engineering techniques for controlling and designing river form and function.
- *Presentational*: restoration is unlikely to succeed without the participation and/or approval of a wide cross-section of interest groups. Restoration itself should not supplant true conservation of undegraded rivers where this is still a possibility.

The 1990s have also seen the emergence of a common set of terms to describe river works relevant to restoration conservation (Boon 1992):

Restoration is recovery obtained by manipulating hydrology, water quality and habitat structure to a pre-disturbance state. This is a relatively infrequently used option. By implication, it is large-scale and longer-term, costly and probably high profile politically, and it requires knowledge of the current and past (pre-disturbance) status of the river system, rather than that simply of a reach or section. Such knowledge is rarely available (particularly in urban contexts), although in the case of river planform characteristics, historical maps as well as remnant channel morphologies may be suitable bases for inferring some characteristics for the redesign for channels, where pre-disturbance states cannot otherwise be defined.

Rehabilitation is partial structural and functional return to a pre-disturbance state. This is the most common objective, representing a compromise between feasible, optimal and ideal solutions. It is appropriate at the reach-scale, where physical changes in the channel can be expected to bring about a partial improvement in flow rates and water quality.

Enhancement describes any improvement of a structural or functional attribute of the channel. Because enhancement is piecemeal, there is always a danger of only limited success or a short lifespan. It may, therefore, be more cosmetic than functional.

Dereliction occurs where a river is so degraded or circumstances are so unfavourable to sustainable improvement that the current state is maintained.

A case study which places these alternatives in the context of urban rivers, and which examines the question of identifying appropriate restoration benchmarks is given by Tapsell (1995).

CONSERVATION MANAGEMENT AND THE PHYSICAL ENVIRONMENT

Successful intervention in rivers requires a firm scientific knowledge of how they work as physical systems, and of the wider context within which the channel system operates. It also requires a good knowledge base or flow of information concerning the present and past state of the environment. These scientific principles and knowledge can then be used to assess the likely impact of development. They can also be used to decide between alternative plans for intervention; as criteria for judging the success or failure of an intervention; or to determine new means of river engineering which satisfy environmental and recreational requirements, as well as traditional concerns such as flood prevention or water supply.

The role of conservation in river management has usually been in three activities: survey, assessment, and ranking. Survey and assessment may be based upon either field or documentary evidence. The key is to place a particular river or river reach in context in order to characterise its actual or potential value for conservation. Once this has been covered, it is then appropriate to consider the degree of present modification from a 'natural' state (either known on the basis of comparison, or assumed), and the likely long-term viability of proposed modification. Assessment must be based upon some kind of model or scheme which relates morphology, flow and sediment characteristics to potential ecological value. Factors considered at this stage depend upon the spatial and temporal extent of the proposed conservation or restoration work.

With respect to the technical aspects of restoration relevant to conservation, there is now a recognition of the need to place works at the channel level into a wider catchment perspective, and to apply principles which stress the spatial, temporal and functional interdependence of hydrological, ecological and geomorphological variables and the importance of heterogeneity and disturbance in maintaining biodiversity (Harper *et al.* 1999). Collectively, these concerns constitute a recognition of the existence of a fluvial hydrosystem (Petts and Amoros 1996) whose conservation management requires a form of holistic streamflow management (Hill, Platt and Beschata 1991).

FLUVIAL HYDROSYSTEMS AND HOLISTIC STREAMFLOW MANAGEMENT

The concepts of fluvial hydrosystems and holistic streamflow management are based upon the tenet that river flows have both direct and indirect controls on the character of riparian ecology. Petts and Amoros (1996:6) listed five features of the fluvial hydrosystem that were keys to the study of rivers:

- Focus on the river corridor, its adjacent floodplain and the underlying alluvial aquifer
- Identification of longitudinal, lateral, vertical and temporal gradients in environmental variables and biological populations, linked by fluxes of energy and materials

- Location of the distribution of biota according to the gradients listed above
- Awareness of the timescale of the controls on the structure and function of the fluvial hydrosystem, principally the legacies of climatic change in the Quaternary and Holocene, and the human impact over the last 200 years or so
- Identification of the relations between flow regime (discharge, water quality and sediment yield), land use and environmental change, and the character of the catchment area.

Awareness of the type and importance of direct controls, the development of techniques to identify these, and latterly, criteria to conserve or even create channels favourable to a particular species has emerged from studies of the North Atlantic salmonids. River depth, flow and gradient have long been identified as significant variables in determining the numbers of salmonid juveniles and the species composition and phenotype of the plant communities on which they depend. A potential to design streams, or to impose substrate and channel morphologies likely to yield favourable habitats for the salmonids was recognised soon after these correlations were confirmed in the 1950s (for a review see Mosley 1985; O'Grady, King and Curtin 1991).

Linking instream flow and habitat may be approached either from 'top-down' or 'bottom-up' perspectives. The top-down approach assumes relations between flow characteristics and species abundance/diversity, and then proceeds to identify biotypes on the basis of physical characteristics, such as surface flow speeds, that are representative of distinctive hydraulic behaviour. The bottom-up approach derives functional habitats on the basis of field sampling and statistical cluster/discriminant analyses (e.g. Kemp, Harper and Crosa 1999; and below on eco-hydraulics). Attention is now focused on the integration of the two approaches (Newson *et al.* 1998b), and on the viability of present databases such as the UK River Habitat Survey to classify and characterise rivers in the necessary detail. The analysis, however, is still disappointingly limited in its success (Newson *et al.* 1998a). More generally, agreed technical standards, protocols and terminologies are required if integrated catchment management is to fulfil its purpose and potential (Raven *et al.* 1998).

Flow character affects ecology indirectly by organising the morphology of rivers and their valleys. For example, field studies of riparian vegetation yield clear associations between species type, substrate type and drainage. These, in turn, reflect proximal/distal relations to the character of the channel, which themselves are related to subsurface drainage and the occurrence of episodic flooding. Channel planform is also implicated, since it affects flooding frequency and magnitude (Birkeland 1996; VanColler, Rogers and Heritage 1997). The type, density and longevity of stands of established vegetation itself affect subsequent sediment movement. As a result, the physical and biological

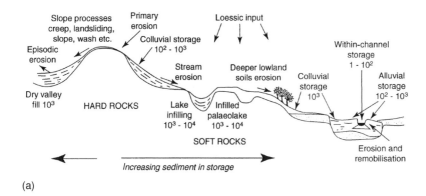

(a)

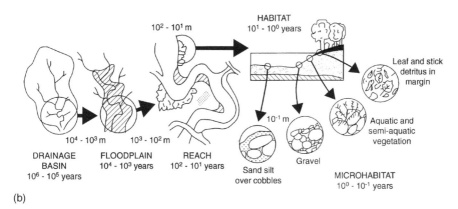

(b)

Figure 4.3. Organisation of stream system and habitat subsystems according to temporal and spatial scales. Sources: (a) from Newson (1992:51); (b) from Boon, Calow and Petts (1992:102) after an original of Frissell *et al.* (1986).

structure of the riverine environment are necessarily interrelated (Gurnell 1997).

Within a drainage basin, habitats are spatially and functionally 'nested', so that the scientific criteria on which conservation management, river restoration or intervention are based are appropriately considered at a variety of levels (Newbury and Gaboury 1993; Figure 4.3):

- Level 1: the morphology (width, depth and slope) of the channel segments between major channel junctions are determined by flows from upstream which are themselves controlled by tributary drainage areas. River segments therefore necessarily reflect basin-scale controls.

- Level 2: reaches may be distinguished within channel segments, and are characterised by a variety of bedforms (e.g. pools and riffles), bed material and channel patterns, each with a differing habitat potential.
- Level 3: within a subreach section, the state and structure of flows must be delineated since this dictates the habitat of a particular organism.
- Level 4: the habitat of a particular organism may be determined from survey, or modelled using information from Levels 2 and 3.
- Level 5: in porous-bed streams, habitats are characterised by interstitial flows through substrates (the hyporheic zone) which, although not observable directly, may be inferred from local piezometric gradients and conductance.

The management of each of these levels depends upon a varying combination of: geomorphological and hydrological knowledge of controlling factors; engineering expertise in achieving designs or outcomes informed by these forms of knowledge; and a complementary knowledge of the ecological significance of existing, planned or consequential outcomes of any intervention. Some of the background to this expertise is outlined below.

GEOMORPHOLOGY AND CONSERVATION OF THE RIVER ENVIRONMENT

Geomorphology may be defined as the scientific study of earth surface processes and landforms. It has a descriptive function, in identifying individual landforms and landform assemblages, and also an explanatory function, in identifying the connections between form and process, and seeking to explain the consequences of these in time and space. Rivers transport both water and sediment. During transport, there are significant storages of sediment, and these storages are controlled by river shape and channel planform change. Whereas engineers tend to focus on shorter-term interactions of flow, sediment dynamics and channel shape, and on the identification of equilibria between these, geomorphologists stress the dynamic relationship over timescales from seconds to thousands of years. To them, the state of river bed sediments may reflect flow–sediment interactions over seconds, minutes and hours, whereas local sediment supplies reflect flooding, channel pattern effects and bank stability operative over years, together with longer-term sediment storage and supply within channels and adjacent hillslopes operating over hundreds or thousands of years. The 'conditioning' of the present environment may be determined by Holocene environmental change. Conservation management of rivers is fundamentally a geomorphological activity, inasmuch as quasi-equilibrium stream channels and functional floodplains promote the greatest aquatic and terrestrial habitat diversity and represent the natural conditions under which riparian ecosystems develop (Morris 1995).

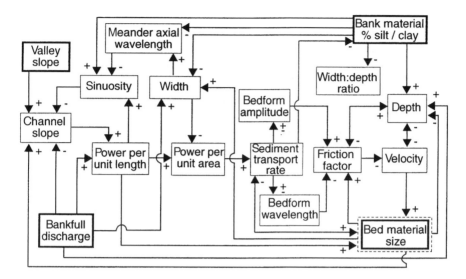

Figure 4.4. The alluvial channel system, showing principal variables linked in a feedback system whose dynamics are further spatially distributed through the operation of geomorphological processes. Heavy outlines refer to 'independent' variables (see also Table 4.1) and + and − refer to positive or negative relationships. After Richards (1982:26).

Channel processes are connected to a wider fluvial system. The variables involved in the dynamics and geometry of alluvial (self-formed) channels are linked in a complex system of interacting feedback relationships (Chorley, Schumm and Sugden 1984; Figure 4.4). Although all the components may be said to be causally interrelated (Lane and Richards 1997), it is common to designate some as 'independent' controls and others as 'dependent' responses (Table 4.1). The distinction depends upon the time- and space-scale of the description or explanation required (Schumm and Lichty 1965).

Using Table 4.1, a set of functional relations can be obtained which link variables on the basis of empirical measurement and/or theoretical

Table 4.1. Components of the fluvial system

Independent variables	Dependent variables
• Valley and channel slope	• Width and depth
	• Velocity
• Discharge	• Flow resistance
• Sediment supply	• Sinuosity

assumptions (e.g. Hey 1986). Recognising the joint importance of water transport and sediment transfers through rivers, basic adjustments may be defined by:

$$LD_{50} \propto SQ \qquad (4.1)$$

i.e. gradient, S, is proportional to amount, L, and calibre, D_{50}, of load and to discharge, Q, and where channel planform is considered via width, w, depth, d, slope, and the meander wavelength, l:

$$Q = wdl/S \qquad (4.2)$$

such that for

$$Q+ \rightarrow w+, d+, l+, S- \qquad (4.3)$$

and for

$$Q- \rightarrow w-, d-, l-, S+ \qquad (4.4)$$

These relations are important for specifying the kind and direction of any impact upon a part of the channel system. They have, therefore, been the subject of research as the underpinning of design criteria for channel restoration, or optimisation to ensure sustainable function (e.g. Hey 2000). Unfortunately, this approach is limited for at least two reasons: first, approached in this manner, the fluvial system is essentially indeterminate — for certain 'forcing' changes and combinations of forcing change which include aspects of sediment transport, response may be either positive (reinforcing) or negative (equilibrating) in any single variable, and involve successively more components of the system (i.e. more potential kinds of adjustment), such as channel sinuosity, P, and the width–depth ratio, F:

$$Q+, L+ \rightarrow w+, d+/-, l+, S+/-, P-, F- \qquad (4.5)$$
$$Q-, L- \rightarrow w-, d+/-, l-, S+/-, P+, F- \qquad (4.6)$$

Second, at best, functional relations must be empirically calibrated to provide 'regime' equations for particular sets of rivers regionally (e.g. Bray 1973) and for particular river reaches, for example:

$$S = 0.956Q^{-0.0334}(D_{50})^{0.586} \qquad (4.7)$$

or at the basin scale (Hack 1957):

$$S = 18(D_{50})0.6/A \qquad (4.8)$$

where A is drainage basin area. However, the coefficients in such equations may persist only for a limited time period.

Approaches based upon physically based numerical simulation may eventually provide better bases for river design, but as yet, full coupling of flow models and sediment models is unsatisfactory. However, it is increasingly possible to utilise models to define the controls on river behaviour, and to identify those parts of the environment that need to be monitored, and which require our control or concern. Used in this way, models provide domains of applicability for particular kinds of river intervention strategy. Underlying these domains are the geomorphological connections between flow and sediment at a variety of scales, as outlined below.

CATCHMENT ENVIRONMENTS: FLOW AND SEDIMENT ROUTING

A useful way to model rivers is to identify zones, or domains, which relate water flow to sediment transport and storage (Figure 4.5). These are: the upland areas of sediment and water production; the intermediate zones of water and sediment transfer; and the lowland areas of sediment deposition. Sediment transfer and storage occur within any of these zones, and it is the amount and stability of these sediment storages and transfers which determine habitat.

One of the most important aspects to consider when determining the conservation potential of rivers (or the need/desirability to restore rivers) is the stability of the water–sediment relation in each zone. Conventionally, this is modelled with respect to the 'effective' discharge, or sediment-transporting flow event. Such models are based upon the joint effect of the magnitude of flow and its frequency of occurrence. Contemporary approaches to river channel dynamics place less emphasis on 'balance', and more on 'continual (re-) adjustment' (Richards 1999). This change in emphasis reflects the fact that rivers are often adjusting in the long term to conditions of sediment supply conditioned by past environments, and that conditions observed at any point, and at any given time, may reflect the influence of short-term but higher-magnitude events, such as floods. These events condition the flow and sediment conditions locally, so that the river is rarely 'stable' for any length of time. Most important is the recognition of the spatially distributed form-process feedback (Ashworth and Ferguson 1986; see Figure 4.4). These concepts allow classification of rivers according to their physical characteristics and environmental regime, and may also be particularly important in defining the reference point to which river-intervention work aspires.

CHANNEL ENVIRONMENTS: THE IMPORTANCE OF BEDFORMS

Within river sections, or reaches (particularly in gravel-bedded rivers), the importance of bedforms must be recognised. Bedforms are associated with

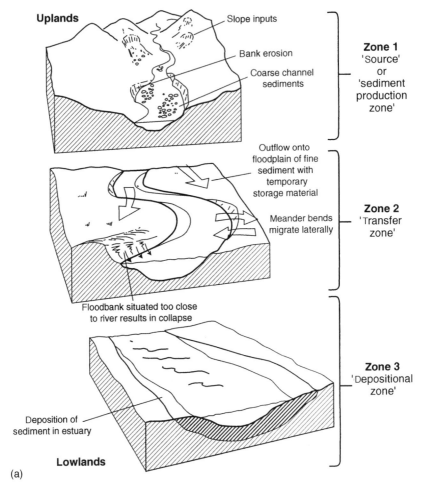

(a)

Figure 4.5. (a) Zones of sediment production, transfer and storage through a river catchment; and (b) channel properties associated with these. Source: RSPB (1994:20, 22). (b) is originally from Calow and Petts (1994).

subreach and small-scale changes in flow and sediment regime, and provide a diversity of environment for fish and invertebrates. Maintaining and recreating a diversity of small-scale habitats is, perhaps, the most important aspect of river conservation management. Figure 4.6 shows a fundamental unit of gravel- and mixed-bedded rivers of low to intermediate slope: the pool–riffle unit. This has received a great deal of research, as a result of the marked spatial differentiation of slow and rapid flow (Figure 4.6.(a)), and the temporal-dependence of these differences as ambient flow rates vary (Figure 4.6(b); for review see Clifford and

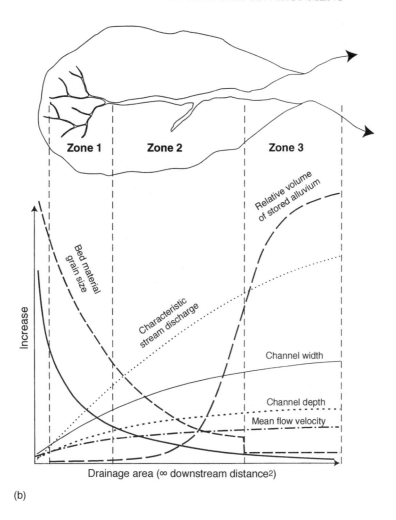

(b)

Figure 4.5. (b)

Richards 1992; Clifford and French 1998). Research is continuing into the possible role of such features in differentiating flows within the cross-section; in controlling the timing and extent of overbank flow; and in regulating water exchanges between the channel and its bed. At present, these processes must be approached empirically, although three-dimensional computational fluid dynamic models capable of at least predicting and/or simulating within-channel eddy circulations are increasingly available. Other research has identified the need to consider flow-dynamics as dependent upon seasonal changes in the

Figure 4.6. (a) Low-flow distinctions of riffle (steep water surface slope, shallow flow and rapid flow) and pool (slight water surface slope, deep water, slow flow). (b) High flow over a riffle and pool, illustrating stage-dependency in the contrast in hydraulic behaviour. Clear spatial distinctions in (a) have now been replaced by a core or 'jet' of high velocity, emerging from the riffle and passing through the downstream pool.

sediment regime of rivers. Bedforms such as riffle–pools are quasi-permanent, but are subject to in-filling and over-passing by fine sediments with consequent habitat loss (Lisle 1982). The amount of in-filling at any given time may also determine subsequent flow behaviour (Lisle and Hilton 1992).

Figure 4.6 (b)

BED STRUCTURE

At a level of much greater detail, a knowledge of fluid flow characteristics may be necessary, for example: in determining the structuring of the bed; sediment intrusion into/evacuation from the bed; and hyporheic temperature profiles (Evans and Petts 1997). Other considerations include the transmission and storage of pollutants, which may be preferentially stored and released in interaction with the sediments of the bed (Bencala *et al.* 1984). Young (1993) pointed out that differentiation of flow regimes in the very near-bed zone of gravel- and mixed-bedded rivers might be of more ecological significance than the dynamics of bedforms such as riffles and pools. Assessing the importance of potential micro-habitats requires fine-scale descriptions of bed geometry and particle composition and texture, and schemes for describing and classifying the detail of associated flow behaviour.

With respect to the specification of near-bed particle geometries, various micro-profiling devices have been employed. These either utilise point-gauge type arrangements in which pins drop vertically from a graduated datum beam onto the surface (Clifford, Robert and Richards 1992; Clifford, in press) or have pins which are forced upwards, mimicking the bed sediment arrangement (Young 1993). Other, non-mechanical, possibilities include acoustic or infra-red devices, but these are costly and can be used only under certain flow conditions. Although there are well-developed classifications of particle arrange-ments in gravel-bedded rivers (e.g. Laronne and Carson 1976; Brayshaw 1985) they have received little attention in the ecological literature. More emphasis is now being placed on the consequences of these arrangements for flow behaviour, which may be approached via flow classification and by direct measurement.

Descriptive classifications of flow 'type' are based upon the visual examination of the water surface behaviour. As such, they are subjective, but quick, and can be related to benthic habitat conditions. Young (1993) reviewed the various schemes that related surface flow disturbance to assumed boundary roughness. Flow types fall between extremes of skimming flow, where the water surface is little disrupted because roughness is low (either particle size is small, or it is small in relation to local depth); wake-interference flow, where roughness is high; isolated roughness flow, where flows are distorted around larger obstacles; and zones of 'dead water' in the shelter of channel banks or large obstacles. As yet, precise relations between the flow types and micro-habitat await research, and this must be informed by refinements to conventional flow measurement.

Direct measurement of flow characteristics relies on the ability to measure velocity fluctuations at relatively high frequencies (fractions of a second), preferably in more than one plane simultaneously. At present, instruments for this are expensive and more difficult to deploy than conventional current metering. Electromagnetic current meters have the longest track record in this respect, although increasing use is being made of acoustic-Doppler veloci-meters. For a review of the instruments and their potential, see Clifford, French and Hardisty (1993).

Once high-frequency flow characteristics have been obtained, there remains the question of deriving physically meaningful parameters. A simple measure of potential activity at or near the bed is the turbulent kinetic energy of the flow. This is associated with the interchange of fine bed sediments and differentially oxygenated water. Kinetic energy is derived from splitting the flow field into three orthogonal planes, U, V and W (streamwise, cross-channel and vertically); and temporally, where the total instantaneous value is the sum of mean and fluctuating parts:

$$U = u + u' \quad V = v + v' \quad W = w + w' \tag{4.9}$$

The total kinetic energy at any instant is:

$$1/2\,\rho(U^2 + V^2 + W^2) \tag{4.10}$$

and the instantaneous turbulent kinetic energy is:

$$1/\rho(u'^2 + v'^2 + w'^2) \tag{4.11}$$

Equations (4.10) and (4.11) should be averaged over some representative time period.

Small-scale flow structures associated with individual particles or particle groupings on the bed can also be determined by analysing the structure of one or more of the velocity time signals. The coherence in the velocity time signal can itself be related to a particular scale of flow structure, which in turn, reflects the influence of a particular bed geometry or particle arrangement (e.g. Clifford 1993). Another approach is to relate higher-frequency flow characteristics to more easily determined 'conventional' time- and/or space-averaged flow measurements. Very few data are available to do this, although Clifford (1997) demonstrated some correlation between mean and fluctuating components. For the future, pressing research questions concern the range of applicability of such correlations, and their incorporation into habitat simulation models, which at present are largely restricted to one-dimensional, time- and space-averaged schemes (see below).

ECOHYDRAULICS

The term 'ecohydraulics' refers to a new form of ecologically relevant hydraulic analysis of rivers. Its components have emerged over a forty-year period. In the 1950s, empirical relationships between flow, current velocity and abundance of salmonids, macroinvertebrates and macrophytes were discovered (see below), and stream ecological studies on energy flows, carbon fluxes and macroinvertebrate life histories were also begun. By the late 1970s and in the 1980s, the role of flow in structuring river ecosystems was clarified in the USA, and deployed practically in the instream flow incremental methodology (IFIM). The approach was introduced to the UK early in the 1990s (Petts and Maddock 1994). Broadly, ecohydraulics is based upon the following considerations connecting river flow and river ecology:

- Flow characteristics define the environmental domains within which biological communities develop.
- Flow rate, timing and nature of sediment transport determine the structure of river corridor morphology.
- River flow and groundwater regime determine the nature of vertical interactions within the hyporheic zone.

Following from these points, the application of ecohydraulics has been characterised first, by the identification of habitat preferences; and second, by the search for appropriate aspects of flow–ecological interactions to incorporate into river design and conservation works. Essentially, these interactions fall into one of two categories: constant, 'normal' or low-flow (particularly the specification of depth and mean flow velocity; and increasingly, consideration of ways to index potential stream dynamism); and episodic 'disturbances', in particular, the effects of floods on ecological functioning.

HABITAT PREFERENCES AND SUITABILITY

Habitat preference and suitability approaches rely on the observation that the energy budget of an organism reflects the relative speed between it and the medium in which it lives. Current velocity, for example, may influence respiration and feeding biology, as well as behaviour such as territoriality. The objective, either by direct survey of species and abundance, and/or by observation of the characteristics of the physical environment, is to specify the range of conditions within which a species is found at an acceptable abundance/productivity, and hence produce a habitat suitability curve. In the ideal case (Figure 4.7), a definite relationship is found, with clear boundaries between suitability, depth and velocity. However, the suitability curve may be life-cycle-dependent (Figure 4.8), so that suitability may be time-dependent unless the local conditions also respond in phase with ecological requirements. Figure 4.9 shows that field data rarely produce such clear-cut criteria of demarcation, and further research is needed to clarify the manner and degree of manipulation appropriate in these cases.

LOW-FLOW DETERMINATIONS AND STREAM DYNAMISM

'Stream power' is a fundamental concept, since it is a means of determining likely sediment movement and the planform stability of the channel. Streams convert potential energy derived from their height above a given reference point, to kinetic energy (a function of the flow velocity) and use this energy in performing work, including the transport of sediment. Stream power, Ω, is the

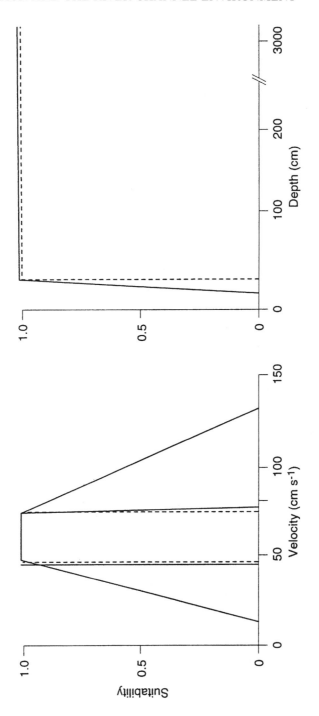

Figure 4.7. Simple habitat suitability curves. Source: Rayleigh *et al.* (1986).

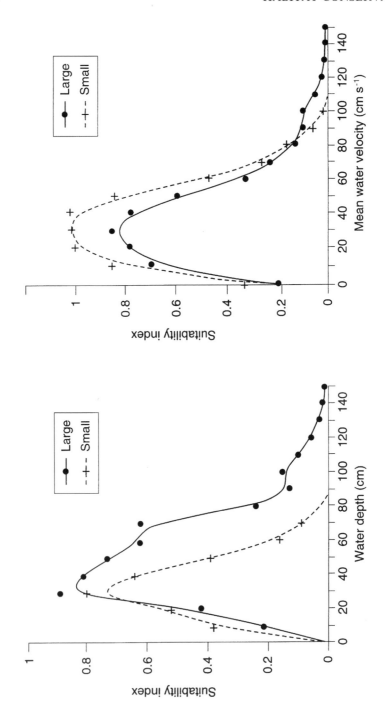

Figure 4.8. Continuous suitability curves with life-cycle dependency for Atlantic salmon parr. Source: Heggenes (1990).

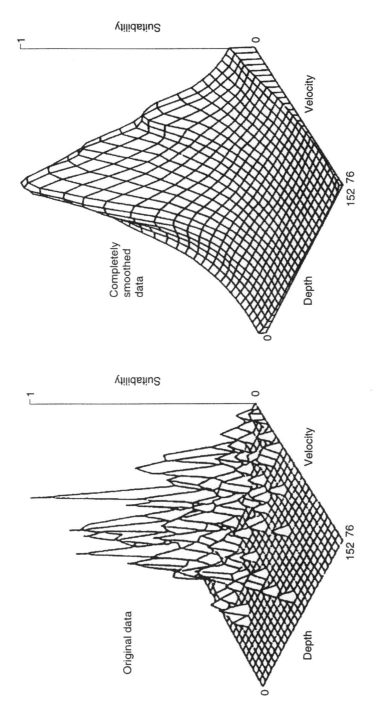

Figure 4.9. Raw and smooth data for habitat suitability approach based upon a response surface. Source: Lambert and Hanson (1989).

rate at which work is done, or the rate of potential energy expenditure per unit length of channel:

$$\Omega = \rho g Q S \tag{4.12}$$

where ρ is the water density, g is gravitational acceleration, Q is the stream discharge and S the channel slope.

The unit stream power, ω (power per unit bed area):

$$\rho g d V S \tag{4.13}$$

(where d is flow depth and V mean stream velocity) is an alternative power criterion which is more easily related to the transport of coarse sediment, and hence to controls on river stability.

An alternative measure relating to sediment transport is the boundary shear stress (τ_0), which is the force exerted upon the bed by the flowing water and which provides the tractive force for sediment transport. Under conditions of uniform flow (no flow acceleration or deceleration along the channel or through time), unit power, velocity and shear stress are related in the following manner:

$$\omega = \tau_0 V \tag{4.14}$$

Power criteria are increasingly of interest in delimiting conditions of potential channel dynamics, and hence the potential applicability of channel stabilisation measures.

More crudely, sediment transport and other aspects of the status of the bed may be indexed by the flow velocity, which is determined by the efficiency of the channel, itself a function of channel shape and boundary resistance or friction. For conservation purposes, increasing use is being made of linked hydraulic-habitat models. These allow flexibility in the design of restoration schemes and in the assessment of river conservation potential, since they allow the computation of river flows over a wider range of discharges than commonly measured in channels, and then link them to potential sediment transport and/ or habitat. PHABSIM is a popular model in this respect (Milhous, Updike and Schneider 1989). Models, however, reflect the assumptions used in the algorithms that are employed, and simple models are restricted to one-dimension (i.e. they represent cross-section-averaged conditions), and to uniform flow conditions. Given these limitations, a particular issue is the representation of channel flow resistance, which reflects some combination of three groups of parameters: (1) the velocity, depth of flow and channel morphology (both along the reach, around bends and bedforms, and at a section); (2) the size and arrangement of material on the bed and banks; and (3) the presence/absence and characteristics of bed and bank vegetation. Determinations of mean velocity used in most conservation applications rely largely on the entirely empirical Manning formula:

$$v = kR^{2/3}s^{1/2}n^{-1} \qquad (4.15)$$

where the constant k depends upon the units of measurement ($= 1$ in SI units), and R is the hydraulic mean depth (usually approximated by the average water depth). Appropriate values of n ('channel roughness') may be obtained from look-up tables, but should be corrected for depth, since as depth increases, the effective roughness of the channel is reduced. Corrections should also be applied for the effects of channel sinuosity. An accessible and comprehensive introduction to open channel hydraulics may be found in French (1986).

A more complete formulation relating channel characteristics to mean velocity is:

$$\frac{1}{\sqrt{f}} = 2.03 \ \log \frac{aR}{3.5D_{84}} \qquad (4.16)$$

Here, a is a shape constant (see below), D_{84} is a measure of particle size (84th percentile of the particle size distribution), R is the hydraulic radius, and f is the Darcy–Weisbach friction factor:

$$f = 8gRS/V^2 \qquad (4.17)$$

The benefit of equation (4.16) is that it incorporates the effect of channel shape into the constant a (defined as the ratio of hydraulic depth to maximum flow depth), and the effect of particle micro-structures on the bed via the multiplier of particle size (Clifford, Robert and Richards 1992). Nevertheless, the equation still systematically underperforms, particularly in pool-like river environments, where a degree of ponding-back of the flow occurs.

Borschardt (1993) stresses that simplistic formulae developed for low flows miss the importance of periodic (hydraulic) disturbances. Disturbances may be forced, as in flood events, or cyclic, as when there is a gradual accumulation and subsequent flushing of woody debris in the low-flow channel. Of these, the ecological importance of floods has received most attention. In semi-arid basins, for example, the lowest species diversity occurs in highly stable flow regimes, and the highest in streams with intermediate flood magnitudes. Models of seed germination also demonstrate the significance of the various components of the annual streamflow hydrograph (Shafroth et al., 1998). The flood-pulse concept is, therefore, the next essential consideration for conservation management which seeks to produce productive, varied and sustainable ecological performance.

THE FLOOD-PULSE CONCEPT

The flood-pulse concept is based upon the idea that the (natural) allocation of water in and around river channels performs specific, seasonal and irregular

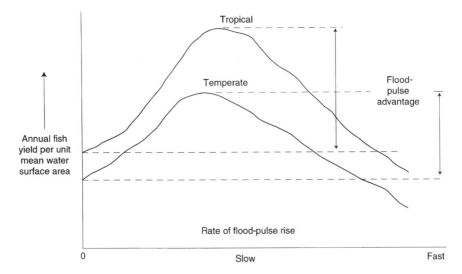

Figure 4.10. Schematic representation of the flood-pulse advantage. Source: Junk, Bayley and Sparks (1989).

functions in the maintenance and structure of riparian ecology (Junk, Bayley and Sparks 1989). Figure 4.10 shows this allocation which has four basic components:

- In-channel flows maintain sedimentary character, bedforms and critical fish and invertebrate needs.
- Overbank flows inundate riparian and floodplain areas (see Hughes and Rood, this volume).
- Flood flows form floodplain and valley features.
- Surface–groundwater interactions sustain the hyporheic system.

At least for rivers with extensive floodplains, lateral exchanges between the channel and its floodplain, and nutrient recycling within the floodplain are likely to be important ecological processes (Hughes and Rood, *this volume*). The river margin is a valuable water–land ecotone, where the biota are adapted to the alternation of distinct aquatic and terrestrial phases determined principally by the hydrological (flood) regime. Bankfull flows, with return periods of between one and three years, are the most effective in transporting sediments and reordering them into bed forms, and possibly in maintaining channel width. 'Flushing flows', which are larger than bankfull flows, are required to maintain quality of substrate by preventing the encroachment of

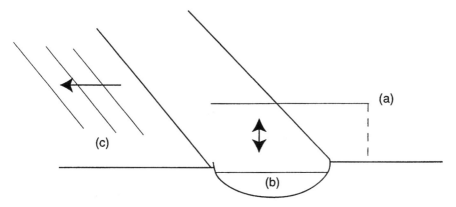

Figure 4.11. Aspects of the operation of the flood-pulse advantage.

vegetation and by removing fine inorganics from the bed. Even larger 'floodplain-maintenance flows' are responsible for channel planform dynamics (such as meander cut-offs), over-bank sedimentation and the disturbance regimes for pioneer species and possibly species diversity (discussed further in Hughes and Rood, *this volume*). Thus, the occurrences, life cycles and abundance of primary and secondary producers are determined by the duration, magnitude, frequency and timing of floods.

The ecological importance of flood pulses is shown in more detail in Figure 4.11. They are important because: (a) aquatic organisms colonise the flood-plain at rising and high water levels (feeding/spawning); (b) terrestrial organisms occupy unflooded habitats, and exploit low water levels; (c) the 'moving littoral', or the inward edge of aquatic environment, traverses the floodplain. As a result of the operation of the flood pulse, environments benefit in the form of an environmental subsidy or 'flood-pulse advantage' (Junk, Bayley and Sparks 1989). This is the degree to which annual multispecies yield exceeds that from a constant water level system. With respect to conservation management, the principal concern is to sustain a sufficient range of flows to create a flood-pulse advantage. Over decades and less, this may be achieved by manipulating hydrological response within and around the channel. Floods over longer timescales, however, are more difficult to predict or manipulate, and do not fit easily into engineering or management options. Perhaps the best example of an attempt to integrate engineering, geomorphology, hydrology and ecology using these ideas is occurring in the restoration of the Kissimmee river and wetlands, in central Florida, USA. Ecological appraisals of the scheme are given in Dahm *et al.* (1995); and Toth *et al.* (1995).

MANAGING RIVER CONSERVATION AND RESTORATION
PROJECTS—THE IMPORTANCE OF THE APPRAISAL PROCESS

Project appraisal is the newest, and perhaps most important, component in environmental policy relating to the river environment. The appraisal process is designed to ensure a cost-effective outcome, in which resources are used accountably and consistently distributed between projects of a similar nature (Gardiner 1991). Appraisal is also important to ensure that, for future projects, those with the greatest potential can be identified. In the past, poor records of methods and costs and poor data on the subsequent performance of river 'improvements' have been a barrier to satisfactory evaluation. In the future, particularly as regulatory frameworks become tighter and better enforced, post-project monitoring will be vital. In addition, the creation of databases with which to compare projects and project options will be necessary if the appraisal process can truly said to be working. For example, in reviewing practice in California over two decades, Kondolf (1998) emphasised the need for some prior (rather than retrospective) ability to see how projects might or might not succeed in a given location. Highly modified systems where water or sediment impoundment has occurred and/or where species diversity and abundance have been constrained for long periods should not be taken as precedents. A scheme of 'best practice' cannot, therefore, be based upon the mimicry of conditions that are assumed to be 'natural' and applicable in different catchments or regions.

The appraisal process is one in which intervention, implementation and assessment of a project is broken down into a series of 'project phases' each of which has a particular activity associated with it. Both the phases and activities need to be coordinated in some way, and each may involve different sets of interested parties, professionals and others, in a consultative, advisory, or operative role. Successful projects require the following aspects of management (Gardiner 1991):

- Survey of basic information
- Classification and ordering of information
- Identification of problems and possible solutions
- Assessment of likely impacts and costs
- Organisation of project team
- Project implementation
- Post-project appraisal.

As a project progresses from consultation/design through implementation and evaluation, there will be a series of technical inputs. They may come from an engineer who advises on the design of a particular bank protection; an economist who assesses short- and longer-term costs; a politician who should be consulted for planning consents and budgetary approval; or others. The role of conservation scientists is in survey, assessment and ranking. Usually, these inputs are advisory, since the coordinating team generally seeks answers to specific questions. Non-technical inputs come, for example, from local residents, local environmental groups, landowners, or river users who may have a statutory right to express an opinion/reaction or who simply wish to make their views and concerns known. These contacts are part of 'project exposure', in which the project is identified, characterised and judged by interested parties. The exposure process is the least clearly identified and most difficult to manage part of the appraisal process, but arguably the most important to the success of the various objectives (Boon 1998).

Most recently, the requirement for dedicated monitoring of the shorter- and longer-term effects of the intervention have been recognised as essential in appraisal (Brookes and Shields 1996). When evaluating the results of a rehabilitation or restoration project from a conservation management standpoint, two basic considerations should be borne in mind:

- First, allowance should be made for the different components of the project to show significant benefit (or otherwise) at varying rates.
- Second, while initial changes are likely to be rapid, different components of change in the ecological and geomorphological system will be characterised by differing timescales of variation. Thus, fish stocks may not change for several years, whereas vegetation may quickly become dominated by exotic species which can reduce potential.

Crucially, the appearance of success in the shorter term does not guarantee successful modification to hydrological or ecological functioning which are required to underpin sustained performance. Conversely, the appearance of failure in the short term may simply reflect a very transient phase of dynamism, which may settle to a more stable and desirable outcome.

CONCLUSIONS

The problems and issues in the conservation management of rivers are changing and becoming more complex. This reflects the wider use of rivers and catchments, and the increased availability of information concerning environmental damage. Environmental legislation now spans local activist groups, national statutory bodies, up to Earth Summits. Most rivers are already affected by human modification of the environment, and are, therefore, already partly 'managed'. While many of the traditional forms of management have been successful by their own standards, the focus of attention is now upon sustainable, multi-use environments, and the restoration of degraded environments. Since it may not be possible to integrate management goals and approaches which satisfy all interest groups, river management must now reflect a choice between many technical alternatives for changing the appearance and operation of the river environment, each exercised in the light of prevailing and anticipated social and economic imperatives. Because impacts upon fluvial systems occur at many scales, management goals and strategies must also be seen in a time- and space-dependent context. The type of expertise required will change as the problems and required/desired solutions themselves change. Strengthening the case for river conservation and rehabilitation needs more basic and applied science, international and local collaboration, better long-term environmental monitoring, and a strong programme of research.

Some limits still apply to our ability to identify the preferred course of river-conservation management. Ideal (technical) solutions may, for example, ignore current land use, water rights and flood defence, and recovery of the physical environment does not ensure that the original species and community structures are recreated. Pre-disturbance conditions are seldom well documented, and fundamental questions remain. What 'past' are we seeking to recreate? What is 'natural'? Nevertheless, principles for future best practice, based upon competent science which is also operationally feasible, have emerged quickly over the last two decades. The physical underpinnings of conservation management in the river environment are to be found within the evolving discipline of eco-hydraulics, which must itself be guided by an appreciation of the geomorphological context and operation of the river system. Together with increasing codification of project management and appraisal, these constitute a good basis for holistic river channel management. As more attempts at conservation management and restoration of rivers are made, experience grows and this should lead to greater success.

ACKNOWLEDGEMENTS

This chapter was completed during receipt of NERC Award GR3/11909 'Identification of physically-based design criteria for riffle–pool sequences in river rehabilitation'. This award is held jointly with Professors G. E. Petts and A. M. Gurnell of the University of Birmingham. Photographs used in Figure 4.6 taken by a field assistant from the University of Birmingham.

REFERENCES

Ashworth PJ & Ferguson RI (1986) Interrelationships of channel processes, changes and sediments in a proglacial braided river. *Geografiska Annaler* **68A**, 361–71.

Balogh SJ, Meyer ML & Johnson DK (1997) Mercury and suspended sediment loadings in the lower Minnesota River. *Environmental Science & Technology* **31**, 198–202.

Banks J (1991) Assessment and practicability. In D Mills (ed.) *Strategies for the rehabilitation of salmon rivers.* London, Linnean Society, 19–24.

Bencala KE, Kennedy VC, Zellinger GW, Jackson AP & Avazino RJ (1984) Interaction of solutes and streambed sediment. 1. An experimental analysis of cation and anion transport in a mountain stream (California). *Water Resources Research* **20**, 1797–1803.

Birkeland GH (1996) Riparian vegetation and sandbar morphology along the lower Little Colorado River, Arizona. *Physical Geography* **17**, 534–53.

Boon PJ (1992) Essential elements in the case for river conservation. In: PJ Boon, P Calow, & GE Petts, (eds) *River conservation and management.* Chichester, John Wiley, 11–33.

Boon PJ (1998) River restoration in five dimensions. *Aquatic Conservation: Marine & Freshwater Ecosystems* **8**, 257–64.

Boon PJ, Calow P & Petts GE (eds) (1992) *River conservation and management.* Chichester, John Wiley.

Borchart D (1993) Effects of flow and refugia on drift loss of benthic macroinvertebrates: implications for habitat restoration in lowland streams. *Freshwater Biology* **29**, 221–7.

Bray DI (1973) Regime relations for Alberta gravel-bed rivers. In: *Fluvial processes & sedimentation.* Research Council of Canada, 440–52.

Brayshaw AC (1985) Bed microtopography and entrainment thresholds in gravel-bed rivers. *Geological Society of America Bulletin* **96**, 218–23.

Brookes A (1985) River channelization: traditional engineering methods, physical consequences and alternative practices. *Progress in Physical Geography* **9**, 44–73.

Brookes A (1995) Challenges and objectives for geomorphology in U.K. river management. *Earth Surface Processes & Landforms* **20**, 593–610.

Brookes A & Shields DF (eds) (1996) *River channel restoration.* Chichester, John Wiley.

Calow P & Petts GE (eds) (1994) *The rivers handbook: hydrological and ecological principles,* Vol. 2. Oxford, Scientific Publications.

Chorley RJ, Schumm SA & Sugden DE (1984). *Geomorphology.* London, Methuen.

Clifford NJ (1993) The analysis of turbulence time series from geophysical boundary layers: statistical and correlation approaches using the MINITAB package. *Earth Surface Processes & Landforms* **18**, 845–54.

Clifford NJ (1997) A comparison of flow intensities from alluvial rivers: characteristics and implications for modelling flow processes. *Earth Surface Processes & Landforms* **23**, 109–21.

Clifford NJ (in press) Statistical properties of the near-bed flow field over a coarse-grained fluvial surface with heterogeneous microtopography, *Geografiska Annaler.*

Clifford NJ & French JR (1998) Restoration of channel physical environment in smaller, moderate gradient rivers: geomorphological bases for design criteria. In: B Sherwood (ed.) *United Kingdom floodplains.* Otley, Westbury, 63–82.

Clifford NJ, French JR & Hardisty J (eds) (1993) *Turbulence: perspectives on flow and sediment transport.* Chichester, John Wiley.

Clifford NJ & Richards KS (1992) The reversal hypothesis and the maintenance of riffle-pool sequences. In: PA Carling & GE Petts (eds) *Lowland rivers: geomorphological perspectives.* Chichester, John Wiley, 43–70.

Clifford NJ, Robert A & Richards KS (1992) Estimation of flow resistance in gravel-bedded rivers: a physical explanation of the multiplier of roughness length. *Earth Surface Processes & Landforms* **17**, 111–26.

Dahm CN, Cummins KW, Valett HM & Coleman RL (1995) An ecosystem view of the restoration of the Kissimmee river. *Restoration Ecology* **3**, 225–38.

Downs PW & Thorne CR (1998) Design principles and suitability testing for rehabilitation in a flood defence channel: The River Idle, Nottinghamshire, UK. *Aquatic Conservation: Marine & Freshwater Ecosystems* **8**, 17–38.

Elliott JG & Parker RS (1997) Altered streamflow and sediment entrainment in the Gunnison gorge. *Journal of the American Water Resources Association* **33**, 1041–54.

Evans EC & Petts GE (1997) Hyporheic temperature patterns within riffles. *Hydrological Sciences Journal* **42**, 199–213.

French RH (1986) *Open-channel hydraulics.* New York, McGraw Hill.

Frissell CA, Liss WJ, Warren CE & Hurley MD (1986). A hierarchical framework for stream classification: viewing streams in a watershed context. *Environmental Management* **10**, 199–214.

Gardiner JL (ed.) (1991) *River projects & conservation: a manual for holistic appraisal.* Chichester, John Wiley.

Gore JA, Crawford DJ & Addison DS (1998). An analysis of artificial riffles and enhancement of benthic community diversity by physical habitat simulation (PHABSIM) and direct observation. *Regulated Rivers: Research & Management* **14**, 69–77.

Graf WL (1987) Late Holocene sediment storage in canyons of the Colorado Plateau. *Geological Society of America, Bulletin* **99**, 261–71.

Gurnell A (1997) The hydrological and geomorphological significance of forested floodplains. *Global Ecology & Biogeography Letters* **6**, 219–29.

Hack JT (1957) Studies of longitudinal stream profiles in Virginia and Maryland. *United States Geological Survey Professional Paper* **294-B**, 45–97.

Hadley RF & Emmett WW (1998) Channel changes downstream from a dam. *Journal of the American Water Resources Association*, **34**, 629–37.

Harper DM, Ebrahimnezhad M, Taylor E, Dickinson S, Decamp O, Verniers G & Balbi T (1999) A catchment-scale approach to the physical restoration of lowland UK rivers. *Aquatic Conservation: Marine & Freshwater Ecosystems* **9**, 141–57.

Harris SC, Martin TH & Cummins KW (1995) A model for aquatic invertebrate response to Kissimmee river restoration. *Restoration Ecology* **3**, 181–94.

Heggenes J (1990) Habitat utilization and preferences in juvenile Atlantic Salmon (*Salmo salar*) in streams. *Regulated Rivers* **5**, 341–54.

Hey RD (1986) River mechanics. *Journal of the Institution of Water Engineers & Scientists* **40**, 139–58.

Hey RD (2000) River processes and management. In: T O'Riordan (ed.) *Environmental science for environmental management* (2nd edn). Harlow, Prentice-Hall, 323–47.

Hill MT, Platt S & Beschata RL (1991) Ecological and geomorphological concepts for instream and out of channel flow requirements. *Rivers* **2**, 198–210.

Howe CW (1990) Technology, institutions, and politics: still out of balance. *Water Resources Research* **26**, 2249–50.

Hughes FMR & Rood SB (2001) Floodplains. In: A Warren & JR French (eds) *Habitat conservation: managing the physical environment.* Chichester, John Wiley, 105–121.

International Union for Conservation of Nature and Natural Resources (1980) *World Conservation Strategy.* Gland, IUCN.

Junk WJ, Bayley PB & Sparks RE (1989) The floodpulse concept in river–floodplain systems. *Special Publication of the Canadian Journal of Fisheries & Aquatic Sciences* **106**, 110–27.

Kemp JL, Harper DM & Crosa GA (1999) Use of 'functional habitats' to link ecology with morphology and hydrology in river rehabilitation. *Aquatic Conservation: Marine & Freshwater Ecosystems* 9, 159–78.

Kondolf GM (1998). Lessons learned from river restoration projects in California. *Aquatic Conservation: Marine & Freshwater Ecosystems* 8, 39–52.

Lambert TR and Hanson DF (1989) Development of habitat suitability criteria for trout in small streams. *Regulated Rivers* 3, 291–304.

Lane SN & Richards KS (1997) Linking river channel form and process: time, space and causality revisited. *Earth Surface Processes & Landforms* 22, 249–60.

Laronne JB & Carson MA (1976) Interrelationships between bed morphology and bed material transport for a small gravel bed channel. *Sedimentology* 23, 67–85.

Lisle TE (1982) Effects of aggradation and degradation on riffle–pool morphology in natural gravel channels, northwestern California. *Water Resources Research* 18, 1643–51.

Lisle TE & Hilton S (1992) The volume of fine sediment in pools: an index of sediment supply in gravel-bed streams. *Water Resources Bulletin* 28, 371–83.

Loughran RJ, Campbell BL & Elliott, GL (1986) Sediment dynamics in a partially-cultivated catchment in New South Wales, Australia. *Journal of Hydrology* 83, 285–97.

Maybeck M, Chapman D & Helmer R (1980) *Global environment monitoring system global freshwater quality: a first assessment.* Oxford, Basil Blackwell.

Miller JR, Lechler PJ & Desilets M (1998) The role of geomorphic processes in the transport and fate of mercury in the Carson River basin, west-central Nevada. *Environmental Geology* 33, 249–62.

Milhous RT, Updike MA, & Schneider DM (1989) Physical Habitat Simulation System Reference Manual — Version II. Instream Flow Information Paper No. 26. *US Fish and Wildlife Service Biological Report* 89(16), Washington, DC.

Morris SE (1995) Geomorphic aspects of stream-channel restoration. *Physical Geography* 16, 444–59.

Mosley MP (1985) River channel inventory, habitat and instream flow assessment. *Progress in Physical Geography* 9, 494–523.

Newbury R (1995) Rivers and the art of stream restoration. In JE Costa, AJ Miller, W Potter and PR Wilcock (eds) Natural and anthropogenic influences in fluvial geomorphology. *Geophysical Monograph 89*. American Geophysical Union, Washington, 239 pp.

Newbury R & Gaboury M (1993) Exploration and rehabilitation of hydraulic habitats in streams using principles of fluvial behaviour. *Freshwater Biology* 29, 195–210.

Newson MD (1992) *Land water and development.* London, Routledge.

Newson MD, Clark MJ, Sear DA & Brookes, A (1998a) The geomorphological basis for classifying rivers. *Aquatic Conservation: Marine & Freshwater Ecosystems* 8, 415–30.

Newson MD, Harper DM, Padmore CL, Kemp JL & Vogel B (1998b) A cost-effective approach for linking habitats, flow types and species requirements. *Aquatic Conservation: Marine & Freshwater Ecosystems* 8, 431–46.

O'Grady MF, King JJ & Curtin J (1991) The effectiveness of two physical in-stream work programmes in enhancing salmonid stocks in drained Irish lowland river system. In: D Mills (ed.) *Strategies for the rehabilitation of salmon rivers.* London, Linnean Society, 154–78.

Ongley ED (1992) Environmental quality: changing times for sediment programmes. *International Association of Hydrological Sciences Publication* 210, 209–18.

Osbourne LL, Bayley PB, Higler LWG, Statzner B, Triska F & Moth Iversen T (1993) Restoration of lowland streams: an introduction. *Freshwater Biology* 29, 187–94.

Petts GE & Amoros C (eds) (1996) *Fluvial hydrosystems.* London, Chapman & Hall.

Petts GE & Calow P (eds) (1996) *River restoration.* Basil Blackwell, Oxford, 231 pp.

Petts GE & Maddock I (1994) Flow allocation for in-river needs. In: P Calow & GE Petts (eds) *Rivers handbook*, Volume 2. Oxford. Blackwell Scientific, 289–307.

Pollard P & Huxham M (1998) The European water framework directive: a new era in the management of aquatic ecosystem health? *Aquatic Conservation: Marine & Freshwater Ecosystems* **8**, 773–92.

Raleigh RF, Miller WJ & Nelson PC (1986) Habitat suitability index models and instream flow suitability curves: Chinook Salmon. *Biological Report 82* (10.122). US Fish and Wildlife Service, Fort Collins, Colorado.

Raven PJ, Boon PJ, Dawson FH & Ferguson AJD (1998). Towards and integrated approach to classifying and evaluating rivers in the UK. *Aquatic Conservation: Marine & Freshwater Ecosystems* **8**, 383–93.

Richards KS (1979) Channel adjustment to sediment pollution by the china clay industry in Cornwall, England. In: D Rhodes & G P Williams (eds) *Adjustments of the fluvial system*. London, Allen and Unwin, 309–31.

Richards KS (1982) *Rivers: form & process in alluvial channels*. London, Methuen.

Richards KS (1999) The magnitude–frequency concept in fluvial geomorphology: a component of a degenerating research programme? *Zeitschrift für Geomorphologie*, NF, Supp. **115**, 1–18.

Rigg J (1995) In the fields, there is dust. *Geography* **80**, 23–32.

River Restoration Centre (1999) *Manual of river restoration techniques*. Silsoe, RRC.

Rosgen D (1996) *Applied river morphology*. Wetland Hydrology, Colorado.

Royal Society for the Protection of Birds (RSPB) (1994) *The new rivers & wildlife handbook*. Sandy, RSPB.

Schumm, SA & Lichty RW (1965) Time, space and causality in geomorphology. *American Journal of Science* **263**, 110–19.

Shafroth PB, Auble GT, Stromberg JC & Patten DT (1998) Establishment of woody riparian vegetation in relation to annual patterns of streamflow, Bill Williams River, Arizona. *Wetlands* **18**, 577–90.

Tapsell SM (1995) River restoration: what are we restoring to? A case study of the Ravensbourne river, London. *Landscape Research* **20**, 98–111.

Toth LA, Arrington DA, Brady MA & Muszick DA (1995) Conceptual evaluation of factors potentially affecting habitat restoration structure within the channelized Kissimmee river ecosystem. *Restoration Ecology* **3**, 160–80.

Trimble SW (1997) Streambank fish-shelter structures help stabilize tributary streams in Wisconsin. *Environmental Geology* **32**, 230–4.

VanColler AL, Rogers KH & Heritage GL (1997) Linking riparian vegetation types and fluvial geomorphology along the Sabie River within the Kruger National Park, South Africa. *African Journal of Ecology* **35**, 194–212.

Walling DE (1983) The sediment delivery problem. *Journal of Hydrology* **65**, 209–37.

Whelan KF (1991) An overview of techniques used in Atlantic salmon restoration and rehabilitation programmes. In: D Mills (ed.) *Strategies for the rehabilitation of salmon rivers*. London, Linnean Society, 6–18.

Woolhouse CH (1989) Managing the effects of urbanisation in the Upper Lee. British Hydrological Society National Symposium, Sheffield.

Wohl EE, Vincent KR & Merritts DJ (1993) Pool and riffle characteristics in relation to channel gradient. *Geomorphology* **6**, 99–110.

Young WJ (1993) Field techniques for the classification of near-bed flow regimes. *Freshwater Biology* **29**, 377–83.

Zalewski M, Bis B, Lapinski M, Frankiewicz P & Puchalski W (1998) The importance of the riparian ecotone and river hydraulics for sustainable basin-scale restoration scenarios. *Aquatic Conservation: Marine & Freshwater Ecosystems* **8**, 287–307.

5 Floodplains

F. M. R. HUGHES[1] AND S. B. ROOD[2]
[1]*University of Cambridge, UK*
[2]*University of Lethbridge, Canada*

INTRODUCTION

Floodplains are unique linear landscapes. They are among the most important ecosystems on the planet because they are highly productive and have a high species diversity. They also play important roles in flood holding and attenuation and in rerouting pollutants between surrounding areas and the river.

Floodplains are complex physical features. They border rivers in areas of low gradients where the river channel is usually meandering, braided or anastomosed and highly mobile. The river's mobility is driven by the frequency and size of floods which bring sediments and nutrients to the floodplain as well as causing erosion. Floodplains therefore process large fluxes of energy and materials from upstream areas. Because of these inputs during floods, floodplains have a very varied microtopography with many depositional features composed of characteristic sediment suites. The duration, frequency, depth and season of flooding experienced on a floodplain are collectively referred to as the hydroperiod and this will vary across the floodplain during any flood event because of variations in microtopography and sediment porosity. The hydroperiod is especially critical in determining species composition of the vegetation in different parts of the floodplain because species which typically live on floodplains have well-defined tolerance limits to flooding depth and duration and are distributed at higher or lower elevations accordingly. It follows that species diversity on a floodplain is at least in part related to the physical complexity.

Floodplains are naturally very dynamic physical environments because they receive floods, usually on an annual or biannual basis. Floods vary from small annual-return period floods to extreme floods with return periods of 100 to 200 or more years. All floods carry out geomorphological work at either a small or a large spatial scale, causing disturbance or perturbations to the floodplain ecosystems. Disturbance by flooding has a number of effects. For example, short-term rearrangement of sediments through the processes of erosion,

Habitat Conservation: Managing the Physical Environment. Edited by A. Warren and J. R. French.
© 2001 John Wiley & Sons Ltd.

transport and deposition causes destruction of part of the floodplain ecosystem while simultaneously providing new sites suitable for the regeneration of vegetation. These highly disturbed environments therefore feature productive vegetation communities which have evolved at one level to tolerate disturbance and at another to depend on disturbance for regeneration opportunities. They have been described as r-landscapes (Warren 1993), characterised by rapid recovery following disturbance and r-selected species which have many adaptations to survive in a disturbed environment. In the case of many floodplain species, these adaptations include prolific seed production combined with an ability to spread vegetatively. Because floodplains are highly disturbed environments, successful regeneration depends on a window of opportunity providing just the right conditions for establishment following germination. This window of opportunity is very variable in its physical parameters depending on the bioclimatic zone. For example, in semi-arid areas, the window of opportunity consists of the availability of bare, newly deposited, moist sediments, coincident with seed dispersal and a water table which recedes sufficiently slowly through the first growing season that the seedlings do not die from drought. On the other hand, in a humid zone where floods can last longer, the window of opportunity consists of a long enough period without flooding for seedlings to become established and grow to a height which will take them above the level of the next flood. In a natural river system, regeneration occurs patchily over space and through time because of this dependence on very specific regeneration conditions provided by floods. In a river system which has been managed, it is frequently the case that suitable regeneration conditions occur far less frequently or never and this has many repercussions for floodplain biodiversity.

CONSERVATION ISSUES IN FLOODPLAINS

The small-scale spatial mosaic of geomorphological features which repetitive flood events promote and rearrange on a floodplain provides many ecotones and a highly diverse range of regeneration niches for plants. Consequently plant diversity per spatial unit and both yearly and seasonal use of a floodplain by wildlife may be high (Brinson et al. 1981). In parts of the Amazon basin it has been suggested that high species diversity in the rainforest is linked to the long-term history of disturbance through flooding and channel migration (Salo et al. 1986). Floodplains frequently act as biological corridors from one part of a landscape to another through terrain which would otherwise be inhospitable to both plant and animal species (Malanson 1993).

Their high species diversity and landscape-corridor function place flood-plains high on the conservation agenda. They are also among the most abused of ecosystems because of competing interests for resources. Dam construction, channelisation, water removal, gravel and sand extraction and forest clearance all affect the timing and quantity of water and sediment delivery to floodplains. These activities have progressively altered or destroyed the integrity of most floodplain ecosystems. The present-day conservation challenge is to reinstate the geomorphological dynamics of floodplains through the water-delivery vehicle of floods.

CONSERVING SPECIES DIVERSITY THROUGH REGULAR FLOOD DISTURBANCE

Floods promote regeneration of plants on floodplains by creating open, moist sites necessary for many riparian species. River damming and flood attenuation have generally led to loss of early successional species and to decline in floodplain forests in many regions. Reduced flood flows have also favoured asexual over sexual reproduction, leading to loss of genetic diversity (Rood et al. 1994). Unseasonal floods can have a similar effect by physically removing recently established seedlings and favouring vegetative propagation (Barsoum 1998).

Increases in species diversity and in the area occupied by floodplain wood-land have been noted downstream of some dams (for example, Johnson 1994, 1997; Stevens et al. 1995), although short-term increases over several decades may be followed by a decrease in diversity over ensuing decades in association with reduced physical disturbance and reduced site turnover (Hughes 1997). In this situation floodplains could become K-landscapes (Warren 1993), occupy-ing a similar spatial location but characterised by generally less opportunistic (K-selected) species, strong biotic controls and long return periods between floods. Such floodplains would presumably also be used by different animal species.

Thomas (1996) questioned whether this new, less geomorphologically dynamic floodplain landscape was any less valuable than the previous one or just different but equally valuable. The value of a floodplain can be measured in terms ranging from agricultural to wildlife, recreational to aesthetic (Yon and Tendron 1981). Perhaps changes in levels of alpha (within-habitat) and gamma (regional) diversity on floodplains through time could provide a biological measure of relative value. There is evidence that both types of biological diversity decrease in association with reduced geomorphological diversity (for example, Nilsson et al. 1991a).

If we view the downstream effects of dams on a long enough timescale (10^3 years) they become superimposed, often unidirectionally, on changes in the abiotic inputs to the floodplain ecosystem that have occurred naturally within the present inter-glacial period. Logically and intuitively, reintroduction of abiotic inputs where they have been removed offers the best chance of dynamic and sustainable recovery for degraded floodplains. However, taking decisions about restoring or rehabilitating floodplain ecosystems, where they have been degraded, is extremely complex in both physical and human terms (Adams and Perrow 1999).

CONSERVING FLOODPLAINS AS LANDSCAPE CORRIDORS

Floodplains are important linear links within the landscape matrix, important for both conservation of landscape and of nature and its diversity (Ward, Tockner and Schiemer 1999). Even if we retain flooding patterns can a fragmented river corridor still function as well for wildlife as a continuous one? Fragmentation of river corridors by dams and channelisation can adversely affect plant species dispersal and use by wildlife (Johansson, Nilsson and Nilsson 1996; Dynesius and Nilsson 1994). Riparian zones are also favoured regions for human settlement and agriculture. In Europe substantial forest clearing has occurred since the early nineteenth century, while in North America, clearing has been both more recent and more rapid (Petts 1990). This deliberate clearing, along with livestock trampling and browsing of seedlings and saplings, has led to the collapse of riparian woodlands in many regions worldwide and has reduced floodplain habitats to a series of disconnected patches. Loss of floodplain forests has received recent attention because they provide exceptionally rich habitats for terrestrial wildlife and birds (Finch and Ruggiero 1993). Considerations of species range and the connectivity and size of the remaining floodplain fragments are important to the assessment of their potential to maintain high levels of biodiversity. In some semi-arid areas, floodplain forests are remnants of formerly continuous rainforest, maintained by the shallow water tables adjacent to active channels. A good example is the floodplain forest along the Tana River in Kenya. This is an evergreen forest with high diversity: 175 woody plant species, more than 250 bird species and at least 57 mammal species have been recorded, including several endemics such as *Populus ilicifolia* and two species of primates. The forest occupies a narrow belt several hundred kilometres long and approximately 2–5 kilometres wide in an otherwise semi-arid area dominated by *Acacia-Commiphora* bush (Medley and Hughes 1996). Biogeographically, therefore, floodplain forests can be important refuges and centres of biodiversity maintained by flooding regimes and the accessibility of the water table. In some locations, floodplain bio-diversity is high because, on an evolutionary timescale, fragmentation of some elements of the floodplain ecosystem have permitted development of endemism. A good example of this is seen in the fish fauna of floodplain lakes on the Phongolo floodplain of South Africa (Heeg, Breen and Rogers 1980).

Floodplain woodlands also contribute substantially to the adjacent aquatic ecosystems through shading, nutrient contributions with leaf litter (Wallace *et al.* 1997), and the removal of agricultural pollutants such as phosphorus and nitrogen which are associated with surface runoff and groundwater that enter the stream (Naiman and Décamps 1997).

THE PHYSICAL PROCESSES IMPORTANT IN CONSERVING FLOODPLAINS

While the physical conditions of the floodplain are largely formed by flood events, the associated riparian ecosystems are also adapted to and dependent upon the physical disturbance that is provided by periodic floods (Cordes, Hughes and Getty 1997; Friedman, Osterkamp and Lewis 1996; Hughes 1994; Johnson, Burgess and Keammerer 1976; Scott, Friedman and Auble 1996; Scott, Auble and Friedman 1997; Stromberg, Fry and Patten 1997; Stromberg, Patten and Richter 1991; Van Splunder *et al.* 1995). Floodplains are periodically inundated when the stream rises following rainfall or rapid snowmelt. The floods produce high stream velocities that are competent to erode and deposit materials, enabling bar formation and channel migration. These dynamic geomorphological patterns are essential to create new, barren areas for recruitment of riparian (floodplain) vegetation. The riparian species are ecological 'pioneers', plants that have prolific reproductive potential, but are non-competitive with established vegetation (Braatne, Rood and Heilman 1996). Thus, cottonwoods (*Populus* spp.), willows (*Salix* spp.) and other riparian phreatophytes require moist and barren nursery sites that are provided after flood peaks; the receding flood waters progressively expose freshly scoured or deposited areas that are suitably saturated for seedling establishment.

The pioneering riparian tree species are relatively short-lived; for example, cottonwoods seldom survive into their second century (Braatne, Rood and Heilman 1996). For their survival as a species the riparian zone must be systematically reworked by the meandering channel that should traverse the woodland floodplain within the trees' lifespan. Flood flow attenuation diminishes the rate of stream migration and can result in channel entrenchment that dissociates the stream from the floodplain (Scott, Auble and Friedman 1997; Rood and Mahoney 1995). This leads to a deficiency in replenishment of the pioneer species. In arid and semi-arid areas in which precipitation is insufficient to support trees, the failure of riparian replenishment will lead to a progressive (or abrupt) loss of the riparian woodland (Rood and Mahoney 1990). In wetter areas, woodland succession would progress and the pioneer trees such as the cottonwoods would be replaced by later successional species, and often conifers. These climax species generally grow more slowly and produce sparser wildlife habitats than the deciduous pioneer trees.

The role of flood-associated disturbance in riparian woodlands parallels the role of fire in some upland forests. In North America, forest fires were historically considered undesirable and park management policies in the mid-1900s often imposed rigorous fire suppression. It is now understood that fire in upland forests provides an essential and natural disturbance that triggers a replenishment cycle that favours many plant species and creates rich wildlife habitat. Similarly, floods are known to erode banks and topple trees and in the immediate term, their impacts may seem unfavourable. However, in the longer timescale of decades to centuries, flood disturbance is essential to maintain the ecological richness and diversity of riparian woodlands (Malanson 1993; Hughes 1997).

Thus floods fulfil an essential role in floodplain ecosystems. They drive the geomorphic processes that underlie the physical structure of the floodplain and also directly and indirectly enable the development of the biological components. The mutual impact on physical and biological processes provides a persistent record of historical flood events. Riparian trees such as cottonwoods are often organised in arcuate bands, rows of even-aged trees positioned at uniform elevations above the stream edge, the rows curving to match the meandering stream channel (Braatne, Rood and Heilman 1996). Each band originated with a recruitment event that was linked to a particular flood that was suitable in timing and pattern for vegetation establishment (Table 5.1). The progressively older bands of trees provide a flood history and demonstrate channel position at the time of their establishment (Everitt 1968; Sigafoos 1964). Through dendrochronological age analyses (tree ring studies), the floodplain history can be determined for the past century or longer (Figure 5.1).

Because floodplains are frequently disturbed and rapidly changing environments it could be argued that if their disturbance regime is reinitiated (for example, by the removal of a dam or by planned flood releases downstream of a dam) then their recovery to some form of previous dynamic equilibrium state should occur rapidly because they are r-adapted landscapes. However, Walker and Thoms (1993) in the Lower Murray argue that the new hydrological regime created below a dam is outside the tolerance range of disturbance events that a floodplain normally experiences and therefore it cannot adjust to the new regime except through permanent change.

CONSERVATION AND RESTORATION OF FLOODPLAIN/RIPARIAN ECOSYSTEMS

LOCAL CONSERVATION OF FLOODPLAIN ECOSYSTEM COMPONENTS (CONSERVATION OF FORM)

With the recognition of the collapse of riparian woodlands, especially in semi-arid regions, and the growing appreciation of their environmental and

Table 5.1. Flood return periods significant to poplar regeneration on selected floodplains. (Modified from Mahoney and Rood 1998)

River	Flood return period (years)	Significance for regeneration	Authors
Milk River, Alberta, Canada	5	Regeneration of *Populus deltoides* on active floodplain edge	Bradley and Smith (1986)
Animas River, Colorado, USA	3	Modelled frequency of good seedling establishment years for *Populus angustifolia*	Baker (1990)
Animas River, Colorado, USA	10–15	Measured frequency of initiation of stands of *Populus angustifolia*	Baker (1990)
Tana River, Kenya	1–2	Regeneration of *Populus ilicifolia* on active channel margins	Hughes (1990)
Bow River, Alberta, Canada	10–20	Regeneration of stands of *Populus deltoides*	Cordes (1991)
Rio Grande, New Mexico, USA	10	Regeneration of *Populus fremontii*	Howe and Knopf (1991)
Milk River, Alberta, Canada	> 5	Regeneration of *Populus deltoides*	Reid (1991)
Hassayampa, Arizona, USA	7	Regeneration of *Populus fremontii*	Stromberg, Fry and Patten (1991)
Red Deer River, Alberta, Canada	10	Regeneration of *Populus deltoides*	Marken (1993)
Colorado River, Utah, USA	10	Regeneration of *Populus fremontii*	Rood, Kalischuk and Polzin (1997)
Missouri River, USA	9	Regeneration of *Populus deltoides*	Scott, Auble and Friedman (1997)
Red Deer River, Alberta, Canada	> 10	Regeneration of *Populus deltoides* on channel margins	Cordes, Hughes and Getty (1997)
Red Deer River, Alberta, Canada	100	*Populus deltoides* regeneration across 50% of the floodplain	Cordes, Hughes and Getty (1997)
Bow River, Alberta, Canada	5–10	Regeneration of *Populus balsamifera*	Rood, Bradley and Taboulchanas 1998a

Figure 5.1. Aerial photograph of the lower Red Deer River near Dinosaur Provincial Park in southern Alberta showing parallel bands of *Populus deltoides* in a meandering floodplain. On either side of the floodplain are badlands which restrict the width of the floodplain.

ultimately economic value, various conservation and restoration efforts have been initiated. Such efforts have commonly involved physical intervention such as re-excavation of meanders in straightened reaches (for example, The River Restoration Project in Europe; Holmes 1998) and vegetation planting to re-establish riparian species such as poplars and willows. Regrettably, exotic species have sometimes been deliberately introduced for streambank stabilisation and riparian 'restoration'. For example, the salt cedar, *Tamarix* spp., was originally imported to the American south-west for streambank revegetation. The plant has subsequently thrived along virtually all waterways in the warm regions of the American west and has excluded the native willows and cottonwoods, often influencing channel-bar morphology, creating monotypic riparian stands that are impoverished with respect to plant and animal species (Birkeland 1996). In a study of eight rivers in south-western France the highest numbers of invading exotics were found in floodplain reaches which were the least physically disturbed (Planty-Tabacchi 1997).

Biodiversity has been affected in a more subtle way in parts of Britain where recent planting of willows in order to stabilise river banks has deliberately favoured male over female stock in order to avoid what is viewed as 'undesirable' production of the cotton-encased seeds. At the present time there is particular interest in the conservation of Black Poplar (*Populus nigra* subsp. *betulifolia*) in Europe, an early successional species in riparian zones which has greatly decreased in abundance because of river-control measures and forest clearance. Plans are being made to conserve the species by creating clone banks from material collected throughout its range in Europe and by planting it in restoration schemes using local stock (Turok *et al*. 1996). Where it still occurs along rivers, Tabbush (1997) distinguishes between static stands of Black Poplars, which might require irrigation for their continued health, and dynamic stands which have remained in some localities where geomorphological processes and natural regeneration are still active ecosystem processes, though often reduced in extent by hydroelectric installations (Pautou and Girel 1994).

Replanting of floodplain forests has implications for sedimentation patterns and flood control, especially in many European rivers where such forests have not existed for several centuries. The feasibility is therefore contentious (Peterken and Hughes 1998) but is being seriously addressed in some countries such as Holland where modelling of the hydraulic roughness of floodplains with different vegetation cover and calculations of critical extents and strategic locations of planting programmes is taking place (Van Splunder 1997).

SYSTEMIC CONSERVATION OF FLOODPLAIN ECOSYSTEMS (CONSERVATION OF FUNCTION)

In the last decade a new approach to riparian restoration has gained momentum. Since the breakdown of floodplain processes along many streams has been partially caused by flood-flow attenuation and artificial, and sometimes arbitrary, patterns of stream-flow regulation, efforts are underway to recover more natural instream flow patterns that will enable recovery of the natural processes (Gourley *et al*. 1997; Rood, Kalischuk and Mahoney 1998b). Some level of flooding is permitted and even encouraged (Collier, Webb and Andrews 1997), to re-establish the geomorphic processes of erosion and deposition that create the barren nursery sites for recruitment of native riparian species. The timing of the high flows is important, since the peak should precede seed dispersal — germinants established prior to the flood peak will be scoured away by the rising stream. Flood-peak timing is naturally somewhat variable, but within a region, the native plants are generally adapted to the natural hydrograph and seed dispersal is often cued by photoperiod (daylength) and modified by temperature and is thus synchronised to follow the typical flood peak (Mahoney and Rood 1998). Along dammed rivers, the timing of peak flows is often substantially altered. This unnatural scheduling may promote the encroachment of invasive exotics that compete and often exclude native plants. Conversely, informed dam operation would involve consideration for the seasonality of hydrograph patterns that could serve as a management tool to favour the native plants and prevent further proliferation of undesirable exotics (Shafroth *et al.* 1995).

Coincidental but independent studies have revealed that the recession limb of the flood hydrograph is particularly important for the successful recruitment of floodplain vegetation (Mahoney and Rood 1992; Segelquist, Scott and Auble 1993). Gradual recession 'ramping' rates are required in which stream stages and corresponding riparian water table levels decline at rates that can be matched by root growth of the seedlings that are dependent on the moisture associated with the capillary fringe above the riparian water table and are especially vulnerable to drought stress (Mahoney and Rood 1998). After initial recruitment of riparian vegetation, stream flows during the subsequent hot and dry periods must be sufficient to provide adequate water for the riparian phreatophytes and thus minimum instream flows are required for floodplain ecosystems (Rood and Mahoney 1990).

The combined requirement for appropriately timed high flows to create and saturate suitable floodplain sites and subsequent gradual flow recession to permit seedling survival have been quantitatively described in the floodplain 'Recruitment Box Model' (Figure 5.2; Mahoney and Rood 1998). This model incorporates the elevation of the floodplain inundation zone in which riparian seedlings could successfully establish and requires that the stream stage recedes through this zone during the period of seed dispersal. The subsequent stage recession should be about 2.5 cm per day or less, although this decline rate is influenced by floodplain substrate texture, plant species and the ambient weather conditions related to water demand, particularly temperature, rainfall events, wind and sunshine (Mahoney and Rood 1992, 1998; Hughes *et al.* 1997; Barsoum and Hughes 1998).

Studies are underway in the semi-arid regions of western North America where riparian cottonwood forests provide the only native trees. These riparian woodlands have been severely degraded by human impacts including direct clearing and the effects of river damming and stream flow regulation. The conservation and restoration of these woodlands has developed as a major environmental priority and the current and imminent relicensing of many of these dams is dependent on reconsideration of operations, practices and environmental impacts. With a general appreciation of the importance of dynamic stream flows for riparian ecosystems, hydrologists, geomorphologists and riparian ecologists are working towards the development of instream flow 'prescriptions' that will permit conservation of remaining riparian woodlands and restoration of degraded riparian ecosystems.

Two important aspects related to the applicability of instream flow prescriptions deserve consideration: (1) the generalisability of instream flow

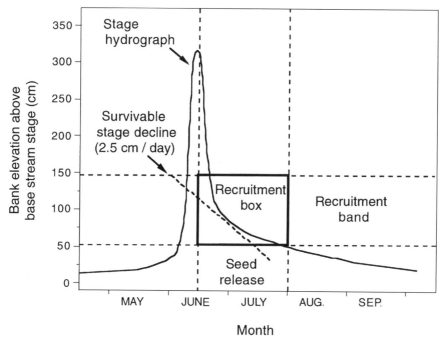

Figure 5.2. The relationship between a generalised hydrograph for southern Alberta and the 'Recruitment Box' including the survivable rate of water table decline for cottonwood seedlings. (From Mahoney and Rood 1998.)

prescriptions for riparian restoration along different rivers and in different regions; and (2) the applicability of common instream flow prescriptions for different riparian trees and shrubs (Shafroth *et al.* 1998). The current 'flow' prescription strategies are actually based on river and floodplain stage (elevation) rather than river discharge (flow). This is appropriate for riparian processes since it is the floodplain stage elevation rather than the amount of water passing through that primarily determines the physical and ecological impact (Mahoney and Rood 1998). The conversion of all data to stage data produces quantitative consistency across a broad range of stream sizes. For example, seedlings of riparian cottonwoods are established in bands at specific floodplain elevations that parallel the stream channel. Across western North America, these bands occur at relatively consistent elevations ranging from about 60 to 200 cm above the base stream stage (the low river elevation that usually occurs towards the end of the plants' growing season). This elevational range is relatively consistent across about 20 degrees of latitude and for streams ranging from relatively small creeks to large rivers. However, the generalisability of flow prescriptions becomes more difficult where there is an element of bedrock control along a river channel as described for the Sabie River of South

Africa by Moon *et al.* (1997) and Van Coller, Rogers and Heritage (1997). Here models at the reach scale, where each reach was composed of different combinations of morphological units, were found to be more useful for predicting changes in geomorphological processes, channel form and dependent vegetation types.

The second consideration relates to the applicability of common instream flow prescriptions for different riparian trees and shrubs. Studies along the Truckee River, Nevada (Gourley *et al.* 1997); the Oldman, St Mary and Bow rivers, Alberta (Rood, Kalischuk and Mahoney 1998b; Rood *et al.* 1995); and the Bill Williams River, Arizona consistently indicate that native willows and cottonwoods and the exotic salt cedar and Russian Olive have generally similar stream stage requirements (Shafroth *et al.* 1998). With respect to control of exotic encroachment, differences in phenology and particularly the timing of seed release, could provide the basis for deliberate management to favour the desirable native trees and shrubs and disfavour the undesirable exotics. The situation is more complicated along rivers whose vegetation types and associated flow needs change along their lengths as described by Nilsson *et al.* (1991b) for rivers in northern Sweden, by Tabacchi and Planty-Tabbachi (1996) for rivers in south-western France and Spain and by Cordes, Hughes and Getty (1997) for the Red Deer River in Alberta, Canada. In these cases, since the flow needs for regeneration of riparian species in an unmanaged river system are satisfied in only certain years (see Table 5.1), flow prescriptions for regeneration need only be implemented in selected years. Different flow prescriptions can, therefore, be recommended in different years to satisfy overall demands of the particular riparian ecosystem.

Through the 1990s, these instream flow prescriptions have actually been implemented for various streams in North America. The deliberate release of a moderate flood flow from the Glen Canyon Dam led to significant sediment remobilisation and beach-building in the Grand Canyon of the Colorado River in Arizona (Collier, Webb and Andrews 1997). Unfortunately the beaches were rapidly eroded following the flood 'pulse' and the long-term geomorphic and ecological impacts are likely to be slight.

More sustained alterations to dam-operation practices have been imposed to recover floodplain processes and conserve and/or restore riparian ecosystems along the Truckee River in Nevada (Gourley *et al.* 1997), the Oldman and St Mary rivers in Alberta (Rood, Kalishuk and Mahoney 1998b) and the Bill Williams River in Arizona (Shafroth *et al.* 1998). Similar adjustments to dam operations to re-establish dynamic floodplain processes for the benefit of riparian and aquatic ecosystems are being considered for numerous other rivers such as the Boise and Snake rivers in Idaho, the Marias River in Montana, the Green River in Wyoming and Utah, and various other cottonwood-dominated rivers in semi-arid regions of western North America.

In the cases of the Truckee, Oldman and St Mary rivers, flood flows have persisted after damming and will continue to occur periodically because the reservoirs have insufficient capacity significantly to attenuate major floods. Additionally, the operational objectives of maximising water storage for irrigation and leaving vacant reservoir capacity for flood flow capture are contradictory management objectives. With the recognition that flooding will occur, refinements to operations patterns for the dams along these rivers were directed towards the delivery of gradual flow recessions after flood peaks. This operational strategy was intended to promote the recruitment of riparian vegetation although in cases such as the Boise River in Idaho, the gradual flow recession is also anticipated to favour stream bank stabilisation — abrupt river stage decline would expose saturated stream banks, which are vulnerable to slumping.

While it is too early to appraise the long-term 'success' of these riparian restoration efforts, initial responses have been both more dramatic and more immediate than anticipated. In the case of the Truckee River, over a twenty-year period of revised flow regulation directed towards the recovery of an endangered fish, the cui-ui sucker (*Chasmistes cujus*), and restoration of the Fremont cottonwood (*Populus fremontii*) groves, the stream channel has become narrower and deeper and focused into a single channel rather than the shallow and broad, braided channels that had followed a century of diversion, damming and channelisation. The stream water became cooler and aquatic conditions vastly more favourable for both the endangered cui-ui and the Lohantan cutthroat trout. The riparian vegetation was successfully recruited with extensive bands of Fremont cottonwoods and coyote willows (*Salix exigua*) and these shrubs and trees further defined and confined the stream channel. This has led to further narrowing and deepening of the stream channel. The shading provided by the riparian vegetation also contributed to cooler water temperatures. The recovery of recruitment of riparian vegetation has resulted in a more complex woodland structure, which has already led to substantial recovery in both diversity and abundance of bird species. Thus, the Truckee River case study demonstrates that stream flow patterns provide the physical foundation for the whole floodplain system, with modifications to instream flow patterns having rapid and dramatic impacts on stream channel geometry and both aquatic and riparian ecosystems.

CONCLUSIONS

There are many options for the conservation of floodplain ecosystems involving local or reach-scale activities which may be practical to implement but are probably not sustainable. The real key to conservation of floodplain ecosystems is maintenance or reinstatement of hydrological pathways and geomorphological processes. However, this systemic approach to floodplain conservation requires whole-catchment planning and is much more difficult to implement because the inputs of water and sediment to a floodplain are derived upstream of the floodplain and are influenced by human activities at many points in the catchment.

REFERENCES

Adams WM & Perrow M (1999) Scientific and institutional constraints on the restoration of European floodplains. In: S Marriot, J Alexander & R Hey (eds) *Floodplains: interdisciplinary approaches*. Geological Society of London Special Publication **163**, 89–97.

Baker WL (1990) Climatic and hydrologic effects on the regeneration of *Populus angustifolia* Lames along the Animas River, Colorado. *Journal of Biogeography* **17**, 59–73.

Barsoum N (1998) *A comparison of vegetative and non-vegetative regeneration strategies in* Populus nigra *and* Salix alba. PhD thesis, University of Cambridge, UK.

Barsoum N & Hughes FMR (1998) The response of vegetative and non-vegetative propagules of *Populus nigra* var *betulifolia* to simulated water table management in different sediments. In: WH Wheater, C Kirby, R Harding & D Gilvear (eds) *Hydrology in a changing environment*. Chichester, John Wiley, 397–412.

Birkeland GH (1996) Riparian vegetation and sandbar morphology along the lower Little Colorado River, Arizona. *Physical Geography* **17**, 534–53.

Braatne JH, Rood SB & Heilman PE (1996) Life history, ecology and conservation of riparian cottonwoods in North America. In: RF Stettler, JHD Bradshaw, PE Heilman & TM Hinckley (eds) *Biology of Populus and its implications for management and conservation*. Ottawa, NRC Research Press, 57–85.

Bradley CE & Smith DG (1984) Meandering channel response to altered flow regime: Milk River, Alberta and Montana. *Water Resources Research* **20**, 1913–20.

Bradley CE & Smith DG (1986) Plains cottonwood recruitment and survival on a prairie meandering river floodplain, Milk River, southern Alberta and northern Montana. *Canadian Journal of Botany* **64**, 1433–42.

Brinson MM, Swift BL, Plantico RC & Barclay JS (1981) *Riparian ecosystems: their ecology and status*. Washington DC, USA Fish and Wildlife Service OBS-81/17.

Collier MP, Webb RH & Andrews ED (1997) Experimental flooding in Grand Canyon. *Scientific American* **276**(1), 82–9.

Cordes LD (1991) The distribution and age structure of cottonwood stands along the lower Bow River, Alberta. In: SB Rood & J Mahoney (eds) *Proceeedings of the biology and management of southern Alberta's cottonwoods conference*. University of Lethbridge, Lethbridge, Alberta, Canada, 13–24.

Cordes LD, Hughes FMR & Getty M (1997) Factors affecting the regeneration and distribution of riparian woodlands along a northern prairie river: the Red Deer River, Alberta, Canada. *Journal of Biogeography* **24**, 675–95.

Dynesius M & Nilsson C (1994) Fragmentation and flow regulation of river systems in the northern third of the world. *Science* **266**, 753–62.

Everitt B (1968) Use of the cottonwood in an investigation of the recent history of a flood plain. *American Journal of Science* **266**, 417–39.

Finch DM & Ruggiero LR (1993) Wildlife and biological diversity in the Rocky Mountains and north Great Plains. *Natural Areas Journal* **13**, 191–203.

Friedman JM, Osterkamp WR & Lewis WM (1996) Channel narrowing and vegetation development following a Great Plains flood. *Ecology* **77**, 2167–81.

Gourley C, Rood SB, Klotz J & Swanson S (1997) Instream flows and the restoration of river and riparian ecosystems along the lower Truckee River, Nevada. In: *Wetlands Heritage and Stewardship. Abstracts of the Society of Wetland Scientists 18th Annual Meeting*. Bozeman, MN, USA, 66.

Heeg J, Breen CM & Rogers KH (1980) The Pongolo Floodplain: A unique ecosystem threatened. In: MN Bruton & KH Cooper (eds) *Studies on the ecology of Maputaland*. Capetown, Cape and Transvaal Printers for Rhodes University and the Natal Branch of the Wildlife Society of Southern Africa, 374–81.

Holmes NTH (1998) Floodplain restoration. In: RG Bailey, PV Jose & BR Sherwood (eds) *United Kingdom floodplains*. Westbury, Samara, 423–36.

Howe WH & Knopf FL (1991) On the imminent decline of Rio Grande cottonwoods in central New Mexico. *Southwestern Naturalist* **36**(2), 218–24.

Hughes FMR (1990) The influence of flooding regimes on forest distribution and composition in the Tana River floodplain, Kenya. *Journal of Applied Ecology* **27**, 475–91.

Hughes FMR (1994) Environmental change, disturbance and regeneration in semi-arid floodplain forests. In: AC Millington & K Pye (eds) *Environmental change in drylands: biogeographical and geomorphological perspectives*. Chichester, John Wiley, 321–46.

Hughes FMR (1997) Floodplain biogeomorphology. *Progress in Physical Geography* **21**, 501–29.

Hughes FMR, Harris T, Richards K, Pautou G, El Hames A, Barsoum N, Girel J, Peiry JL & Foussadier R (1997) Woody riparian species response to different soil moisture conditions: laboratory experiments on *Alnus incana* (L.) Moench. *Global Ecology and Biogeography Letters* **6**, 247–56.

Johansson ME, Nilsson C & Nilsson E (1996) Do rivers function as corridors for plant dispersal? *Journal of Vegetation Science* **7**, 593–8.

Johnson WC (1994) Woodland expansion in the Platte River, Nebraska: patterns and causes. *Ecological Monographs* **64**, 45–84.

Johnson WC (1997) Equilibrium response of riparian vegetation to flow regulation in the Platte River, Nebraska. *Regulated Rivers: Research and Management* **13**, 403–15.

Johnson WC, Burgess RL & Keammerer WR (1976) Forest overstory vegetation and environment on the Missouri River floodplain in North Dakota. *Ecological Monographs* **46**, 59–84.

Mahoney JM & Rood SB (1992) Response of a hybrid poplar to water table decline in different substrates. *Forest Ecology and Management* **54**, 141–56.

Mahoney JM & Rood SB (1998) Streamflow requirements for cottonwood seedling recruitment — an integrative model. *Wetlands* **18**, 634–45.

Malanson GP (1993) *Riparian landscapes*. Cambridge Studies in Ecology series. Cambridge, Cambridge University Press.

Marken SL (1993) *Plains cottonwodds and riparian plant communities on the lower Red Deer River*. MSc thesis, University of Calgary, Alberta, Canada.

Medley K & Hughes FMR (1996) Riverine Forests. In: McClanahan TR & Young TP (eds) *East African ecosystems and their conservation*. New York, Oxford University Press, 361–84.

Moon BP, van Niekerk AW, Heritage GL, Rogers KH & James CS (1997) A geomorphological approach to the ecological management of rivers in the Kruger National Park: the case of the Sabie River. *Transactions of the Institute of British Geographers* **NS22**, 31–48.

Naiman RJ & Décamps H (1997) The ecology of interfaces: Riparian Zones. *Annual Review of Ecological Systematics* **28**, 621–58.

Nilsson C, Ekblad A, Gardfjell M & Carlberg B (1991a) Long-term effects of river regulation on river margin vegetation. *Journal of Applied Ecology* **28**, 963–87.

Nilsson C, Grelsson G, Dynesius M, Johansson M & Sperens U (1991b) Small rivers behave like large rivers: effects of post-glacial history on plant species richness along riverbanks. *Journal of Biogeography* **18**, 541–33.

Pautou G & Girel J (1994) Interventions humaines et changements de la végétation alluviale dans la vallée de l'Isère de Montmelian à Port St Gervais. *Revue de Géographie Alpine* **RXXXII**, 27–46.

Peterken GF & Hughes FMR (1998) Limitations and opportunities for restoring floodplain forest in Britain. In: RG Bailey, PV Jose & BR Sherwood (eds) *United Kingdom floodplains*. Westbury, Samara, 423–36.

Petts GE (1990) Forested river corridors: A lost resource. In: D Cosgrove & GE Petts (eds) *Water, engineering and landscape transformation in the modern period*. London, Belhaven Press, 12–33.

Planty-Tabacchi AM (1997) Invasions des corridors fluviaux du sud-ouest par des espèces végétales exotiques. *Bulletin Français de Pêches et Pisciculture* **344/345**, 427–39.

Reid DE (1991) Cottonwoods of the Milk River. In: SB Rood & J Mahoney (eds) *Proceeedings of the biology and management of southern Alberta's cottonwoods conference*. University of Lethbridge, Alberta, Canada, 35–42.

Rood SB, Hillman C, Sanche T & Mahoney JM (1994) Clonal reproduction of riparian cottonwoods in southern Alberta. *Canadian Journal of Botany* **72**, 1766–74.

Rood SB, Bradley C & Taboulchanas K (1998a) *Instream flows and riparian vegetation along the middle Bow River, Alberta*. Report prepared for Alberta Environmental Protection, Edmonton, Alberta, Canada.

Rood SB, Kalischuk AR & Mahoney JM (1998b) Initial cottonwood seedling recruitment following the flood of the century along the Oldman River, Alberta, Canada. *Wetlands* **18**, 557–70.

Rood SB, Kalischuk AR & Polzin ML (1997) *Canyonlands cottonwoods: mortality of Fremont cottonwoods in the Matheson Wetlands Preserve along the Colorado River at Moab, Utah*. Submission to The Nature Conservancy, Moab, UT. University of Lethbridge, AB, 38 pp.

Rood SB & Mahoney JM (1990) Collapse of riparian poplar forests downstream from dams in Western Prairies: Probable causes and prospects for mitigation. *Environmental Management* **14**, 451–64.

Rood SB & Mahoney JM (1995) River damming and riparian cottonwoods along the Marias River, Montana. *Rivers* 5, 195–207.

Rood SB, Mahoney JM, Reid DE & Zilm L (1995) Instream flows and the decline of riparian cottonwoods along the St Mary River, Alberta. *Canadian Journal of Botany* **73**, 1250–60.

Salo J, Kalliola R, Hakkinen I, MakinenY, Niemala P, Puhakka M & Coley P (1986) River dynamics and the diversity of Amazon lowland forest. *Nature* **322**, 254–8.

Scott ML, Auble GT & Friedman JM (1997) Flood dependency of cottonwood establishment along the Missouri River, Montana, USA. *Ecological Applications* **7**, 677–90.

Scott ML, Friedman JM & Auble GT (1996) Fluvial process and the establishment of bottomland trees. *Geomorphology* **14**, 327–39.

Segelquist CA, Scott ML & Auble GT (1993) Establishment of *Populus deltoides* under simulated alluvial groundwater declines. *American Midland Naturalist* **130**, 274–85.

Shafroth PB, Auble GT & Scott ML (1995) Germination and establishment of the native plains cottonwood (*Populus deltoides* Marshall subsp. *monilifera*) and the exotic Russian-olive (*Eleagnus angustifolia* L.). *Conservation Biology* **9**, 1169–75.

Shafroth PB, Auble GT, Stromberg JC & Patten DT (1998) Establishment of woody riparian vegetation in relation to annual patterns of streamflow, Bill Williams River, Arizona. *Wetlands* **18**, 577–90.

Sigafoos RS (1964) Botanical evidence of floods and flood-plain deposition. *United States Geological Survey Professional Paper* **485A**.

Stevens LE, Schmidt JC, Ayers TJ & Brown BT (1995) Flow regulation, geomorphology and Colorado River marsh development in the Grand Canyon, Arizona. *Ecological Applications* **5**, 1025–39.

Stromberg JC, Fry J & Patten DT (1997) Marsh development after large floods in an alluvial, arid-land river. *Wetlands* **17**, 292–300.

Stromberg JC, Patten DT & Richter BD (1991) Flood flows and dynamics of Sonoran riparian forests. *Rivers* **2**, 221–35.

Tabacchi E, Planty-Tabbachi AM, Salinas MJ & Décamps H (1996) Landscape structure and diversity in riparian plant communities:A longitudinal comparative study. *Regulated Rivers: Research and Management* **12**, 367–90.

Tabbush P (1997) Dynamic processes in riparian ecosystems: implications for *Populus nigra* gene conservation strategies. Unpublished paper, EUFORGEN and Forest Research, UK.

Thomas DHL (1996) Dam construction and ecological change in the riparian forests of the Hadejia–Jama'are floodplain, Nigeria. *Land Degradation and Rehabilitation* **7**, 279–95.

Turok J, Lefévre F, Cagelli L & de Vries S (eds) (1996) Report of the second meeting of the European Forest Genetics Resources Programme (EUFORGEN) *Populus nigra* network. International Plant Genetics Resources Institute (IPGRI).

Van Coller AL, Rogers KH & Heritage GL (1997) Linking riparian vegetation types and fluvial geomorphology along the Sabie River within the Kruger National Park, South Africa. *African Journal of Ecology* **35**, 194–212.

van Splunder I (ed.) (1997) *Floodplain forest, willows and poplars along rivers*. RIZA Report 97.030.

van Splunder I, Coops H, Voesnek LACJ & Blom CWPM (1995) Establishment of alluvial forest species in floodplains: the role of dispersal timing, germination characteristics and water level fluctuations. *Acta Botanica Neerlandica* **44**, 269–78.

Walker KF & Thoms MC (1993) Environmental effects of flow regulation on the lower Murray River, Australia. *Regulated Rivers:Research and Management* **8**, 103–19.

Wallace JB, Eggert SL, Meyer JL & Webster JR (1997) Multiple trophic levels of a forest stream linked to terrestrial litter inputs. *Science* **277**, 102–4.

Ward JV, Tockner K & Schiemer F (1999) Biodiversity of floodplain river ecosystems:ecotones and connectivity. *Regulated Rivers: Research and Management* **15**, 125–39.

Warren A (1993) Naturalness: a geomorphological approach. In: FB Goldsmith & A Warren (eds) *Conservation in progress*. Chichester, John Wiley, 15–24.

Yon D & Tendron G (1981) *Alluvial forests of Europe*. Nature and Environment Series 22. Strasbourg, Council of Europe.

6 Lakes

LAURENCE CARVALHO[1] AND N. JOHN ANDERSON[2]
[1]*University College London, UK*
[2]*University of Copenhagen, Denmark*

INTRODUCTION: CONSERVATION ISSUES IN LAKES

The focus of conservation in lake environments is currently on preserving sites that illustrate the 'best' of nature. As with many habitats, selection of these priority sites for nature reserves has used criteria such as rarity, diversity, representativeness and typicalness with certain biological groups being particularly favoured in conservation legislation. Sites with diverse or sizeable bird populations, because of their visible public appeal, are particularly valued and are the main concern of international legislation such as the Ramsar Convention. Lake-classification schemes developed specifically to ensure adequate coverage of typical or representative lake types or for monitoring environmental change have similarly been based upon favoured biological assemblages (Palmer, Bell and Butterfield 1992), with little consideration of physical features or processes.

An alternative perspective is to view the biological assemblage simply as a response to the environment. Physical features and processes within the environment could then be considered highly relevant. Faith and Walker (1996) argued that to protect maximum biodiversity it might be better to survey sites for major environmental gradients than to carry out extensive biological surveys. Site selection along environmental gradients may be more objective and protect greater genetic diversity than selection that is based on a limited survey of favoured biological groups.

The naturalness and the fragility or stability of a site are two additional criteria often used in designating protected areas. In the UK, the naturalness of the catchment is a specific criterion considered in the choice of lake sites for protection (NCC 1989). The concept has also been specifically highlighted in EC conservation legislation with 'natural eutrophic lakes with *Magnopotamion*- or *Hydrocharition*-type vegetation' being listed for special attention in Annex I of the EC Habitats Directive. Naturalness is also implicit in any notion of ecological restoration in conservation management.

Habitat Conservation: Managing the Physical Environment. Edited by A. Warren and J. R. French.
© 2001 John Wiley & Sons Ltd.

The concept of 'naturalness' is a relative measure of how greatly sites have been modified by human activities. It has largely been assessed subjectively on relatively recent historical knowledge. The assumption that lakes have a stable natural status does, however, conflict with the accepted view that ecosystems are inherently dynamic (Harris 1980; Reynolds 1990). Lakes exhibit considerable inter-annual variability. While they do show *natural systematic change* over long timescales (thousands of years) in relation to changes in climate and catchment vegetation, they are also variable over shorter-term timescales (10–100 years). This dynamism may make it difficult, when managing for conservation, to identify natural baselines for lakes and their biological communities.

Fragility and stability both reflect the degree of sensitivity to environmental change. Because of strong seasonal changes, the stability of lakes is often gauged in terms of annual totals or averages (e.g. annual mean chlorophyll *a* concentrations are often used as a measure of primary production). Environmental change can then be considered as any systematic change, or trend, in these annual measures of lake state (Sas 1989).

When dealing with lake ecosystems, we are in the fortunate situation of having many components of past lake communities preserved in the sediments (Anderson 1993; Anderson and Battarbee 1994). These not only provide objective restoration targets (Bennion, Juggins and Anderson 1996; Flower, Juggins and Battarbee 1997) but can also be used to explore theories on lake dynamics, stability and naturalness (Jones, Stevenson and Battarbee 1989). The assessment of naturalness using relatively objective palaeoecological techniques has been reviewed thoroughly by Birks (1996).

This chapter aims to explore the physical features and processes that are most pivotal in shaping lake communities, with an emphasis on environmental change, in particular natural-dynamic versus systematic-anthropogenic change. Much of the discussion will centre on whether physical processes result in certain states that are intrinsically more natural in some lake basins than in others and whether naturalness is a useful criterion for conservation management in general. Before we focus on these physical features and processes, two issues need to be highlighted: first, our need to understand the spatial and temporal scales operating within lake ecosystems, and second, the major processes of environmental change that affect the conservation interests of lakes.

SPATIAL AND TEMPORAL SCALES

In addition to shifting our perceptions on the spatial scale in management we also need to understand what temporal scales are involved (Reynolds 1990; Harris 1980). Light and nutrients are clearly dynamic at diurnal, seasonal and inter-annual scales. Many lacustrine organisms have very short generation times and so their populations contain a high degree of temporal sensitivity with respect to environmental change. Sommer *et al.* (1993) described how one year covers hundreds of generations of planktonic organisms and is, therefore, comparable to century-scale change in woodland tree communities. They pointed out that, in terms of planktonic communities, variations in environmental conditions should be viewed on a relatively short timescale, such that 'weather', which may be stable over several planktonic generations, should be considered equivalent to our perception of 'climate'.

Fish and plant communities, on the other hand, show responses to disturbance at a variety of temporal scales that reflect their lifespan (Carpenter and Kitchell 1987). When change is observed in one community, it may be the result of many years of disturbance at lower (or higher) trophic levels. This has important consequences for restoration as it frequently means that there are no short-term management solutions.

Similarly, hydrological changes affecting catchment sources of nutrients, wind-induced sediment resuspension reducing light, and thermal stratification affecting light and chemical gradients are all essentially physical processes that interact over a multitude of spatial and temporal scales. Management timescale must encompass the different timescales of biological and physical variability within lakes. Likewise, monitoring for change or recovery will often have to be carried out over long timescales and the responses reviewed over several generations (Lancaster *et al.* 1996). Essentially lake managers need to decide what the appropriate scale of observation should be for their specific conservation question.

Given suitable monitoring records, the earliest signs of change to a lake will probably occur at the microscopic scale in the algal communities. Prolonged or frequent disturbances may then be transferred to higher trophic levels. When change is observed at the macroscopic scale in the more visible plant and animal communities, it is often tempting to focus management on these visible components: enhancing habitats for particular rare fish or birds, such as the vendace or bittern, or protecting botanically rich bays within a large lake. In many situations, however, practical management has to focus on primary productivity and associated nutrient inputs from the catchment. From this perspective, a fundamental issue is the cause of changes in plant and algal resource availability, in particular light resources and limiting nutrients.

ENVIRONMENTAL CHANGE IN LAKE ENVIRONMENTS

The principal processes of environmental change affecting the conservation status of lakes are eutrophication, acidification and climatic change. Irregular fluctuations in physical processes (or, more fundamentally, in climate) may obscure or exaggerate deleterious systematic trends such as eutrophication or acidification, or the outcome of management designed to deal with them.

Eutrophication

Eutrophication is one of the most widespread problems affecting lake environments. It is the process of the enrichment in plant nutrients, generally phosphorus and nitrogen, that are washed into lakes from their catchments, via surface streams and rivers, drainage or deeper groundwater sources. Increased loadings are principally associated with inputs from sewage effluent and agriculture. Nutrients may also be gained from, or lost to, internal sediments dependent on interactions between physical, chemical and biological processes at the sediment–water interface (Marsden 1989).

In lakes, the symptoms of eutrophication are primarily reflected in increased macrophyte (initially) and algal growths. However, as enrichment proceeds to a hypertrophic state, increased algal growths can result in loss of macrophytes and dominance of phytoplankton. The Norfolk Broads illustrate clearly the impact of eutrophication on conservation interest. Hoveton Great Broad developed from a clear-water lake rich in charophytes in the nineteenth century through a phase of more competitive plants such as rigid hornwort (*Ceratophyllum demersum*) and fennel pondweed (*Potamogeton pectinatus*). After further urbanisation and agricultural intensification following the Second World War, submerged macrophytes were largely lost and a turbid phytoplankton-dominated state was established (Moss 1988).

In addition to these changes in aquatic plants, associated changes in macroinvertebrate, fish and bird communities occur as lakes become enriched. The vendace (*Coregonus albula*) and the schelly (*Coregonus lavaretus*), two of the UK's rarest fish, are both represented in deep oligo-mesotrophic waters and are particularly sensitive to enrichment (see section on stratification below). In conservation terms, the implications of enrichment may, therefore, be most significant at low nutrient concentrations. A recent audit of lake sites in England that had Site of Special Scientific Interest (SSSI) protection status revealed that over 75% of the 102 sites surveyed showed symptoms of eutrophication and could be classed as eutrophic or hypereutrophic. Very few were classed as oligotrophic or mesotrophic (Carvalho and Moss 1995). In the UK and Europe, mesotrophic lakes have been highlighted as an increasingly rare habitat (HMSO 1994). By definition they support a high macrophyte biomass and generally have a large number of species including a high proportion of nationally scarce and rare aquatic plants, such as the slender naiad (*Najas flexilis*) and ribbon-leaved water-plantain (*Alisma gramineum*).

Acidification

The primary cause of acidification is acid deposition resulting from the burning of fossil fuels. Acidification is most commonly associated with lakes in regions of base-poor geology, for these have little buffering capacity. The biological impacts of acidification are wide-ranging. In terms of conservation interest the most publicised are the effects on fish communities. It has been estimated that in Norway an area of $13\,000\,km^2$ became devoid of fish due to acidification (Sevalrud 1980) and in about 9000 lakes in Sweden the pH declined to below 5.5, severely affecting fish stocks (Johanson and Nyberg 1981). The decline of the natterjack toad (*Bufo calamita*) in shallow pools on heathland in southern England has also been attributed to acidification in the twentieth century (Beebee *et al.* 1990).

Climate Change

Climate, both directly and indirectly, affects lake physics (flushing, stratification), chemistry (DOC, pH, nutrients) and biology. Lake managers need to be aware of how sensitive their sites are to climate variability and in particular, how climate-driven change can obscure or exaggerate the processes of acidification or eutrophication. Sommaruga-Wögrath *et al.* (1997) described the relationship between increased pH and warmer years in a large series of high-altitude, meltwater-fed, alpine lakes, while Webster *et al.* (1990) reported the opposite trend, rapid acidification following five years of drought, in a low-altitude, seepage lake in the USA.

There are very few studies showing direct climate effects on conservation interests. One is a long-term study of the Experimental Lakes Area in northwest Ontario in Canada, a region of numerous small glacial lakes. Increased water temperatures and more intense thermal stratification recorded since 1969 have resulted in a decline in the number of species that inhabit summer coldwater habitats, such as lake trout and opossum shrimp (Schindler *et al.* 1990). Magnuson *et al.* (1990) predicted the response of fish communities to climate change in the North American Great Lakes using Global Climatic Model (GCM) climate scenarios. They concluded that fish populations in general may benefit from global warming, but certain cold-water species might be reduced while warmer waters might also benefit exotic introductions. Lake-level lowering could also seriously reduce breeding and nursery areas for fish and wildlife.

THE PHYSICAL ENVIRONMENT AND CONSERVATION OF LAKES

This is considered in six sections: location in landscape, lake morphology, lake-catchment hydrology, thermal processes (stratification and mixing), water-level change and infilling. The last four can be considered dynamic. On the timescale of lake management, location and lake morphology can be considered static.

LOCATION IN LANDSCAPE—THE PHYSICAL SETTING

Much of the intrinsic conservation value of lakes, from a public perspective, derives from their location in the landscape—islands of water in a physical setting. The interplay between a lake and its surroundings is fundamental to lake management and any worthwhile attempt to conserve lake communities, or species, will be ineffectual if the lake is viewed separately from its surrounding catchment. Many important and highly relevant management decisions actually relate to the catchment rather than to the lake basin itself. Although it is tempting to treat lakes as partially closed systems, reflecting the links with their catchments, the importance of viewing them as part of extended atmospheric–terrestrial systems is indicated by the problem of lake acidification. Long-term amelioration of lake acidification due to deposition of industrially derived acidity requires wide-reaching agreement on levels of atmospheric pollution, often with an international component. Liming of a lake or its catchment is only a short-term management response to a long-term chronic problem.

Although lakes can be viewed as aquatic islands in a terrestrial sea, the rapid colonisation ability of many aquatic species also indicates the extent to which lakes are more subtly connected across a landscape. Many lakes are, of course, highly connected by streams and rivers with the result that species migrations can be very rapid indeed (Lodge 1993). This can be particularly detrimental in the case of introduced species (exotics) which may expand their range quickly resulting in the local extinction of the native species (for example, the introduction of North American crayfish into the UK), sometimes with profound impacts on the whole foodweb structure. For some conservationists, the isolation of lakes from disturbance by exotics is one of the most critical management problems. If the aim of conservation is to protect the *status quo*, it is, therefore, not necessarily important how the present-day species pool was assembled, more that it is protected from exotics.

Landscape position is critical to a lake's hydrology and chemistry (Kratz *et al*. 1997). Position determines the relative importance of inputs of groundwater and precipitation. Lakes high in the local hydrological landscape receive a greater proportion of their water from precipitation than lakes lower in the hydrological landscape. Any changes in the proportions of catchment-derived minerals compared with those received in direct precipitation are therefore more obvious at high landscape sites than at low landscape sites, where most hydrological inputs are via groundwater. As direct precipitation is a significant hydrological component in many of these high-landscape sites, many span a slightly acid pH gradient over which major biological change can occur and they should, therefore, be considered intrinsically more sensitive to climate change and acidification.

Low alkalinity, high-landscape lakes may be most sensitive to the hydrological impacts of climate change, but as later sections show, many other lake types have more subtle responses to climate change. In particular, the sensitivity of deep lakes to changes in the seasonal stratification pattern is discussed below. Those high-altitude lakes studied to date (Psenner and Schmidt 1992; Sommaruga-Wögrath *et al.* 1997) appear to produce a different pH response from that of high-landscape, low-altitude lakes because of the overwhelming impact of temperature on snow and ice cover in their catchments. In warmer years, alpine catchments are more susceptible to weathering, resulting in increased run-off of catchment-derived minerals.

LAKE-CATCHMENT HYDROLOGY

Water can enter and leave lakes by various pathways: groundwater, subsurface drainage, overland flow, direct precipitation and evapotranspiration. Each of these varies both spatially and temporally with respect to climate, catchment soils and geology and the characteristics of the lake itself. Lakes can be functionally separated into two main hydrological classes: (1) closed-basin lakes which have no significant outflow and whose water loss is dominated by evapotranspiration; and (2) open-basin lakes where water loss is largely by surface drainage or by seepage into groundwater. Closed-basin saline lakes are often considered the most sensitive to climate change as they lose water by evaporation so that climatic changes cause large fluctuations in lake level and salinity (Street 1980). Because much smaller-scale fluctuations in lake level and salinity occur in hydrologically open lakes, they are generally considered much less sensitive to short-term (< 10 years) climatic variations.

The flushing rate of a lake, or its 'inverse retention time', is another hydrological parameter that is important in estimating its sensitivity to environmental change, and is particularly relevant to eutrophication and acidification. Flushing rate (ρ) is defined as the volume of water flowing out of a lake per unit time (V_{out}) divided by the volume of the lake itself (V_{lake}) (Vollenweider and Kerekes 1982):

$$\rho = V_{out}/V_{lake} \qquad (6.1)$$

Flushing rate determines the rate of output or retention of material (dissolved nutrients, metals, phytoplankton, etc.) and hence the time available for in-lake processes to occur. It is, therefore, an important component of management models used to predict the effects of changing nutrient loading on in-lake nutrient concentrations and algal standing crop (Vollenweider 1975; Dillon and Rigler 1974; Vollenweider and Kerekes 1982). The lower the flushing rate, the more sensitive a site will be to eutrophication because the longer the residence time of nutrients in the site, the more available they are for primary production. Conversely, high flushing rate systems may be resilient to increased nutrient loadings as algae (and nutrients) are washed out of the lake more quickly than they can be replenished.

One 18-year case study that illustrates the importance of flushing rate was carried out on Loch Leven, Scotland. It showed that variation in flushing rate appeared to have 'a considerable effect on temperature regimes and the supplies and in-loch dynamics of nutrients; through such changes, flushing rate controls major features of phytoplankton succession . . . and as a consequence, in some cases, of the animals preying on them' (Bailey-Watts et al. 1990:85–6).

Flushing rate clearly fluctuates naturally with respect to climate (Rippey, Anderson and Foy 1996), but it is increasingly being affected also by human activity in the catchment. Land management (deforestation or afforestation) may have a profound influence on catchment hydrology. The increasing channelisation of rivers has almost certainly resulted in flashier responses to rainfall and wider variations in flushing rate throughout the year. The significance of this for lake ecology is largely unexplored. Increased groundwater abstraction may also reduce the amount of water reaching a lake and hence reduce the flushing rate, and this could be an important contributory factor to increased incidences of the symptoms of eutrophication. In the UK, the Habitat Action Plan for Mesotrophic Lakes (HMSO 1994) proposed that water-resource use affecting current mesotrophic lakes with SSSI status be reviewed to limit potential reduced flushing effects of over-abstraction.

This section has highlighted a range of sensitivities to hydrological change, often driven by climate variability. Large, deep, groundwater-dominated lakes may show a sensitivity to hydrological change over much longer timescales and these are only identifiable by using palaeoecological approaches. Their biological communities can, however, be sensitive in other ways to climatically driven physical processes, and respond at a much finer temporal scale. One such example is the onset and timing of thermal stratification, which is discussed in a later section.

EXAMPLE 1: EUTROPHICATION OF LOWLAND LAKES

Within the area covered by the last ice sheet in north-west Europe and North America, and particularly close to the ice margin where there are extensive morainic deposits, lakes are common landscape features. Because of the fertility of the surrounding glacial deposits, the lakes are often surrounded by intensive farmland, which has been utilised for thousands of years. The present-day management of these lakes is intrinsically linked to both catchment activity, groundwater processes and their cultural history. Many thermally stratify in the summer and have anoxic hypolimnia with a substantial internal phosphorus load (Marsden 1989) resulting in major restoration problems. In this section, we provide two examples (Northern Ireland and the Shropshire–

Cheshire Meres) where stratification and internal phosphorus loads are associated with conservation management problems.

At White Lough, a small inter-drumlin lake in County Tyrone (Northern Ireland), rapid eutrophication in the mid-1970s resulted in massive *Oscillatoria* (a blue-green alga) crops and concern about its fishery status (Foy and Fitzsimons 1987). Initially thought to be related to land-management, later work (Rippey, Anderson and Foy 1996) demonstrated that this eutrophication event resulted from the internal load being retained in the lake due to reduced flushing during two dry winters. In more typical winters, the lake is flushed and the phosphorus released from the sediment is lost from the lake and not available for use by the phytoplankton. This shows that not only is the resilience of a system to eutrophication related to its flushing rate, but also that the recovery, or restoration, is also affected. Importantly, however, it is clear that the internal phosphorus load which caused the rapid eutrophication at White Lough was also a result of long-term eutrophication caused by farming and land management changes.

Naturally low flushing rates appear to be of conservation significance in the debate over naturally eutrophic lakes in the Shropshire–Cheshire Meres in England. Many of the deeper meres have exceptionally high phosphorus concentrations and have a long history of blue-green algal blooms. This led to them being described as 'Britain's naturally eutrophic lakes' (Reynolds and Sinker 1976). Palaeolimnological studies reinforce a picture both of a long history of cyanobacterial abundance (McGowan *et al.* 1999) and reveal at other sites long-term enrichment which has proceeded in several waves associated with the development of agriculture in the catchment (Nelms 1984; Anderson 1995).

This does not, however, fully explain the excessively high nutrient concentrations in some of the deeper meres. Many of these have insignificant or no surface outflow and water is lost slowly by seepage into the groundwater. The high retention time (low flushing rate), in combination with other factors (stratification-related sedimentary P-release and migrating algae), appear to provide an enhanced opportunity for the internal recycling of nutrients (Moss *et al.* 1997). The deeper meres may, therefore, be intrinsically or naturally much more sensitive to enrichment than more highly flushed systems. The importance of the internal nutrient pool in the Meres reflects, however, their location within a cultural landscape. The transfer of nutrients from the catchment to the lake has exacerbated internal nutrient loading. Even limited catchment disturbance at these sites may result in highly eutrophic conditions. For conservation purposes, a eutrophic state dominated by planktonic cyanobacterial biomass may be a realistic baseline for these sites, although the process of eutrophication has clearly been severely exacerbated in many sites through inputs from sewage and intensive agriculture (Carvalho, Beklioglu and Moss 1995).

LAKE MORPHOLOGY: SIZE, AREA, SHORELINE AND DEPTH

Lake morphology is a major influence on many of the biological, physical (light and hydrodynamics) and chemical processes in lakes and, as such, has a considerable influence on lake restoration and conservation projects. There are no definitive morphological definitions of shallow, deep, small or large lakes. In general, large lakes tend to be deeper than small lakes.

Until recently, large lakes were assumed to be more stable and consequently more resistant to perturbations than small ones, but the sensitivity of large lakes is being increasingly realised. Palaeolimnological studies of some large, deep lakes have indicated that change occurs in algal communities before standard monitoring programmes are able to detect it (Jones et al. 1997). Likewise, palaeolimnological analyses of the American Great Lakes clearly show that changes in their biota were associated with European settlement, forest clearance and the start of agriculture (Stoermer, Wolin and Schelske 1993). When changes do occur in large, deep lakes, the changes can be very resilient and more difficult to reverse (Sas 1989). Examples of greater and lesser resilience in large deep lakes have been highlighted in sites such as Lake Baikal (Mackay et al. 1998) and the North American Great Lakes (Stoermer, Wolin and Schelske 1993).

Large lakes with complex shapes may demand various management approaches. Each basin may require its own management plan, depending on the degree of physical connectivity in the system (i.e. transfer of nutrients and pollutants between the different basins). The Lough Erne system in Ireland is a good example, having a clear nutrient gradient down the system from the shallow and very eutrophic Upper Lough to the deep, mesotrophic Lower Lough (Gibson, Foy and Fitzsimmons 1980). Although large lakes do contain much of the world's freshwater resources, small shallow lakes are more typical and can be more important from a conservation perspective (Biggs et al. 1994).

Shorelines are the interface between a lake and its catchment, and edge effects are greatest for small lakes with complex shorelines. Edge effects are important in terms of their influence on chemical transfers to a lake from its catchment. Wetlands around the lake can have major impacts on both the nutrient and acidity status of the lake (Lazerte 1992; Yan et al. 1996). Shoreline developments, such as the construction of housing, can create multiple point-source nutrient inputs, whereas farming up to the edge of a lake can facilitate nutrient transfer from fields. Heavy recreational use can have detrimental effects on the diversity of the littoral zone, through physical disturbance and trampling.

All lakes are subject to mixing by winds at a variety of scales and frequencies. The extent of shoreline exposure is dependent upon a lake's morphology and landscape position and the dominant local wind direction. Wind speed and lake area influence the extent of energy input to the system, which in turn governs the area of the lake from which bottom sediments can be resuspended. Within-lake distributions of macrophytes are related to the distribution of wave energy (Keddy 1985; Weisner 1987). Exposure can have implications for management of lake-macrophyte communities. For example, if attempts are made to reseed macrophyte beds, wind waves can uproot seedlings and/or erode the shoreline, disturbing both the future seedbank source and the growth mediums. Physical resuspension of sediment can also be important for nutrient release processes as well as the reduction of ambient light for macrophyte growth.

Depth has a strong influence on the biological communities owing to the scale of environmental gradients in the water column. Light, temperature and pressure gradients in a metre or so of water are equivalent to hundreds of metres in the atmosphere. Nutrient and oxygen concentrations may also vary dramatically over a few metres, or even a few millimetres below the sediment–water interface.

From a lake management perspective, lake morphology is very important. In deep lakes phytoplankton dominate productivity while in shallower lakes there are extensive littoral zones and macrophyte–periphyte-dominated productivity (Tilzer 1990). Although of limited extent in deep lakes, littoral plants may still be extremely important for fish breeding. Shallow, littoral bays should, therefore, be primary sites for conservation within large deep lakes.

The actual depth at which light limits plant growth varies according to the colour of the water, the concentration of suspended particles (affected by turbulence, resuspension and inflowing waters), and the densities of planktonic organisms. A depth increase of only 10 m increases hydrostatic pressure by 1 atmosphere. Submerged bryophytes and macroalgae which do not have intercellular gas systems are less affected by increases in pressure, and as a result can grow to much greater depths in lakes (Hannon and Gaillard 1997).

THERMAL STRATIFICATION AND MIXING

Thermal stratification and mixing is strongly influenced by depth. In shallow lakes, wind energy is frequently sufficient to mix the water column throughout the year. In deep lakes, where wind-induced mixing has less of an impact, thermal and chemical stratification can develop. A lake can be considered as a functionally deep, incompletely mixed lake if its mean depth (h) is larger than its growth-season mixing depth (h_{mix}) (Sas 1989). Thermal stratification develops when the surface waters are heated more rapidly than the heat is distributed by mixing. Stratification is, therefore, a fairly permanent feature in deep, low-altitude, tropical lakes but may only develop during summer in temperate lakes (Figure 6.1(a)). As water is densest at $4\,°C$, inverse stratification can also develop during winter under ice-cover.

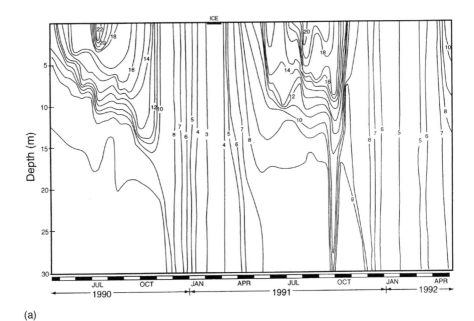

(a)

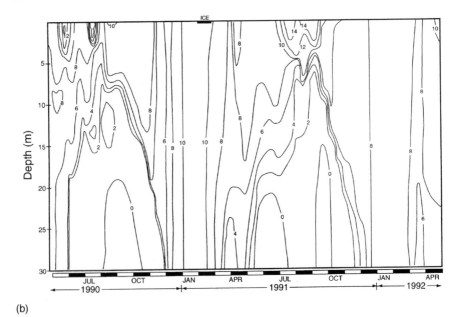

(b)

Figure 6.1. Depth–time diagrams for (a) temperature (°C) and (b) dissolved oxygen (mg l^{-1}) in a temperate lake (Rostherne Mere, Cheshire).

During summer stratification, the water column can be divided into an upper layer of warm water, the epilimnion, a lower layer of cold water, the hypolimnion, and a sharp transitional layer between, the thermocline. Stratification has strong effects on water chemistry. In the epilimnion, respiration and photosynthesis can lead to temporary changes in oxygen and carbon dioxide concentrations and the depletion of limiting nutrients such as phosphorus and nitrogen. In the hypolimnion, photosynthesis is negligible, yet respiration of bacteria associated with sedimenting detritus continues, resulting in lower oxygen concentrations (Figure 6.1(b)). The more productive the lake, the more severe the depletion of oxygen. Phosphorus and nitrogen bound in lake sediments are transformed and released in anaerobic environments into the hypolimnion, enhancing the enrichment process.

The timing of the onset of stratification in spring appears to have particular biological significance. Long periods of stable stratification promote buoyant or motile algae such as blue-green algae, while delayed or interrupted stratification promotes heavy non-motile algae such as diatoms with their heavy silica skeletons (Reynolds and Bellinger 1992; IFE 1998). This interannual variability associated with stratification has been linked to large-scale climatic phenomena such as the El Niño Southern Oscillation in the Pacific (ENSO) or the North Atlantic Oscillation (NAO). In Esthwaite Water in the Cumbrian Lake District a highly significant relationship was shown between the late summer phytoplankton standing crop and the position of the Gulf Stream in the Atlantic, recorded as the Gulf Stream Index (IFE 1998). Years with more northerly Gulf Stream activity are characterised by more prolonged stratification, while 'southerly' Gulf Stream years are characterised by more frequent mixing. The magnitude of the peak summer phytoplankton biomass is influenced by the quantity of nutrients transferred from the hypolimnion to the epilimnion by this mixing: windier summers result in higher epilimnetic nutrient concentrations and greater phytoplankton standing crops. George *et al.* (1990) show how windier years in Esthwaite Water also encouraged algae that were edible to *Daphnia* zooplankton, whereas calmer years encouraged unpalatable cyanobacteria, preventing the development of large *Daphnia* populations. This illustrates how variations in primary productivity and community composition can be transmitted through the zooplankton community and have impacts throughout the lake food web, particularly on the fecundity of zooplanktivorous fish (Carpenter and Kitchell 1987).

The research at Esthwaite Water not only illustrates how variations in stratification can affect the conservation interest it also highlights how changes in stratification pattern can obscure or exaggerate trends in eutrophication. Equally, the process of eutrophication may modify the impact of stratification on the conservation interest of a lake through influences on the extent of hypolimnetic deoxygenation and nutrient recycling. As stratification and deoxygenation are more stable in deep lakes, they may, therefore, intrinsically be more prone to blue-green algal blooms and be more efficient at recycling sediment sources of nutrients than shallow mixed lakes (Moss *et al.* 1997). This response to physical processes has been exploited in reservoir management where aerators have been installed to encourage destratification and oxygenation and discourage the development of large populations of blue-green algae.

EXAMPLE 2: CONSERVATION ISSUES IN LARGE DEEP LAKES: FISH COMMUNITIES

The conservation issues relating to large, deep lakes are frequently centred on the fish communities (trout, salmon and coregonids) and the associated highly specialised bird species (e.g. divers). Depth is important in relation to fish because of its role in habitat partitioning. The narrow depth tolerance of the different cichlid species in many of the African Great Lakes (e.g. Tanganikya and Malawi) has been strongly implicated in the rapid allopatric speciation that has occurred there, with over 200 races/species present in Lake Malawi alone (Martens 1997). Similarly, Lough Melvin in north-west Ireland, a large relatively deep lake, has three highly specialised varieties of lake trout, which occupy different depth zones within the lake and do not interbreed (Fergusen and Mason 1981).

The magnitude and severity of hypolimnetic deoxygenation is closely associated with the intensity of stratification in combination with the productivity of the water body. Deoxygenation of the hypolimnion can exclude many animals from inhabiting a large volume of the lake; fish are particularly susceptible. Cyprinid fish can tolerate higher water temperatures and lower oxygen concentrations than salmonids and coregonids. Whitefish (coregonids) are particularly sensitive as they prefer cooler, darker waters. In many of the larger lakes of the west of Ireland, the numbers of Arctic Char, a fish associated with oligotrophic lakes with stony bottoms, are declining because of land management-induced eutrophication.

The loss of some of the UK's rarest fish is believed to be due to stratification-related deoxygenation events (Maitland and Lyle 1992). Simple deoxygenation of the bottom waters alone can have an impact on egg survival of whitefish in mesotrophic lakes (Muller 1992) and can result in a switch from a salmonid and coregonid assemblage to a cyprinid fish assemblage. Sites containing rare whitefish species, such as Bassenthwaite Lake and Llyn Tegid in the UK, therefore need particular protection from nutrient enrichment. For deeper, stratifying lakes the effects of global warming may have implications for fish that occupy the deeper, cooler hypolimnetic waters during the summer. Warmer air temperatures will strengthen stratification, which in conjunction with enrichment may lead to reduced oxygen availability, thereby reducing habitat availability for coregonids and char (Magnuson, Meisner and Hill

1990; Tippel, Eckmann and Hartman 1991). The response of fish to climate change is complex, however, because of the possible positive effects of warming on food availability (zooplankton), thereby increasing fish growth and reproductive success (Magnuson, Meisner and Hill 1990).

WATER-LEVEL CHANGE

In most lakes the water level varies at a variety of scales, varying from seasonal drops of a few centimetres during the summer to massive drops of hundreds of metres in deep lakes over tens to hundreds of years associated with climate change (Street and Grove 1976). Some ephemeral lakes dry out altogether, seasonally, such as the turloughs in the west of Ireland (Reynolds 1998), or in drier years, as in playa lakes in semi-arid regions (Hutchinson 1957). The causes of lake-level changes, over both long- and short-term timescales, are varied. They include (1) long-term shifts in climate (precipitation–evaporation ratios); (2) long-term shallowing resulting from infilling; (3) short-term, seasonal variability (increased evaporation during the summer); (4) anthropogenic lake-level alterations, which include land drainage and damming for water supply, and finally (5) biotic factors, such as beaver dams.

There is widespread evidence for substantial lake-level lowering in northwest Europe during the sub-Boreal (Harrison and Digerfeldt 1993). In more temperate zones, long-term lake-levels have stabilised during the late Holocene, a situation that contrasts to that in the drier parts of the world. The implications of future global warming may, however, alter this situation. Moreover, superimposed on these natural patterns is a variable degree of anthropogenic interference. In many parts of Europe lake levels were increased by damming during the seventeenth century to provide a reliable head of water for local mills. In the late nineteenth century in Denmark, many lakes suffered lake-level drops following efforts to drain the surrounding farmland, and in the twentieth century major abstraction for irrigation caused great drops in water levels, the Aral Sea probably being one of the most catastrophic examples (Williams and Aladin 1991).

The large fluctuating water levels associated with drawdown zones in reservoirs may result in low aquatic plant diversity, yet may provide an important habitat for several specialist rare plants such as the mudwort (*Limosella aquatica*). In this respect, reservoirs may actually offset the introduction of water-level controls on many smaller (but not necessarily natural) lakes and ponds where management has had the opposite effect of reducing seasonal fluctuations in water level.

INFILLING

The long-term changes in water depth that are brought about by infilling reflect the close relationship between lakes and their catchment. Infilling of lakes is a natural process that can be altered substantially by anthropogenic activity within the catchment. The main processes, increased sediment yield and nutrient transfer reflect disturbance of the catchment and illustrate the extent to which management cannot ignore the location of a lake within its catchment. Lakes with a large catchment to lake area ratio are subject to a higher rate of input of terrestrial material from the catchment. Danish lakes today have a mean depth of little more than 1 m. Many, however, contain over 20 m of sediment, a substantial proportion of which has accumulated since the start of agriculture around 6000 years BP. With such high catchment erosion rates following deforestation, it is no wonder that some lakes disappeared from the landscape very quickly. Eventually, lakes shallow sufficiently to permit rapid colonisation by both submerged macrophytes and fringing emergents (Figure 6.2). The macrophyte communities hold sediment efficiently and themselves input more refractory organic matter, with the result that the littoral zone expands outwards and infilling accelerates.

This aspect of lake ontogeny, the development of hydroseres, creates a dilemma for conservationists: to protect the *status quo* or not. Hydroseres are clear examples of natural ecological succession, and preserving the *status quo* is in reality attempting to arrest this succession and freeze the lake–fen ecosystem in a point in time (Tallis 1973). In some places the repeated cutting of marginal reeds for thatching, over centuries, may have sustained habitat diversity against this natural process of encroachment. An alternative management option for preserving this diversity is to increase the water level, which may reduce the rate of encroachment (Reynolds 1987). When viewed from a longer-term perspective, however, efforts like these could be considered futile. The natural situation is for the lake to infill and turn into a fen, but given our ability to dam streams, to create new artificial lakes as well as to drain wetlands, controlling lake levels is one of the few physical management actions that can be undertaken.

EXAMPLE 3: CONSERVATION ISSUES IN SHALLOW LAKES: SUBMERGED MACROPHYTES

In contrast to deep, pelagic-dominated systems, productivity in shallow lakes is dominated by littoral macrophytes, epiphytes and the associated benthic microbiological complex (Scheffer 1998; Stevenson, Bothwell and Lowe 1996). Dense macrophyte beds represent an important food source for aquatic birds, such as coots and swans. In recent decades, however, many shallow lakes have completely lost their submerged macrophyte communities and productivity is now dominated by phytoplankton. Shallow lakes have generally been categorised into two relatively stable states: a clear-water, macrophyte-dominated state; and a turbid, phytoplankton-dominated state (Figure 6.2; Scheffer *et al.* 1993). Shifts between the two can result in major ecosystem

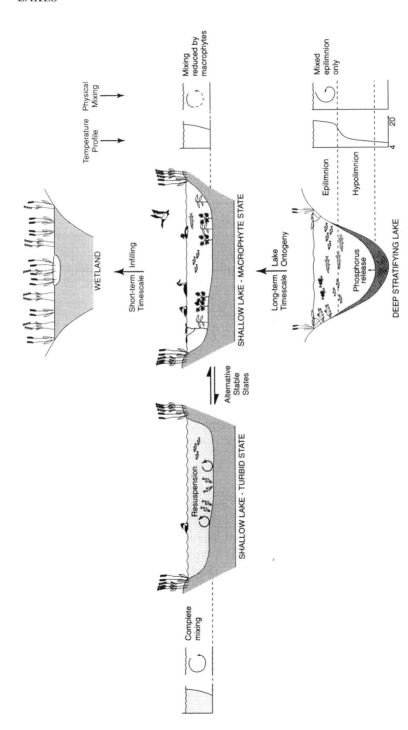

Figure 6.2. Diagramatic representation of lake structure over time from a deep, stratifying lake to a terrestrialised wetland basin. Shallow lake stage represented by two alternative stable states (Scheffer *et al.* 1993): a turbid state (of low conservation value and fully mixed) and a macrophyte-dominated state (of high conservation value with macrophytes reducing mixing depth).

restructuring, although equally shifts within a 'stable state' can be immensely significant in terms of specific conservation interests.

Conservation and management of shallow lakes in Europe tends to centre on the restoration and maintenance of the clear-water phase with its luxuriant submerged macrophyte growth. This primarily stems from the greater aesthetic appeal of this 'state' compared with the turbid, phytoplankton-dominated stage, but also from concern for waterfowl and other associated wildlife.

Lakes are known to fluctuate between the clear-water macrophyte-dominated stage and the turbid phytoplankton stage on a variety of timescales (Scheffer 1998; Blindow 1992; Blindow et al. 1993). The cause of these switches is, however, a matter of considerable debate (Scheffer et al. 1993). Although the primary causal factor is high nutrient loadings, the situation is more complex and depends upon both physical and internal biological interactions. For example, the stocking of benthivorous fish, such as carp and bream, for angling, makes biologically mediated sediment resuspension an increasingly common threat to submerged aquatic plant communities (Carvalho and Moss 1995) and can be an important factor in the inability of a lake to switch back to a clearwater stage. Losses at protected sites of conservation importance in England include two nationally rare plants: eight-stamened waterwort (*Elatine hydropiper*) and starfruit (*Damasonium alisma*). Lake managers are now turning to fish removals/manipulations in an attempt to speed up restoration of submerged plant communities containing these species.

The main physical factors affecting the switch between the alternative states in shallow lakes are light availability, sediment resuspension and water-level change. Shallow lakes thermally stratify infrequently, if ever, and much of their bottom zone is within the wave-influenced depth of even moderate wind speeds. As a result, resuspended matter in the water column reduces light penetration.

In nutrient-enriched waters light availability is more often reduced by increased phytoplankton biomass, but as macrophytes reduce water movements and increase sediment stability, their loss creates a positive feedback where more sediment is resuspended by physical processes thereby reducing light penetration even more. Although sediment resuspension may not often be the primary reason for the switch between the clear-water and turbid stages in lakes, it may be an important factor in inhibiting the recovery of the macrophytes. Resuspension and siltation can additionally create problems for fish spawning.

If too extreme, water-level fluctuations can cause a switch from a macrophyte-dominated system to a turbid one with high phytoplankton biomass. At two shallow Swedish lakes (Tåkern and Krankesjön), Blindow et al. (1993) reported how water-level fluctuations resulted in the loss of macrophytes in this way. In Lake Tämnaren in Sweden, intentional lowering of water levels has resulted in dense macrophyte beds and the influx of large numbers of migratory birds including about 500 mute swans. Raising of water levels again by as little as 0.3 m led to the disappearance of much of the submerged vegetation (Wallsten and Forsgren 1989).

It is, however, becoming increasingly clear that luxuriant submerged plant communities do not always represent natural baseline conditions. In part, they reflect the shallowing of the lakes associated with long-term changes, increased soil erosion and cultural eutrophication (Figure 6.2). Protecting them for nature conservation may be acceptable, but one should be aware that what is being protected is not necessarily the 'natural state'.

THE IMPLICATIONS FOR CONSERVATION MANAGEMENT

Conservation is often concerned with maintaining the *status quo*, in part because of the assumption that if ecosystems are not in equilibrium with their surrounding environment, they might be able to reach one. That lakes are seldom in equilibrium, however, being dynamic at a variety of different timescales has been made apparent by long-term monitoring and palaeolimnological studies. Lakes are changing continually, both seasonally, from year to year and over much longer time periods (decades, centuries). This dynamism has major implications for the lake conservationist in terms of defining background conditions (as restoration targets) and what the natural community composition of a lake was, and what it should be restored to. Should a background community be chosen that dates from the 1930s, prior to the major changes accompanying post-war industrialisation, or conditions that pre-date the start of early agriculture, up to 6000 years ago? Clearly, the latter is not feasible, but maintaining lakes in a macrophyte-dominated state ignores the fact that this developed in many temperate lakes, as a function of lake shallowing and enrichment. Both of these factors have a substantial anthropogenic component. Moreover, the macrophyte state is only transitory in terms of the lake's past history and its future. Ecologists have become increasingly aware of what Shugart (1998) refers to as the 'omnipresence of change', in lakes as in the other habitats discussed in this book. This dynamic aspect of lakes and their relationship to the physical environment has important implications for how their conservation should be approached.

It may be an acceptable management decision to maintain a dense, submerged macrophyte flora in shallow lakes as a suitable means of maximising biodiversity of specific biological groups. It is, however, wrong to present this as the lake's 'natural' condition. Management for conservation should have an understanding of natural dynamics and how human activity is affecting patterns of natural disturbance, as in the interrelationship between stratification and eutrophication described earlier. For most temperate lakes, the natural state is not readily definable because of natural forcing nor can it be regained because it was lost thousands of years ago with the start of agriculture and catchment deforestation. Rather than specifically trying to conserve a lake's 'naturalness' (natural state), conservationists need to address the central questions: 'what are we trying to preserve and why?'; and 'is it species conservation, habitats, or genetic diversity?'. When these questions have been answered management plans can be initiated.

From the perspective of a lake and its physical setting, management options are however, relatively limited. The physical processes shaping lake communities are largely determined by the physical setting of the lake and its position in the catchment and are primarily climatic. Many aspects of a lake's physical environment are, of course, difficult to change (such as size and physical setting) while others, such as wind and rainfall, which have important implications for hypolimnetic anoxia and nutrient loading respectively, are not readily controllable by managers.

Management can, however, aim to reduce the impact of anthropogenic environmental change. Hydrological processes, for example, have been greatly affected by human activity and some remediation can be envisaged (such as river-restoration projects to reduce channelisation; see Clifford 2001). Further catchment management options include the encouragement of lower nutrient loading to reduce the impact of stratification-related deoxygenation events or the reduction of sediment loads to slow infilling. The use of buffer strips and wetland fringes as sediment and nutrient traps may provide limited management options to counter these processes.

The fact that many of the most realistic management options for lake restoration and conservation actually involve control over land use and water abstraction highlights the absolutely essential need for protected lakes to have protected catchments. Simply looking at the water body in isolation from its physical setting offers very little hope for conservation management.

REFERENCES

Anderson NJ (1993) Natural versus anthropogenic change in lakes: the role of the sediment record. *Trends in Ecology & Evolution* **8**, 356–61.

Anderson NJ (1995) Naturally eutrophic lakes: reality, myth or myopia. *Trends in Ecology & Evolution* **10**, 137–8.

Anderson NJ & Battarbee RW (1994) Aquatic community persistence and variability: a palaeolimnological perspective. In: PS Giller, AG Hildrew & D Raffeli (eds) *Aquatic Ecology: Scale, Pattern and Process*. Oxford, Blackwell Scientific Publications, 233–59.

Bailey-Watts AE, Kirika A, May L & Jones DH (1990) Changes in phytoplankton over various time scales in a shallow, eutrophic lake: the Loch Leven experience with special reference to the influence of flushing rate. *Freshwater Biology* **23**, 85–111.

Beebee TJC, Flower RJ, Stevenson AC, Patrick ST, Appleby PG, Fletcher C, Marsh C, Natkanski J, Rippey B & Battarbee RW (1990) Decline of the Natterjack Toad *Bufo calamita* in Britain: palaeoecological, documentary and experimental evidence for breeding site acidification. *Biological Conservation* **53**, 1–20.

Bennion H, Juggins S & Anderson NJ (1996) Predicting epilimnetic phosphorus concentrations using an improved diatom-based transfer function and its application to lake eutrophication management. *Environmental Science & Technology* **30**, 2004–7.

Biggs J, Corfield A, Walker D, Whitfield M & Williams PJ (1994) New approaches to the management of ponds. *British Wildlife* **5**, 273–287.

Birks HJB (1996) Contributions of quaternary palaeoecology to nature conservation. *Journal of Vegetation Science* **7**, 89–98.

Blindow I (1992) Long- and short-term dynamics of submerged macrophytes in two shallow eutrophic lakes. *Freshwater Biology* **28**, 15–27.

Blindow I, Andersson G, Hargeby A & Johansson S (1993) Long-term pattern of alternative stable states in two shallow eutrophic lakes. *Freshwater Biology* **30**, 159–67.

Carpenter SR & Kitchell JF (1987) The temporal scale of variance in limnetic primary production. *American Naturalist* **129**, 417–33.

Carvalho L, Beklioglu M & Moss B (1995) Changes in a deep lake following sewage diversion — a challenge to the orthodoxy of external phosphorus control as a restoration strategy? *Freshwater Biology* **34**, 399–410.

Carvalho L & Moss B (1995) The current status of a sample of English Sites of Special Scientific Interest subject to eutrophication. *Aquatic Conservation* **5**, 191–204.

Clifford NJ (2001) Conservation and the river channel environment. In: A Warren & JR French (eds) *Habitat conservation: managing the physical environment.* Chichester, John Wiley, 67–103.

Dillon PJ & Rigler FH (1974) The phosphorus–chlorophyll relationship in lakes. *Limnology and Oceanography* **19**, 767–73.

Faith DP & Walker PA (1996) Environmental diversity: on the best-possible use of surrogate data for assessing the relative biodiversity of sets of areas. *Biodiversity and Conservation* **5**, 399–415.

Ferguson A & Mason FM (1981) Allozyme evidence for reproductively isolated sympatric populations of brown trout *Salmo trutta* L. in Lough Melvin, Ireland. *Journal of Fish Biology* **18**, 629–42.

Flower RJ, Juggins S & Battarbee RW (1997) Matching diatom assemblages in lake sediment cores and modern surface sediment samples: the implications for lake conservation and restoration with special reference to acidified systems. *Hydrobiologia* **344**, 27–40.

Foy RH & Fitzsimons AG (1987) Phosphorus inactivation in a eutrophic lake by the direct addition of ferric aluminium sulphate: changes in phytoplankton populations. *Freshwater Biology* **17**, 1–13.

George DG, Hewitt DP, Lund JWG & Smyly WJP (1990) The relative effects of enrichment and climate change on the long-term dynamics of *Daphnia* in Esthwaite Water, Cumbria. *Freshwater Biology* **23**, 55–70.

Gibson CE, Foy RH & Fitzsimmons AG (1980) A limnological reconnaissance of the Lough Erne system, Ireland. *Internationale Revue gesmaten Hydrobiologie* **66**, 641–4.

Hannon GE & Gaillard M-E (1997) The plant-macrofossil record of past lake-level changes. *Journal of Paleolimnology* **18**, 15–28.

Harris GP (1980) Temporal and spatial scales in phytoplankton ecology. Mechanisms, methods, models and management. *Canadian Journal of Fisheries & Aquatic Science* **37**, 877–900.

Harrison SP & Digerfeldt G (1993) European lakes as palaeohydrological and palaeoclimatic indicators. *Quaternary Science Reviews* **12**, 233–48.

HMSO (1994) *Biodiversity. The UK Action Plan*. London, HMSO.

Hutchinson GE (1957) *A Treatise on Limnology Volume I: Geography, Physics and Chemistry*. New York, John Wiley.

IFE (1998) Climate and stratification in Esthwaite Water. *Institute of Freshwater Ecology Annual Report* 1996–97.

Johanson K & Nyberg P (1981) *Acidification of surface waters in Sweden — effects and extent 1980*. Drottingholm, Sweden, Institute Freshwater Research.

Jones VJ, Battarbee RW, Rose NL, Curtis C, Appleby PG, Harriman R & Shine A (1997) Evidence for the pollution of Loch Ness from the analysis of its recent sediments. *Science of the Total Environment* **203**, 37–49.

Jones VJ, Stevenson AC & Battarbee RW (1989) Acidification of lakes in Galloway, south west Scotland: a diatom and pollen study of the post-glacial history of the Round Loch of Glenhead. *Journal of Ecology* **77**, 1–23.

Keddy PA (1985) Wave disturbance on lakeshores and the within-lake distribution of Ontario's Atlantic coastal plain flora. *Canadian Journal of Botany* **63**, 656–60.

Kratz TK, Webster KE, Bowser CJ, Magnuson JJ & Benson BJ (1997) The influence of landscape position on lakes in Northern Wisconsin. *Freshwater Biology* **37**, 209–17.

Lancaster J, Real M, Juggins S, Monteith DT, Flower RJ & Beaumont WRC (1996) Monitoring temporal changes in the biology of acid waters. *Freshwater Biology* **36**, 179–201.

Lazerte BD (1992) The impact of drought and acidification on the chemical exports from a minerotrophic conifer swamp. *Biogeochemistry* **18**, 153–75.

Lodge DM (1993). Biological invasions: lessons for ecology. *Trends in Ecology & Evolution* **8**, 133–7.

McGowan S, Britton G, Haworth E & Moss B (1999) Ancient blue-green blooms. *Limnology and Oceanography* **44**, 436–9.

Mackay AW, Flower RJ, Kuzmina AE, Granina LZ, Rose NL, Appleby PG, Boyle JF & Battarbee RW (1998) Diatom succession trends in recent sediments from Lake Baikal and their relation to atmospheric pollution and to climate change. *Philosophical Transactions Royal Society London* B **353**, 1011–55.

Magnuson JJ, Meisner JD & Hill DK (1990) Potential changes in the thermal habitat of Great Lakes fish after Global Climate Warming. *Transactions of the American Fisheries Society* **119**, 2554–640.

Maitland PS & Lyle AA (1992) Conservation of freshwater fish in the British Isles: the current status and biology of threatened species. *Aquatic Conservation* **1**, 25–54.

Marsden MW (1989) Lake restoration by reducing external phosphorus loading: the influence of sediment phosphorus release. *Freshwater Biology* **21**,139–62.

Martens K (1997) Speciation in ancient lakes. *Trends in Ecology and Evolution* **12**, 177–82.

Moss B (1988) The palaeolimnology of Hoveton Great Broad, Norfolk: clues to the spoiling and restoration of Broadland. In: P Murphy & C French (eds) *Exploitation of wetlands*. British Archaeological Reports, International Series, 163–91.

Moss B, Beklioglu M, Carvalho L, Kilinc S, McGowan S & Stephen D (1997) Vertically-challenged limnology: contrasts between deep and shallow lakes. *Hydrobiologia* **342**, 257–67.

Muller R (1992) Trophic state and its implications for natural reproduction of salmonid fish. *Hydrobiologia* **243/244**, 261–8.

NCC (1989) *Guidelines for Selection of Biological SSSIs*. Peterborough, Nature Conservancy Council.

Nelms R (1984) *Palaeolimnological studies of Rostherne Mere (Cheshire) and Ellesmere (Shropshire)*. PhD thesis, Liverpool Polytechnic.

Palmer MA, Bell SL & Butterfield I (1992) A botanical classification of standing waters in Britain: applications for conservation and monitoring. *Aquatic Conservation* **2**, 125–43.

Psenner R & Schmidt R (1992) Climate-driven pH control of remote alpine lakes and effects of acid deposition. *Nature* **356**, 781–3.

Reynolds CS (1987) Lake communities: an approach to their management for conservation. In: IF Spellerberg, FB Goldsmith, & MG Morris (eds) *The Scientific management of temperate communities for conservation.* Oxford, Blackwell Scientific Publications.

Reynolds CS (1990). Temporal scales of variability in pelagic environments and the response of phytoplankton. *Freshwater Biology* **23**, 25–53.

Reynolds CS & Bellinger EG (1992) Patterns of abundance and dominance of the phytoplankton of Rostherne Mere, England: evidence from an 18-year data set. *Aquatic Sciences* **54**, 10–36.

Reynolds CS & Sinker CA (1976) The meres: Britain's eutrophic lakes. *New Scientist* **71**(1007), 10–12.

Reynolds JD (1998) *Ireland's freshwaters.* Dublin, The Marine Institute.

Rippey B, Anderson NJ, & Foy RH (1996) Accuracy of diatom-inferred total phosphorus concentrations and the accelerated eutrophication of a lake due to reduced flushing. *Canadian Journal of Fisheries & Aquatic Science* **54**, 2637–46.

Sas H (ed.) (1989) *Lake restoration by reduction of nutrient loading. Expectation, experiences, extrapolation.* St Augustin, Academia Verlag Richarz.

Scheffer M (1998) *Ecology of shallow lakes.* London, Chapman & Hall.

Scheffer M, Hosper SH, Meijer M-L, Moss B & Jeppesen E (1993) Alternative equilibria in shallow lakes. *Trends in Ecology and Evolution* **8**, 275–9.

Schindler DW, Beaty KG, Fee EJ, Cruikshank DR, Debruyn ER, Findlay DL, Linsey GA, Shearer JA, Stainton MP & Turner MA (1990) Effects of climatic warming on lakes of the central boreal forest. *Science* **250**, 967–70.

Sevalrud IH (1980) Loss of fish populations in southern Norway. Dynamics and magnitude of the problem. In: D Drabløs & A Tollan (eds) *Proc. of International Conference on the Ecological Impact of Acid Precipitation.* Sandefiord, 350–1.

Shugart HH (1998) *Terrestrial ecosystems in changing environments.* Cambridge, Cambridge University Press.

Sommaruga-Wögrath S, Koinig K A, Schmidt R, Sommaruga R, Tessadri R & Psenner R (1997) Temperature effects on the acidity of remote alpine lakes. *Nature* **387**, 64–7.

Sommer U, Padisak J, Reynolds CS & Juhasznagy P (1993) Hutchinson's heritage — the diversity–disturbance relationship in phytoplankton. *Hydrobiologia* **249**, 1–7.

Stevenson RJ, Bothwell ML & Lowe RL (1996) *Algal ecology: freshwater benthic ecosystems.* San Diego, Academic Press.

Stoermer EF, Wolin JA & Schelske CL (1993) Paleolimnological comparison of the Laurentian Great Lakes based on diatoms. *Limnology & Oceanography* **38**, 1311–16.

Street FA (1980) The relative importance of climate and local hydrogeological factors in influencing lake-level fluctuations. *Palaeoecology of Africa* **12**, 137–58.

Street FA & Grove AT (1976) Environmental and climatic implications of late Quaternary lake-level fluctuations in Africa. *Nature* **261**, 385–90.

Tallis JH (1973) The terrestrialisation of lake basins in North Cheshire, with special reference to the development of a 'Schwingmoor' structure. *Journal of Ecology* **61**, 537–67.

Tilzer MM (1990) Specific properties of large lakes. In: MM Tilzer & C Serruya (eds) *Large lakes: ecological structure and function.* Berlin, Springer-Verlag, 39–43.

Tippell EA, Eckmann R & Hartman J (1991) Potential effects of global warming on whitefish in Lake Constance, Germany. *Ambio* **20**, 226–31.

Vollenweider RA (1975) Input–output models; with special reference to the phosphate loading concept in limnology. *Schweizerische Zeitschrift für Hydrologie* **37**, 53–84.

Vollenweider RA & Kerekes JJ (1982) Background and summary results of the OECD cooperative programme on eutrophication. Appendix I in *The OECD cooperative programme on eutrophication*. Canadian contribution (compiled by LL Janus & RA Vollenweider). Environment Canada, Scientific Series **131**.

Wallsten M & Forsgren PO (1989) The effects of increased water level on aquatic macrophytes. *Journal of Aquatic Plant Management* **27**, 32–7.

Webster KE, Newell AD, Baker LA & Brezonik PL (1990) Climatically induced rapid acidification of a softwater seepage lake. *Nature* **347**(6291), 374–6.

Weisner SEB (1987) The relation between wave exposure and distribution of emergent vegetation in a eutrophic lake. *Freshwater Biology* **18**, 537–44.

Williams WD & Aladin NV (1991) The Aral Sea — recent limnological changes and their conservation significance. *Aquatic Conservation* **1**, 3–23.

Yan ND, Keller W, Scully NM, Lean DRS & Dillon PJ (1996) Increased UV-B penetration in a lake owing to drought-induced acidification. *Nature* **381**(6578), 141–3.

7 Freshwater Wetlands

J. R. THOMPSON[1] AND C. M. FINLAYSON[2]

[1]*University College London, UK*
[2]*National Centre for Tropical Wetland Research, Environmental Research Institute of the Supervising Scientist, Jabiru, Australia*

INTRODUCTION: WETLANDS

This chapter outlines the key environmental processes and characteristics of freshwater wetland environments. It shows how the evaluation and under-standing of these factors is central to the conservation of these ecologically, economically and socially important landscapes.

It is commonly asserted that wetlands cover approximately 6% of the Earth's surface or some 8.6 million km^2 (Maltby 1986; Mitsch and Gosselink 1993). However, a recent global analysis of the extent of wetlands has raised many concerns about such figures, based as they are on poorly or inaccurately compiled data, differences in definitions and methodologies (Finlayson and Spiers 1999). Clearly environments this ubiquitous must encompass an enormous diversity of habitats and this is mirrored in the large number of wetland definitions that have been devised. For example, Dugan (1990, 1993) observe that over 50 definitions are in everyday use. Davies and Claridge (1993) further emphasised this diversity noting that wetlands meant different things to different people. The Convention on Wetlands of International Importance especially as Waterfowl Habitat, or the Ramsar Wetlands Convention, introduced what has become one of the most widely employed definitions. It argued that wetlands were:

'areas of marsh, fen, peatland or water; whether natural or artificial, permanent or temporary, with water that is static or flowing, fresh, brackish or salt, including areas of marine waters, the depth of which at low tide does not exceed six metres [and may include] riparian and coastal zones adjacent to the wetlands and islands or bodies of marine water deeper than six metres at low tide lying within the wetlands' (Davis 1994).

Dugan (1990) classified wetlands into 39 categories according to their basic biological and physical characteristics. This was simplified into seven distinct landscape units that were either wetlands or areas of which wetlands formed an

---

Habitat Conservation: Managing the Physical Environment. Edited by A. Warren and J. R. French.
© 2001 John Wiley & Sons Ltd.

important part: estuaries; open coasts; floodplains; lakes; swamp forests; peatlands; freshwater marshes. The focus of this chapter is the last two of these broad landscape units, which are here broadly divided into mires and marshes.

FRESHWATER WETLANDS: TERMINOLOGY AND CLASSIFICATION

A wide range of terms has been applied in different countries and by different authors to freshwater wetlands (Gore 1983; Finlayson and van der Valk 1995). Table 7.1 summarises some of the more popular terms used to describe them (Mitsch and Gosselink 1993; Mitsch, Mitsch and Turner 1994). Figure 7.1 shows the potential confusion in the terminology. North American terminology classifies inland non-forested wetlands into either marshes or peat-forming, low-nutrient acid bogs. In contrast, European terminology, which is older and consequentially more diverse, identifies at least four wetland types from swamps (mineral-rich reed beds) to wet grassland marshes, to fens and subsequently bogs or moors (Mitsch and Gosselink 1993). Within single countries terminology also varies. For example, in their introduction to British wetlands, Hughes and Heathwaite (1995) highlighted a number of classification systems based upon factors including vegetation, chemistry, hydrology and conservation status. As a result, the number of wetland types generated by different classification schemes can be variable and, in some cases, large. Rodwell (1991) listed 38 wetland types separated by their differences in plant communities. In contrast, Heathwaite *et al.* (1993) employed a classification system based upon water status (very wet, wet, fluctuating wet/dry cycles), source of water and water movement (river flow, spring, surface runoff), and water chemistry (eutrophic, mesotrophic, oligotrophic, dystrophic). A broad classification scheme for British wetlands, based predominantly on substrate, produces six principal wetland types (Hughes and Heathwaite 1995; Figure 7.2).

The diversity of wetland types has been responsible for the tremendous difficulties experienced by those attempting to classify wetlands, delineate their boundaries and develop plans for their management and conservation, as reported in the global analyses presented in Finlayson and Spiers (1999). Semeniuk (1987) noted these difficulties and presented a wetland classification scheme that was based fundamentally on the two features — namely, landform and water regardless of the climatic setting, soil type, vegetation cover, or origin. This classification scheme brings out the underlying and unifying features of wetlands that occur across the spectrum of climatic and physiographic settings. Semeniuk and Semeniuk (1995, 1997) extended this classification scheme, pointing out that it also overcame major inconsistencies in other systems that had been primarily based on vegetation characteristics, either by themselves or in association with soil/substrate or inundation patterns. By classifying wetlands initially on the basis of five landform

Table 7.1. Selected common names for freshwater wetlands around the world (adapted from Mitsch and Gosselink 1993 and Mitsch, Mitsch and Turner 1994)

Billabong	Australian term for lagoon.
Bog	A peat-accumulating wetland that has no significant inflows or out-flows and supports acidophilic mosses, particularly sphagnum.
Bottomland	Lowlands along streams and rivers, usually on alluvial floodplains that are periodically flooded. These are usually forested and are called bottomland hardwood forests in south-eastern United States.
Carr	Term used in Europe for forested wetlands, characterised by alders (*Alnus*) and willow (*Salix*).
Fen	A peat-accumulating wetland that receives some drainage from surrounding mineral soil and usually supports marsh-like vegetation.
Lagoon	Term frequently used in Europe to denote deep-water enclosed or partially opened aquatic system, especially in coastal delta regions.
Marsh	A frequently or continually inundated wetland characterised by emergent herbaceous vegetation adapted to saturated soil conditions. In European terminology a marsh has a mineral soil substrate and does not accumulate peat.
Mire	Synonymous with any peat-accumulating wetland (European definition).
Moor	Synonymous with any peatland (European definition). A high moor is a raised bog while a low moor is a peatland in a basin that is not elevated above its perimeter.
Muskeg	Large expanses of peatlands or bogs, particularly used in Canada and Alaska.
Peatland	A generic term of any wetland that accumulates partly decayed plant matter (peat).
Playa	An arid- to semi-arid region wetland that has distinct wet and dry seasons. Term used in south-west United States for marsh-like ponds similar to potholes but of a different geologic origin.
Pothole	Shallow marsh-like ponds, particularly as found in the Dakotas, USA and central Canadian provinces.
Reedswamp	Marsh dominated by *Phragmites* (common reed); term used particularly in Europe.
Riparian system	Ecosystems with a high water table because of proximity to an aquatic system, usually a stream or river. They are given such names as bottomland hardwood forests, floodplain forests, bosques, and streamside vegetation strips.
Slough	A swamp or shallow lake system in the northern and mid-western United States. A slowly flowing shallow swamp or marsh in south-eastern United States.
Swamp	Wetlands dominated by trees or shrubs (US definition). In Europe, forested fens and areas dominated by reed grass (*Phragmites*) are also called swamps.
Tidal fresh-water marsh	Marshes along rivers and estuaries close enough to the coastline to experience significant tides by non-saline water. Vegetation is often similar to non-tidal freshwater marshes.
Vernal pool	Shallow, intermittently flooded wet meadow, generally typical of Mediterranean climate with dry season for most of summer and autumn.
Wet meadow	Grassland with waterlogged soil near the surface but without standing water for most of the year.
Wet prairie	Similar to a marsh but with water levels usually intermediate between a marsh and a wet meadow.

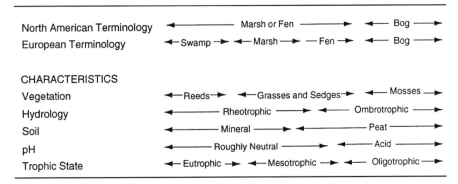

Figure 7.1. A comparison between North American and European terminology for similar inland non-forested freshwater wetlands. (Source: Mitsch and Gosselink 1993.)

attributes and four water characteristics some 13 categories were identified (Table 7.2). These are mutually exclusive categories and provide a more consistent basis for identifying wetlands. The categories used in the geomorphic system have single-word terms that avoid confusion with existing commonly used names for wetlands types, such as bog or marsh, although it is noted that agreement on the names is not a central feature of this system. This approach brings out the underlying similarity of wetlands across a wide range of climatic, geomorphic, soil, and vegetation settings based on the rationale that landform and water characteristics are the dominant and/or common feature for all wetlands, regardless of their setting. This classification scheme can be extended to a further level by the addition of descriptors for the prevailing salinity and its seasonal variability. Descriptors for the shape and size of the wetland can also be provided, while vegetation can be described in terms of the amount of cover and its complexity.

THE IMPETUS FOR CONSERVATION: WETLAND BENEFITS AND WETLAND LOSS

There is a limited number of attributes common to freshwater wetlands which are fundamental to their formation and existence and subsequently their conservation (Hughes and Heathwaite 1995). In many cases the failure to appreciate or understand these attributes has been a major contributory factor in wetland destruction or degradation. Before evaluating these attributes the two principal driving forces that have been largely responsible for stimulating and driving wetland conservation need to be understood. Awareness of wetlands among conservationists, hydrologists, water managers, politicians and indeed the general public has blossomed over the last few decades (e.g.

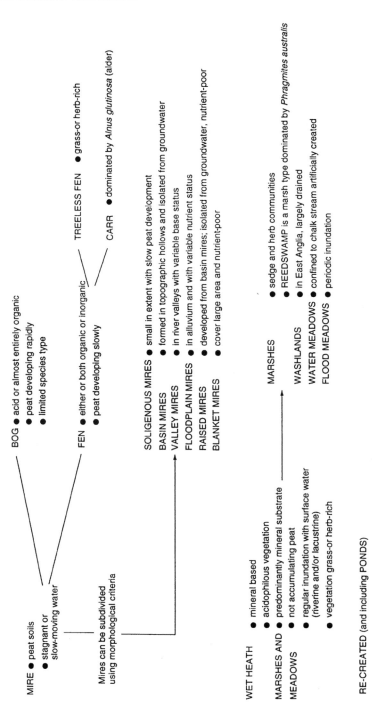

Figure 7.2. Principal British freshwater wetland types. (Source: Hughes and Heathwaite 1995.)

Table 7.2. Schematic presentation of the classification of wetland habitats based on landform and water characteristics (Semeniuk and Semeniuk 1997)

Water longevity	Landform				
	Basin	Channel	Flat	Slope	Highland
Permanent inundation	Lake	River			
Seasonal inundation	Sumpland	Creek	Floodplain		
Intermittent inundation	Playa	Wadi	Barlkara		
Seasonal waterlogging	Dampland	Trough	Palusplain	Paluslope	Palusmont

Table 7.3. Wetlands beyond wildlife habitat: functions, products and attributes associated with wetlands

Functions	Products	Attributes
Groundwater recharge	Agricultural crops	Active recreation
Groundwater discharge	Grazing resources	Passive recreation
Flood storage and desynchronisation	Fish	Roles in global chemical cycles
Shoreline anchoring and dissipation of erosive forces	Fuel	Maintenance of bio-diversity
Sediment trapping	Building materials	Microclimate stabilisation
Nutrient retention and removal		

Williams 1990; Finlayson and Moser 1992; Mitsch and Gosselink 1993; Mitsch 1994; McComb and Davis 1998). Whereas wetlands were all too frequently associated with wastelands of little value, they are now widely recognised as being among the Earth's most productive and important ecosystems (e.g. Mitsch *et al.* 1994).

The first factor that is responsible for raising concern for wetlands, their management and conservation is the recognition that wetland environments provide a host of benefits which secure goods and services for the health, safety and welfare of human communities (e.g. Hollis *et al.* 1988; Hollis 1992a). The literature on these wetland benefits, which are conventionally categorised as functions or environmental services, products and attributes (Table 7.3), is extensive, continually expanding and beyond the scope of this chapter. Mitsch and Gosselink (1993), Thompson and Hollis (1997) and Williams (1990) are some examples from the flourishing body of literature devoted to expounding the virtues of wetlands.

The benefits provided by an individual wetland are the result of the inter-action of its many physical and biological characteristics. Under the Ramsar Wetlands Convention these have been defined as follows (Finlayson 1996):

- *Functions* are the result of the interactions between biological, chemical and physical wetland components such as soils, water, plants and animals, and include: water storage; storm protection and flood mitigation; shoreline stabilisation and erosion control; groundwater recharge; groundwater discharge; retention of nutrients, sediments and pollutants; and stabilisation of local climatic conditions, particularly rainfall and temperature.
- *Products* are generated by the interactions between the biological, chemical and physical components and include: wildlife resources; fisheries; forests; forage; agricultural resources; and water supply.
- *Attributes* have value either because they induce certain uses or because they are valued themselves, and include the following: biological diversity; geomorphic features; and unique cultural and heritage features.

The combination of wetland functions, products and attributes give the wetland benefits and values that make it important to society.

The degree to which wetlands provide different benefits varies dramatically and few wetlands perform all the functions, provide all the products or are associated with all the attributes in Table 7.3 (Hollis 1990). However, as Hollis (1992b) argued, it is possible for small wetlands, which may have traditionally had small conservation interest, to provide valuable benefits. The growing recognition that through these benefits, which often have significant economic values (e.g. Barbier Acreman and Knowler, 1997), wetlands are important beyond mere wildlife habitat, presents an important opportunity for enhancing the political and institutional will for their conservation (e.g. Dugan 1994; Lemly, Kingsford and Thompson 2000; Maltby 1991).

The second major factor that has stimulated wetland conservation is the realization that their destruction and degradation has been taking place at a rapid, and in many cases, accelerating rate (Finlayson and Moser 1992; Finlayson, Hollis and Davis 1992; Dugan 1993). Wetlands are rarely static. Many, through the accumulation of mineral or organic matter, gradually dry out (Gilman 1994), but human activities have been increasingly responsible for hastening the rate of loss. For many wetlands it is too late and in some countries, especially in Europe and North America, natural wetlands are nearly extinct (Mitsch, Mitsch and Turner 1994). Those that remain are severely threatened (Green 1981; Mitsch, Mitsch and Turner 1994). For example, the USA has lost 54%, or 87 million ha, of its original wetlands (Tiner 1984; Dahl 1990). Throughout Europe, wetlands have been destroyed wholesale (Hollis and Jones 1991; Stevenson and Frazier 1999). The magnitude of European wetland loss is illustrated by the national examples shown in Table 7.4 (CEC 1995). Since 1900 the south-west of England has lost 92% of its wet pastures (Denny 1993) while in the area of the Thames Estuary the extent of grazing marshes has fallen by 28 000 ha or 65% since the 1930s (Thornton and Kite 1990). Freshwater meadows, bogs and woods once extended over 1.3 million

Table 7.4. The magnitude of wetland loss in six European countries (CEC 1995)

Country	Period	% loss of wetlands
Netherlands	1950–1985	55
France	1900–1993	67
Germany	1950–1985	57
Spain	1948–1990	60
Italy	1938–1984	66
Greece	1920–1991	63

ha of France. Baldock (1989) estimated that this area has recently been declining by 10 000 ha each year. Similarly, wetlands covered 10% of Italy (3 million ha) in Roman times, yet by 1865 this had declined to only 764 000 ha and by 1920 only 190 000 ha remained (Ramsar Convention Bureau 1990). In Spain, 60% of wetlands in the Castille La Mancha region have been lost with 75% of this loss, some 200 000 ha, taking place since the mid-1960s (Montes and Bifani 1989). Psilovikos (1992) estimated that of the 98 600 ha of fresh-water marshes in pre-1930s Macedonia only 5600 ha, or less than 6%, remain.

In most cases, human-induced wetland losses have resulted from activities, such as drainage and water diversion, that have been aimed at fundamentally modifying the natural features which provide wetlands with their unique characteristics. Frequently, these features are poorly appreciated (Denny 1995). Unexpected and deleterious impacts often occur following interventions within a wetland or within its catchment. Attempts to restore wetlands are often hampered by the lack of knowledge on the causes of damage, the nature of controls upon key environmental variables and the responses of organisms to these variables (Wheeler 1995). The conservation of wetlands requires an understanding and evaluation of the processes operating within them and the nature of the features that furnish them with their particular characteristics. Ecological character is a function of all the components of a wetland and is intricately related to the values and benefits. Hollis (1998) went further and argued that, not only was the hydrological regime the driving force of all wetlands, but it was the key to their conservation within a much larger scenario of water management.

WETLAND HYDROLOGY:
THE KEY TO WETLAND CONSERVATION

The central role of hydrology, the science of the occurrence, movement and distribution of water, within wetland conservation is evident from the dominance of water within the three common components that feature within the multitude of alternative wetland definitions (Mitsch and Gosselink 1993):

- Wetlands are distinguished by the presence of water, either at the surface or within the root zone for at least part of the time. The depth and duration of flooding or soil saturation may vary from year to year and from wetland to wetland.
- Wetlands often have unique soil conditions that differ from adjacent uplands. Hydric soils are saturated, flooded or ponded long enough during the growing season to develop anaerobic conditions in the upper horizons.
- Vegetation in wetlands is adapted to the wet conditions found there (most species are hydrophytes). Conversely, vegetation that is intolerant to surface flooding or soil saturation is characteristically absent.

The hydrological regime is supremely important for maintaining a wetland's structure and functional characteristics. Unique physiochemical conditions distinguish wetlands as ecotones between well-drained terrestrial systems and deepwater aquatic systems. In common with other ecotones, wetlands share some features of the systems they lie between (Lewis 1995). The influence of the hydrological regime upon these characteristics is best summarised using a conceptual model (Figure 7.3).

Figure 7.3 shows that the hydrological regime within a wetland affects many chemical and physical wetland properties such as soil and water salinity, soil anaerobiosis, nutrient availability, pH and sediment properties including deposition rates and texture (e.g. Holland, Whigham and Gopal 1990; Lewis 1995; Mitsch and Gosselink 1993). Wetland biotic characteristics, such as the distribution of vegetation communities, are also directly affected. Factors such as water depth and the period of inundation control the nature of the adaptations required by plants to establish, live and reproduce within a wetland. The chemical and physical properties that are influenced by the hydrological regime also affect biotic characteristics. The substrate, including the availability of nutrients within it, exerts a major influence on the rate of establishment of plants, the range of species that are supported and ultimately their long-term survival, as well as the range of habitat available for invertebrates, fish (in particular for spawning) and other animal life (Weller 1994).

As shown in Figure 7.3, a wetland's chemical and physical properties and its biotic components are also active in influencing and altering the hydrological regime through a feedback mechanism. For example, the growth of vegetation can modify hydrological conditions in a number of ways. Vegetation slows water movement thereby promoting the deposition of sediment (Weller 1994). Plant roots bind sediments and so reduce erosion, a mechanism which is enhanced by the protection vegetation affords against wind and wave action. The accumulation of decaying plant matter combined with trapped sediment reduces the capacity of the wetland over time and may decrease the duration and frequency of inundation (Mitsch and Gosselink 1993). Within bogs, the accumulation of peat raises the elevation of the surface to the point where

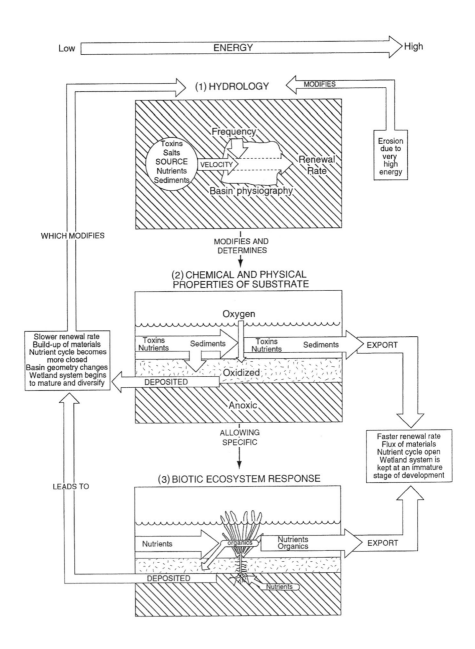

Figure 7.3. Conceptual model of the direct and indirect effects of hydrology on wetlands. (Source: Mitsch and Gosselink 1993.)

water flowing from the surrounding catchment may no longer influence it. In such circumstances bogs become wholly dependent upon precipitation for water. The shading of water surfaces by plants may reduce evaporation rates while transpiration by hydrophytes often represents a significant flux of water from wetlands.

Animals can also modify hydrological conditions within wetlands. In North America, beavers (*Castor canadensis*) create wetlands by building dams on streams. Wetland vegetation can be opened up by the consumption of rootstocks by ducks, geese, fish and aquatic rodents (Weller 1994), or by grazing and trampling, such as that caused by the introduced feral water buffalo (*Bubalis bubalus*) in northern Australian wetlands. Such activities may cause large areas of plants to float to the water surface carrying soil with the roots and subsequently deepening the wetland bed. The creation of other depressions, of various sizes and in different geographical locations, results from the activities of crayfish, alligators, rodents, elephants, hippopotami and other animals. The disturbance of wetland substrate associated with these activities also increases the turbidity of the water within the wetland and can lead to premature drainage of fresh water from wetlands that are flooded seasonally or even less regularly (Storrs and Finlayson 1997).

The strong interconnections between the hydrological regime and other physical properties, chemical properties, and the biotic components of a wetland ensure that even small changes can be responsible for enormous modifications to species composition and diversity and to the productivity of the ecosystem (Mitsch and Gosselink 1993). For example, premature drainage of seasonally inundated freshwater wetlands along the coast of northern Australia can lead to the incursion of saline tidal water and is possibly not inseparable from the effects of global climate change (Bayliss *et al.* 1997; Eliot, Waterman and Finlayson 1999). This sensitivity strengthens the requirement for sound hydrological foundations upon which to build wetland management and conservation strategies. There are several key hydrological attributes common to freshwater wetlands which, following the nature of hydrological systems, are interrelated. The remainder of this chapter outlines each of these attributes and, using examples, shows how an understanding of these attributes is central to wetland conservation.

THE WETLAND WATER-LEVEL REGIME

A wetland's water-level regime, commonly referred to as the hydroperiod in North America, is the seasonal pattern of surface and subsurface water elevation. It is the hydrological signature of each wetland type and the constancy of its pattern on a year-to-year basis ensures reasonable stability (Mitsch and Gosselink 1993). Figure 7.4 shows the water-level regimes of 12 different wetland types (Lewis 1995). It demonstrates the tremendous variety in

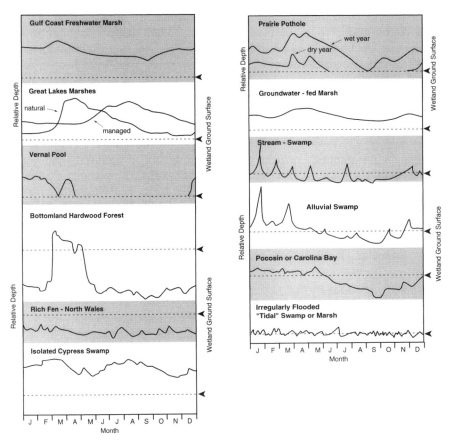

Figure 7.4. Water-level regimes or hydroperiods of different wetland types. (Sources: Mitsch and Gosselink 1993; Lewis 1995.)

the pattern of seasonal rising and falling water levels. This diversity is reflected in the classification of water-level regimes for non-tidal wetlands in Table 7.5.

Both the length of time during which standing water is present and the average number of times this inundation occurs in a given period, the flood duration and flood frequency respectively, vary from wetland to wetland. Figure 7.4 indicates that, whereas standing water is a permanent feature of many marshes, inundation is a seasonal occurrence in wetlands such as the bottomland forests along lowland streams and rivers (see Hughes and Rood, this volume). An increased frequency of inundation is experienced within riverine wetlands along small streams where water levels respond rapidly to local rainfall events (Junk, Bayley and Sparks 1989). Tidally influenced freshwater wetlands may experience semi-diurnal water-level regimes super-imposed upon variations associated with spring and ebb tides. In some

Table 7.5. Definitions of wetland water-level regimes (Cowardin *et al.* 1979)

Waterl-evel regime	Characteristics
Permanently flooded	Flooded throughout the year in all years.
Intermittently exposed	Flooded throughout the year except in years of extreme drought.
Semi-permanently flooded	Flooded in the growing season in most years.
Seasonally flooded	Flooded for extended periods in the growing season but usually no surface water by the end of the growing season.
Saturated	Substrate is saturated for extended periods in the growing season but standing water is rarely present.
Temporarily flooded	Flooded for brief periods in the growing season but the water table is otherwise well below the surface.
Intermittently flooded	Surface is usually exposed with surface water present for variable periods without detectable seasonal pattern.

wetlands, such as the fens and mires discussed by Gilman (1994), water-level changes are confined below the ground surface (Figure 7.4). Open water is limited to channels and ditches that surround or cross the wetlands.

Variations in the duration and frequency of inundation are as diverse as the depth and range of water level. Weller (1994) used the range and mean water depth to classify seven wetland types ranging from lakes through floodplains and marshes to tidally influenced freshwater wetlands (Figure 7.5). Subsurface water-level regimes of fens display only limited seasonal variations (Figure 7.4). In contrast, many wetlands, such as those closely associated with streams and rivers, experience highly variable water levels. The large variations in these cases are associated with seasonal or periodical flooding 'pulses' which are responsible for supplementing wetlands with additional nutrients and removing detritus (Junk, Bayley and Sparks 1989). Another important property of a wetland's water-level regime is the degree of year-to-year variation. In many wetlands, as the example of the prairie pothole in Figure 7.4 indicates, water-level regimes vary considerably from one year to another (Weller 1994). Such variation can be even greater in inland freshwater wetlands that are flooded intermittently or episodically. Many inland wetlands undergo an irregular cycle of inundation and these habitats are far more common, especially in Africa and Australia, than is often realised (Williams 1998)

The water-level regime is a major factor controlling a wetland ecosystem's structure and its ecological processes. Many properties of the water-level regime may impact the growth and survival of plants (Wheeler and Shaw 1995). The effect of high or low water levels upon both plant growth and community composition could be dependent upon their magnitude, duration, frequency and periodicity. The repercussions of these factors are different for

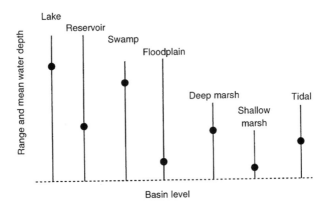

Figure 7.5. Classification of wetland types according to range and mean water depth. (Source: Weller 1994.)

overwintering success, renewal of growth in spring and reproduction in summer. In arid or semi-arid areas the balance between fresh and saline conditions in wetlands is linked with the extent and depth of flooding. Further, in tropical wetlands the establishment, development and dominance of vegetation are seemingly linked with the timing and duration of inundation and water depth (Finlayson 1988; Finlayson, Bailey and Cowie 1989).

The responses of individual species to water-level regime are also variable. Different plants have 'preferred' optimum positions on a gradient reflecting the duration and depth of inundation or soil saturation (e.g. Newbold and Mountford 1997). Rheinhardt and Hershner (1992) evaluated the influence of water-level regime upon canopy composition within tidal freshwater swamps along the Pamunkey River which flows into Chesapeake Bay, USA. They showed that swamps dominated by ash (*Fraxinus* spp.) and swamp blackgum (*Nyssa sylvatica* var. *biflora*) occurred in wetter areas where mean water table depths were 0.17 m. Drier areas, with mean water table depths of 0.21–0.30 m, were dominated by sweetgum (*Liquidambar styraciflua*) and red maple (*Acer rubrum*). Since a difference in mean water table depth of only a few centimetres was shown to influence canopy composition, even slight modifications to the hydrological regime, perhaps associated with relative sea-level rise, would lead to significant structural changes in these tidal swamps. However, for many plant species the importance of water depth has not be ascertained: for example, the *Melaleuca* (paperbark) trees that occur in many tropical freshwater wetlands seemingly have preferred depth tolerances, but the tolerance of individual species or sub-species is not well known. This is well illustrated by *Melaleuca cajaputi* which, in northern Australia, seems to favour 3–6 months flooding (Finlayson, Cowie and Bailey 1993), whereas in Vietnam it commonly grows in areas with 10–11 months of inundation (J. Barzen, personal communication).

The magnitude of fluctuations in wetland soil saturation, and subsequently oxidation and reduction conditions, also exerts a strong influence on nutrient, in particular nitrogen, transformations and cycling (Ross 1995). Permanent or long-term saturation reduces the oxygen content of wetland soils and produces chemical conditions that limit the number of macrophytic plant species that can survive without appropriate adaptations. In general, wetlands that experience long flood duration therefore possess lower species diversities than wetlands that are flooded less frequently and where surface water remains for shorter periods. Primary productivity is frequently higher in wetlands with water-level regimes characterised by large ranges in water level resulting from the flooding pulses described by Junk, Bayley and Sparks (1989) than within wetlands that experience sluggish or still water conditions (e.g. Mitsch 1988). Junk, Bayley and Sparks (1989) also stressed that the predictability of flood pulses influences ecosystem responses. Unpredictable pulses tend to impede adaptations to inundation among organisms. In contrast, where pulses are regular, adaptations and strategies in wetland organisms can develop that make the most efficient use of available resources.

The common misconception that wetlands require year-round standing water has therefore been criticised (Fredrickson and Reid 1990). In many wetlands that are managed there has been a tendency in the past to restrict and stabilise fluctuations in water level and thereby disrupt the natural water-level regime to which wetland flora and fauna have adapted (e.g. Deuver 1988; Brock, Smith and Jarman 1999; Kingsford, 1999). Many interventions within wetlands have resulted in fundamental modifications to water-level regimes that have been responsible for their destruction or degradation. Conservation and management initiatives for wetlands must, therefore, aim to establish water-level regimes that are in tune with those required for the establishment and maintenance of the desired ecosystem structure and vegetation community composition.

Lowland wet grasslands provide examples from the UK in which the establishment of suitable water-level regimes is central to wetland management and conservation. These wetlands, which are subject to periodic flooding and high water tables, are characteristically found in river valleys, areas of impeded drainage and behind sea defences (Jefferson and Grice 1998). In the latter situation, where land has often been reclaimed from salt marshes, wetlands are traditionally termed 'coastal grazing marshes'. Notable examples include the Somerset Levels, the North Kent Marshes, the Pevensey Levels and the Norfolk Broads (e.g. Royal Society for the Protection of Birds (RSPB) *et al.* 1997). Important features of lowland wet grasslands are the often complex networks of drainage ditches which surround and cross them. In the case of coastal grazing marshes these ditches reflect the former salt marsh drainage

channels although many of these have been straightened, widened and
deepened. The ditches are often rich in plant and invertebrate species while
the grasslands, which contain a variety of habitats ranging from drier to wetter
areas, are important breeding and wintering grounds for wading birds and
other waterfowl (e.g. English Nature 1997).

Lowland wet grasslands have evolved under extensive or low-intensity
agricultural management, principally grazing and hay-cutting. Post-Second
World War agricultural intensification saw the conversion of many areas of
wet grassland to arable or improved grassland (e.g. Glaves 1998; Jefferson and
Grice 1998; Jenman and Kitchen 1998). This was facilitated by improvements
in drainage infrastructure such as the installation of pumping stations and field
under-drainage. Al-Khudhairy and Thompson (1997) and Al-Khudhairy et al.
(1999) employed hydrological modelling to investigate the influence of a
pumping station, under-drainage and conversion to arable cropping upon a
catchment within the North Kent Marshes. They showed that, although
drainage did not change the annual peak groundwater levels, the duration of
high water levels was reduced considerably (Figure 7.6). Summer groundwater
elevations were considerably higher under natural conditions than when
pumping, drainage and arable cropping had been imposed.

This kind of change in water-level regime, which has been repeated in many
lowland wet grasslands, has important implications for conservation value.
Gowing, Spoor and Mountford (1998) demonstrated the strong influence of
water table elevation upon wet grassland plant communities in the Somerset

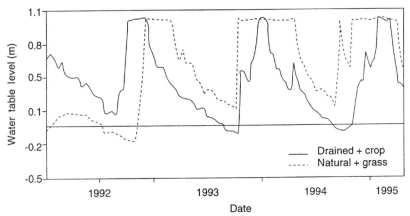

Figure 7.6. Simulated groundwater elevations within a catchment of the North Kent
Marshes under natural grassland conditions and following the imposition of pumping
under-drainage and conversion to arable (Al-Khudhairy and Thompson 1997; Al-
Khudhairy et al. 1999.)

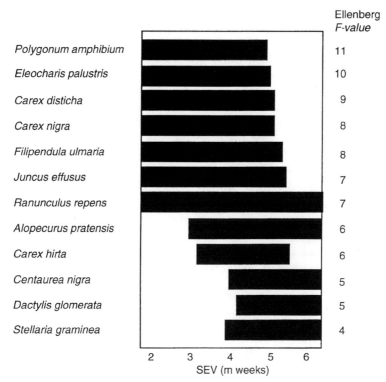

Figure 7.7. Summary of tolerated ranges for a number of plant species found in the Somerset Moors with respect to potential drought stress below a 0.45 cm threshold (SEV = Sum Exceedence Value) (Gowing, Spoor and Mountford 1998.)

Moors. They employed the concept of Sum Exceedence Value (SEV), which expresses the degree to which water tables fall below, or rise above, threshold elevations representing exposure to drought conditions and anoxia caused by high water tables or flooding respectively. Figure 7.7 shows that the tolerance of different species to potential drought stress varied markedly. Changes in water-level regime associated with prolonging periods of lower water table elevations can, therefore, be expected to favour the establishment of plant communities that are more tolerant of dry conditions. Drier conditions as a result of drainage improvements have also been blamed for declining populations of waterfowl normally associated with wet grasslands (e.g. Glaves 1998; Jenman and Kitchen 1998; RSPB *et al.* 1997).

In response to these declines, hydrological management of wet grasslands for nature conservation is now predominantly concerned with maintenance of appropriate water regimes, particularly wet conditions for certain periods of the year (Joyce and Wade 1988). A central feature of Environmentally

Sensitive Area (ESA) schemes, such as those developed for the Somerset Levels and Moors, the North Kent Marshes and the Norfolk Broads as well as the Pevensey Levels Wildlife Enhancement Scheme, is the maintenance of more ecologically sympathetic high ditch water levels (e.g. Armstrong and Rose 1998, Jenman and Kitchen, 1998; Ministry of Agriculture, Fisheries and Food (MAFF) 1989). In their prescriptions for the North Kent Marshes ESA, MAFF (1995) required land owners to maintain ditch water levels at not less than mean field level between 1 December and 30 April. High water levels were acknowledged to be essential for overwintering and breeding birds, wetland flora and other fauna. They also provided 'wet fencing' and water for stock. In addition, the creation of shallow pools by high ditch water levels provided suitable conditions for feeding waders. The ESA prescriptions required the provision of 0.3 m of water in the bottoms of ditches and dykes between 1 May and 30 November. The transition from winter to summer water levels should be a gradual process, occurring naturally, in order to maximise benefits to wildlife. Similar water level requirements feature within the Water Level Management Plans (MAFF 1991) which have been produced for many lowland wet grasslands in the UK. Control over water levels in accordance with these schemes is obtained through the appropriate operation of bunds, dams and sluices installed within wetland ditch networks.

THE WETLAND WATER BALANCE

Rising and falling water levels reflect the balance of inflows and outflows. This water balance or water budget, which is often evaluated on a monthly basis, but which may occasionally be examined using a daily time step, is a crucial component of wetland hydrology (Hollis and Thompson 1998). Unlike wholly terrestrial ecosystems, the water balances of wetlands are dominated by storage in open-water bodies and particularly in the saturated and unsaturated parts of the soil. Soil storage dominates the hydrology of the North Kent Marshes. Depending on the time of year, between 89% and 95% of the water is stored within the soil (Hollis, Fennessy and Thompson 1993). The ditches and infrequent surface inundation constitute a minor proportion of the total water balance.

Gilman (1994) attributed the dominance of storage within wetland water balances to gentle topographic and hydraulic gradients combined with the texture of wetland soils, which frequently provides high storage coefficients, and resistance to flow associated with vegetation. Water stored within wetlands is generally slowly released as surface flow, groundwater discharge or evapotranspiration.

Wetland water balances are conventionally expressed as:

$$V_t = V_{t-1} + (P - I) + Q_i + G_i - E_0 - E_t - Q_0 - G_0 \pm T \pm (H) \qquad (7.1)$$

where V_t is the volume of water in the wetland at time t; and V_{t-1} is the volume of water in the wetland at time interval $t - 1$; and the following are inputs and outputs between time $t - 1$ and t:

P is precipitation directly onto the wetland
I is evaporation of water intercepted by vegetation within the wetland
Q_i are the contributions from rivers and surface runoff to the wetland
G_i is the groundwater inflow to the wetland
E_0 is evaporation from areas of open water
E_t is evapotranspiration from wetland vegetation and soils
Q_0 are the outflows from the wetland carried by streams and rivers
G_0 is the groundwater recharge from the wetland
T are any tidal inputs ($+$) or outputs ($-$) associated with the wetland
H are any human water withdrawals (e.g. for irrigation) or returns of water to the wetland (e.g. effluent discharges).

Reviews of each of the components of the wetland water balance, which are beyond the scope of this chapter, are provided by Gilman (1994), Hollis (1992b), Ingram (1983) and Mitsch and Gosselink (1993). Effective wetland management and conservation depends upon a thorough understanding of how the different elements of the water budget interact to provide the water-level regime, the template on which wetland plant communities develop (Gilman 1994). The identification and evaluation of the components in an individual wetland is a prerequisite in the development of management and conservation initiatives.

Changes in any inflow or outflow of a wetland's water budget can result in significant ecological changes. Redgrave and Lopham Fen in East Anglia is the largest example of a valley head mire in England and is both a Site of Special Scientific Interest (SSSI) and a Ramsar Site (Denny 1993; Harding 1993). In common with other examples of this type, a major feature of the water balance of the fen is irrigation by groundwater (Wheeler and Shaw 1995). However, abstraction via a public-supply borehole located close to the fen lowered the water table and eliminated the seepage of groundwater. The hydrology of the fen became controlled by rainfall patterns and levels within the River Waveney (Harding 1993). These changes were responsible for a decline in floristic quality. A shift from species-rich, soligenous calcareous mire communities to degraded topogenous fen communities occurred. The floristic changes were paralleled by declines in the site's fauna. Invertebrates associated with spring-fed calcareous fens declined, in some cases, catastrophically. The fen raft spider (*Dolomedes plantarius*), which is found at only two wetlands within the UK, faces extinction on the fen (Harding 1993).

Each of the water inflows and outflows represented in Equation (7.1) vary in importance between different wetland types and not all of the terms of the water budget apply to all wetlands (Heathwaite 1995). This variability is demonstrated in Figure 7.8, which shows the annual water balances for some different wetland types. Table 7.6 characterises the pattern of the major components of the water budget and the types of wetland that they affect. Although annual water balances, such as those shown in Figure 7.8, provide a useful means of summarising the relative importance of the different elements of a wetland's hydrological cycle, seasonal and inter-annual variations in water balances can be considerable. Winter and Rosenberry (1995) used wetland and shallow well water level records to identify different patterns in the relationship between prairie potholes and groundwater in North Dakota, USA. They demonstrated that, whereas wetlands in the topographically highest parts of the region recharged groundwater whenever they received water from precipitation, wetlands in intermediate altitudes received groundwater for much of the time but also experienced some seepage when wetland water levels were higher than local water table elevations. The timing, duration and location of this seepage around the perimeter of the wetland varied seasonally and from year to year.

The water balance and the relative importance of the different components within it play a fundamental role in determining other environmental conditions. The interaction of precipitation, evapotranspiration, surface water and groundwater affects hydrochemistry and that in turn influences vegetation (Vardavas 1989; Finlayson, Bailey and Cowie 1989; Gilvear et al. 1993). Ingram (1993) divided mires into two distinct categories according to the nature of their water supplies and their impact on wetland water chemistry. According to this division, wetlands that are recharged mainly by meteoric water (precipitation in all its forms) and are isolated from the influence of surface and groundwater flows are characterised by low nutrient status. These ombrotrophic wetlands are dominated by raised mires and blanket bogs (Figure 7.2; Hughes and Heathwaite 1995). In contrast, wetlands where surface and groundwater flows are major components of the water balance are recharged by water potentially high in nutrients derived from the soils and rocks of the surrounding catchment (Ross 1995). Such minerotrophic wetlands are therefore dependent upon hydrological processes operating beyond their boundaries and cannot be considered in isolation from broader catchment influences. Reduced flooding following drainage improvements within fens in the Jegrznia area of Poland stopped the inflow of nutrients from floodwater (De Mars 1996). This resulted in dramatic shifts in vegetation composition and

---

Figure 7.8. *(opposite)* Annual water balances for different wetland types. P = precipitation, ET = evapotranspiration, I = interception, P_n = net precipitation, S_i = surface inflow, S_0 = surface outflow, G_i = groundwater inflow, G_0 = groundwater outflow, $\Delta V/\Delta t$ = change in storage per unit time, T = tides. All values expressed in $cm\ yr^{-1}$. (Source: Mitsch and Gosselink 1993.)

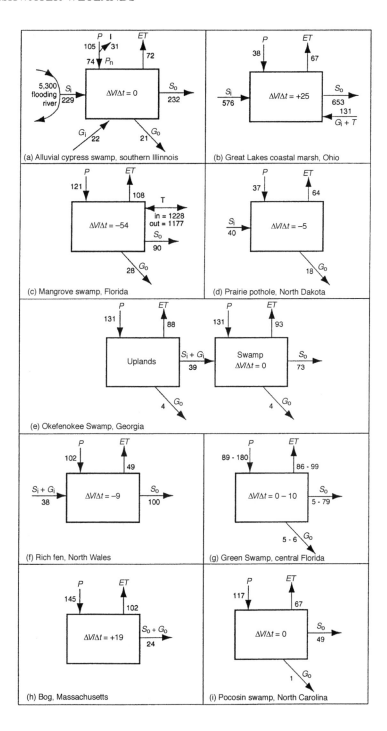

(a) Alluvial cypress swamp, southern Illinnois

(b) Great Lakes coastal marsh, Ohio

(c) Mangrove swamp, Florida

(d) Prairie pothole, North Dakota

(e) Okefenokee Swamp, Georgia

(f) Rich fen, North Wales

(g) Green Swamp, central Florida

(h) Bog, Massachusetts

(i) Pocosin swamp, North Carolina

Table 7.6. Major components of freshwater wetland water balances (Source: Mitsch and Gosselink 1993)

Component	Pattern	Wetlands affected
Precipitation	Varies with climate although many regions exhibit distinct wet and dry seasons.	All
Evapo-transpiration	Seasonal with peaks in summer and low rates in winter. Controlled by meteorological, physical and biological conditions within wetlands.	All
Surface inflows and outflows	Seasonal, often matching precipitation pattern or spring thaw. Can be channeled as stream flow or non channeled as runoff and includes river flooding of alluvial wetlands.	Potentially all wetlands except ombrotrophic bogs. Surface flows are particularly important for riparian wetlands.
Groundwater	Less seasonal than surface flows and not always present.	Potentially all wetlands except ombrotrophic bogs and other perched wetlands.
Tides	One or two tidal periods each day while the flood frequency depends upon wetland elevation.	Tidal freshwater wetlands.

low above-ground biomass. The reliance of many wetlands upon outside hydrological influences means, therefore, that effective wetland management and conservation requires integrated and coordinated plans for water management within whole catchments (Denny 1993).

It may also be necessary to consider intercatchment water management given that groundwater may extend beyond the surface catchment or that floods may cross catchment boundaries. The latter is a common circumstance in northern Australia where annual flooding can provide surface linkages between wetlands that are associated with neighbouring rivers (Storrs and Finlayson 1997). The interrelatedness of wetlands in such environments can provide a particular management challenge when the freshwater floods extend into coastal zones that are also affected by tidal regimes, necessitating consideration of integrated coastal zone management (Eliot, Waterman and Finlayson 1999) in addition to catchment-wide management.

In wetlands where the water balance includes large surface or subsurface flows, especially if they are associated with high amplitude oscillations in water table level, peat development is usually retarded (Hughes and Heathwaite 1995). Providing significant erosion does not take place, high rates of water flow through a wetland can lead to the development of vegetation communities dominated by taller species and, as noted above, higher primary productivity

(Gilman 1994). Moving water can also remove toxic substances, such as ferrous iron, manganese and aluminium, while providing vegetation roots with nutrients even if concentrations are low.

WETLAND LEVEL–AREA–VOLUME RELATIONSHIPS

Relationships between water level, area of inundation and volume of water determine how the water balance influences the water level regime, or hydroperiod. These transformations are essential for understanding how hydrological inputs and outputs impact upon such ecologically vital elements as level of saturation in the root zone, the distribution of different water depths required for various floral and faunal assemblages and the area of open water (Hollis and Thompson 1998). Digital hydrological models developed for wetlands need these relationships to convert the water volumes derived from water balance calculations into depths and extents of inundation (e.g. Sutcliffe and Parks 1989; Thompson and Hollis 1995).

A prime factor that controls the seasonal fluctuations of both the vertical and horizontal extent of a wetland is its topography (Deuver 1988). It determines the volume of water that is stored within depressions before runoff is generated. Where channels and ditches are dominant, as in many European wet grasslands, the relationship between volume of water and its level within these channels determines the elevation of water required to inundate adjacent areas. The area that is inundated is in turn determined by the combined level–area–volume relationships of the channel network and the surrounding wetland.

Gasca-Tucker (in preparation) has produced level–volume–area relationships for a number of catchments within the Pevensey Levels, an internationally important lowland wet grassland in East Sussex, UK (Figure 7.9). The derivation of these relationships, which has been undertaken for both gravity-drained areas and catchments drained by pumps, has important implications for conservation. The volume of water required to satisfy the water-level prescriptions laid down by the Wildlife Enhancement Scheme and Water Level Management Plan can be evaluated. The threshold water levels required to initiate surface inundation and subsequently the gain in the area of flooding for a given increase in ditch water level can also be derived. In addition to providing a larger area of wetland habitat, the flooding of land within the Levels during the winter may provide a means of retaining sufficient water to satisfy spring and summer requirements. The opportunities for inundation are conditioned by the availability of funds to provide financial compensation to landholders whose traditional low-intensity grazing activities would be affected (Jenman and Kitchen 1998). Additionally, the management of raised water levels must be undertaken in coherent, hydrologically discrete areas in which all landholders have agreed to participate (Tidy 1995).

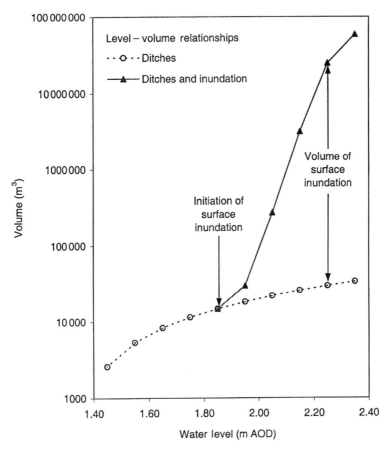

Figure 7.9. Level–area–volume relationships for catchments within the Pevensey Levels. (Source: Gasca-Tucker in preparation.)

Although level–area–volume relationships are most easily applied to wetlands with extensive areas of open water, the same principles can be applied in wetlands where shallow groundwater and sub-surface water storage are the major features. The central parameter in evaluating the relationship between change in water storage (s) and any shift in water elevation (h) is the specific yield (S_y) of the substrate (Gilman 1994). These are related as follows:

$$\Delta s = \frac{S_y}{100} \Delta h \qquad (7.2)$$

The specific yield, which is expressed as a percentage, is a function of the porosity and hence the degree of compaction of the substrate. The extent of

water level change in response to the annual climatic cycle and to any phenomena operating within the wetland such as droughts, short-term de-watering
and artificial drainage is controlled by the specific yield. In peat-dominated
wetlands the specific yield is controlled by both the extent of humification and
the degree of compaction. In fresh, undecomposed peats near the surface of
acid mires, specific yields are characteristically over 50% (Gilman 1994). In
contrast, peats which are derived from reed and sedge remains and those from
deep within acid mires that have been subject to longer periods of humification
have much lower specific yields of between 10% and 20%. The enhanced
compaction and humification with depth in peat wetlands such as ombrotrophic bogs is responsible for the declines in hydraulic conductivity which can
be of two or three orders of magnitude (Proctor 1995).

HYDROLOGICAL EXTREMES

Extreme hydrological events, such as floods and droughts, often play an
influential role in the development of wetlands and in controlling their physical
and ecological characteristics. For example, a level–area–volume relationship
can be modified by erosion by fast-flowing waters during periods of flooding or
by sediment deposition within the wetland as flood waters recede. Alter-
natively, drought periods resulting in the desiccation of a wetland or a major
interlude of enhanced salinity within coastal freshwater wetlands can have a
major bearing on the type of biota that survives. It is often these unpredictable
or catastrophic events that play a significant role in initiating changes in
wetland vegetation communities (Niering 1988).

An example of the potentially enormous impacts of hydrological extremes,
in this case floods, upon wetlands was provided by Stromberg, Fry and Patten
(1997). They documented the fivefold expansion in the area of riverine marsh
dominated by species including southern cattail (*Typha domingensis*), Olney's
bulrush (*Scirpus americanus*) and jointed rush (*Juncus articulatus*) along the
Hassayampa River of arid central Arizona, USA. This expansion was
attributed to large winter floods in January and February 1993 followed by
above-average flows throughout spring and summer of the same year. The
1993 flood enlarged the river's active channel and degraded terraces and
groundwater levels rose by approximately 0.5 m. Overall, the flood resulted in
large changes in surface elevations relative to the water table. Before the flood
less than 20% of the floodplain-channel surfaces were within 1 m of the water
table. In 1994, following the flood, the corresponding figure was 45%. The
growth in extent of areas with saturated soils or shallow water tables promoted
the establishment of wetland vegetation. Another flood two years later in
February 1995 removed most of the wetland vegetation and deposited sediment across the area which, by increasing the distance between the ground
surface and the water table, prevented the re-establishment of wetland

vegetation (Stromberg, Fry and Patten 1997). This example shows that floods can be a double-edged sword for wetland development.

At the other end of the spectrum of hydrological extremes, Niering (1988) described the important role played by drought in controlling the dynamics of prairie pothole wetlands in the American Midwest. These wetlands are characterised by cycles of desiccation and drowning. Drought periods are associated with the absence of wetland vegetation which, through the preservation of a seed bank within the dry marsh substrate, is re-established with the return of normal precipitation. High water, combined with disease, insects and senescence of the vegetation, can lead to the degeneration of the marsh, while muskrats (*Ondatra zibethicus*) may play a major role in stripping the wetland of its vegetation (Weller, 1994). The return of drought leads to desiccation and the replay of this cycle. Brock (1998) summarised information on the role of flooding and drought on seedbanks of temporary wetlands in Australia and South Africa and pointed out that these wetlands were resilient and that many contained species-rich germinable seed banks which enabled revegetation after drying cycles.

CONCLUSIONS

Hydrology is the key to understanding and evaluating the physical, chemical and biological characteristics of freshwater wetlands. These characteristics develop as a result of the interactions between some key hydrological attributes. The spatial and temporal distribution of inflows and outflows of water, when coupled with level–area–volume relationships, determines the water-level regime, hydroperiod or hydrological signature. This seasonal pattern of surface and subsurface water elevation and the associated expansion and contraction of inundation is the template on which the flora and fauna develop. Extreme hydrological events, such as floods and droughts, are also often an important influence. By upsetting prevailing hydrological conditions such events stimulate change.

The dominance of water within wetlands means that success in their management and conservation is dependent upon the establishment of sound hydrological foundations. An evaluation of the water balance is fundamental to understanding how a wetland works and how and when it will respond to climate change, drainage and any other alterations (Clymo *et al.* 1995). The success of management and conservation in wetlands depends upon the development of schemes that reflect the overriding importance of hydrology. Initiatives should include hydrological monitoring in addition to monitoring of flora and fauna. Raingauges, stage boards, automatic water-level recorders, piezometers and dip wells are among the hydrometeorological instruments which should be seen as invaluable tools for the wetland conservationist.

REFERENCES

Al-Khudhairy D & Thompson JR (1997) Modelling the hydrology of an underdrained catchment and the effects of reverting it to its former natural state. In: JC Refsgaard & EA Karalis (eds) *Operational water management*. Proceedings of the International Conference, Copenhagen, Denmark 3–7 September 1997. Balkema, Rotterdam, 330–40.

Al-Khudhairy D, Thompson JR, Gavin H & Hamm NAS (1999) Hydrological modelling of a drained grazing marsh under agricultural land use and the simulation of restoration management scenarios. *Hydrological Sciences Journal* **44**(6), 943–971.

Armstrong A & Rose S (1998) Managing water for wetland ecosystems: A case study. In: CB Joyce & PM Wade (eds) *European wet grasslands: biodiversity, management and restoration*. Chichester, John Wiley, 201–15.

Baldock D (1989) Agriculture and habitat loss in Europe, WWF International CAP Discussion Paper, Number 3, Institute for European Environmental Policy, London.

Barbier EB, Acreman MC & Knowler D (1997) *Economic valuation of wetlands: A guide for policy makers and planners*. Gland, Ramsar Convention Bureau.

Bayliss BL, Brennan KG, Eliot I, Finlayson CM, Hall RN, House T, Pidgeon RWJ, Walden D & Waterman P (1997) *Vulnerability assessment of predicted climate change and sea level rise in the Alligator Rivers Region, Northern Territory, Australia*. Supervising Scientist Report **123**. Canberra, Supervising Scientist Group, Environment Australia.

Brock MA (1998) Are temporary wetlands resilient? Evidence from seedbanks of Australian and South African wetlands. In: AJ McComb & JA Davis (eds) *Wetlands for the future*. Adelaide, Gleneagles Publishing, 193–206.

Brock MA, Smith RG & Jarman PJ (1999) Drain it, dam it: Alteration of water regime in shallow wetlands on the New England Tableland of New south Wales, Australia. *Wetlands Ecology and Management* **7**, 37–46.

Clymo RS, Dawson FH, Bertram BCR, Burt TP, Gilman K, Ingram HAP, James R, Kirkby MJ, Lee JA, Maltby E, Wheeler BD & Wilcock D (1995) Conclusion: Directions for research on wetlands in Britain. In: J Hughes & L Heathwaite (eds) *Hydrology and hydrochemistry of British wetlands*. Chichester, John Wiley, 445–78.

CEC (1995) *Wise use and conservation of wetlands*. Communication from the Commission to the Council of the European Parliament COM (95), Brussels.

Cowardin LM, Carter V, Golet FC & La Roe ET (1979) *Classification of wetlands and deepwater habitats of the United States*. United States Fish and Wildlife Service Publication FWS/OBS-79/31, Washington, DC.

Dahl TE (1990) *Wetland losses in the United States 1780s to 1980s*. US Department of the Interior and US Fish and Wildlife Service, Washington DC.

Davies J & Claridge G (eds) (1993) *Wetland benefits: the potential for wetlands to support and maintain development*. Asian Wetland Bureau Publication No. 87, Kuala Lumpur; IWRB Special Publication No. 27, Slimbridge; Wetlands for the Americas Publication No. 11, Massachusetts.

Davis TJ (1994) *The Ramsar Convention manual: a guide to the Convention on Wetlands of International Importance Especially as Waterfowl Habitat*. Gland, Ramsar Convention Bureau.

De Mars H (1996) *Chemical and physical dynamics of fen hydro-ecology*. Nederlandse Geografische Studies **203**. Utrecht.

Denny P (1993) Water management strategies for the conservation of wetlands. *Journal of the Institution of Water and Environmental Management* **7**, 387–94.

Denny P (1995) Benefits and priorities for wetland conservation: The case for national wetland conservation strategies. In: M Cox, V Straker & D Talyor (eds) *Wetlands: Archaeology and nature conservation*. London, HMSO, 249–74.

Deuver MJ (1988) Hydrologic processes for models of freshwater wetlands. In: WJ Mitsch, M Straškraba & SE Jørgensen (eds) *Wetland modelling*. Amsterdam, Elsevier, 9–39.

Dugan PJ (ed.) (1990) *Wetland conservation: A review of current issues and required action*. Gland, IUCN.

Dugan PJ (ed.) (1993) *Wetlands in danger*. Londo, Mitchell Beazley.

Dugan PJ (1994) Wetlands in the 21st Century: The challenge to conservation science. In: WJ Mitsch (ed.) *Global wetlands: old world and new*. Amsterdam, Elsevier, 75–87.

Eliot I, Waterman P & Finlayson CM (1999) Monitoring and assessment of coastal change in Australia's wet–dry tropics. *Wetlands Ecology and Management* **7**, 63–81.

English Nature (1997) *Wildlife and freshwater: an agenda for sustainable management*. Peterborough, English Nature.

Finlayson CM (1988) Productivity and nutrient dynamics of seasonally inundated flood plains in the Northern Territory. In: D Wade-Marshall & P Lovejoy (eds) *North Australia: progress and prospects, Volume 2 Flood plain research*. Darwin, ANU Press, 58–83.

Finlayson CM (1996) The Montreux Record: A mechanism for supporting the wise use of wetlands. In: *Proceedings of the 6th meeting of the Conference of the Contracting Parties, Technical sessions: Reports and presentations, Technical sessions B and D*. Gland, Ramsar Convention Bureau, 32–8.

Finlayson M & Moser M (eds) (1992) *Wetlands*. Oxford, Facts on File, Oxford.

Finlayson CM & Spiers AG (eds) (1999) *Global review of wetland resources and priorities for inventory*. CD-ROM version. Supervising Scientist Report, Supervising Scientist Group, Environment Australia, Canberra.

Finlayson CM & van der Valk AG (1995) *Classification and inventory of the world's wetlands*. Advances in Vegetation Science 16, Kluwer. Dordrecht, Academic Press.

Finlayson CM, Bailey BJ & Cowie ID (1989) *Macrophyte vegetation of the Magela Floodplain, Alligator Rivers Region, Northern Territory*. Supervising Scientist for the Alligator Rivers Region Research Report No. 5, Canberra.

Finlayson CM, Cowie ID & Bailey BJ (1993) Litterfall in a Melaleuca forest on a seasonally inundated flood plain in tropical northern Australia. *Wetlands Ecology and Management* **2**, 177–88.

Finlayson CM, Hollis GE & Davis TD (eds) (1992) *Managing Mediterranean wetlands and their birds*. IWRB Special Publication **20**, Slimbridge.

Fredrickson LH & Reid FA (1990) Impacts of hydrologic alteration on management of freshwater wetlands. In: JM Sweeney (ed.) *Management of dynamic ecosystems, North Central Section*. West Lafayette, The Wildlife Society.

Gasca-Tucker DL (in preparation) *Hydrology and hydrological management of the Pevensey Levels*. PhD thesis, University College London, London.

Gilman K (1994) *Hydrology and wetland conservation*. Chichester, John Wiley.

Gilvear DJ, Andrews R, Tellam JH, Lloyd JW & Lerner DN (1993) Quantification of the water balance and hydrogeological processes in the vicinity of a small groundwater-fed wetland, East Anglia, UK. *Journal of Hydrology* **144**, 311–34.

Glaves DJ (1998) Environmental monitoring of grassland management in the Somerset Levels and Moors Environmentally Sensitive Area, England. In: CB Joyce & PM Wade (eds) *European wet grasslands: biodiversity, management and restoration*. Chichester, John Wiley, 73–94.

Gore AJP (1983) Introduction. In: AJP Gore (ed.) *Ecosystems of the World 4A. Mires: swamp, bog, fen and moor. General studies*. Amsterdam, Elsevier, 1–34.

Gowing JG, Spoor G & Mountford O (1998) The influence of minor variations in hydrological regime on grassland plant communities: Implications for water management. In: CB Joyce & PM Wade (eds) *European wet grasslands: biodiversity, management and restoration.* Chichester, John Wiley, 218–27.

Green B (1981) *Countryside conservation.* London, Allen and Unwin.

Harding M (1993) Redgrave and Lopham Fens, East Anglia, England: A case study of change in flora and fauna due to groundwater abstraction. *Biological Conservation* **66**, 33–45.

Heathwaite AL (1995) Overview of the hydrology of British wetlands. In: JMR Hughes & AL Heathwaite (eds) *Hydrology and hydrochemistry of British wetlands.* Chichester, John Wiley, 11–20.

Heathwaite AL, Göttlich Kh, Burmeister EG, Kaule G & Grospietsch Th (1993) Mires: Definitions and form. In: AL Heathwaite & Kh Göttlich (eds) *Mires: process, exploitation and conservation.* Chichester, John Wiley, 1–75.

Holland MM, Whigham DF & Gopal B (1990) The characteristics of wetland ecotones. In: RJ Naiman & H Décamps (eds) *The ecology and management of aquatic– terrestrial ecotones.* United Nations Educational, Scientific and Cultural Organisation, Paris and Parthenon, Carnforth, UK and New Jersey, 171–98.

Hollis GE (1990) Environmental impacts of development on wetlands in arid and semi-arid lands. *Hydrological Sciences Journal* **35**(4), 411–28.

Hollis GE (1992a) The causes of wetland loss and degradation in the Mediterranean. In: CM Finlayson, GE Hollis & TD Davis (eds) *Managing Mediterranean wetlands and their birds*, IWRB Special Publication 20, Slimbridge, 83–92.

Hollis GE (1992b) The hydrological functions of wetlands and their management. In: PE Gerakis (ed.) *Conservation of Greek wetlands.* Proceedings of a Greek wetlands workshop, Thessaloniki, Greece, 17–21 April 1989. IUCN, Gland, 9–60.

Hollis GE (1998) Future wetlands in a world short of water. In: AJ McComb & JA Davis (eds) *Wetlands for the future.* Adelaide, Gleneagles Publishing, 5–18.

Hollis GE, Fennessy S & Thompson JR (1993) *A249 Iwade to Queenborough: wetland hydrology.* Report to Ove Arup and Partners, Wetland Research Unit, Department of Geography, University College London, London.

Hollis GE, Holland MM, Maltby E & Larson JS (1988) Wise use of wetlands. *Nature and Resources* **24**(1), 2–13.

Hollis GE & Jones TA (1991) Europe and the Mediterranean Basin. In: CM Finlayson & M Moser (eds) *Wetlands.* International Waterfowl and Wetlands Research Bureau. Oxford, Facts on File, 27–56.

Hollis GE & Thompson, JR (1998) Hydrological data for wetland management. *Journal of the Chartered Institution of Water and Environmental Management* **12**, 9–17.

Hughes JMR & Heathwaite AL (1995) Introduction. In: JMR Hughes & AL Heathwaite (eds) *Hydrology and hydrochemistry of British wetlands.* Chichester, John Wiley, 1–8.

Ingram HAP (1993) Hydrology. In: AJP Gore (ed.) *Ecosystems of the world 4A. Mires: swamp, bog, fen and moor. General studies.* Amsterdam, Elsevier, 67–158.

Jefferson RG & Grice PV (1998) The conservation of lowland wet grassland in England. In: CB Joyce & PM Wade (eds) *European wet grasslands: biodiversity, management and restoration.* Chichester, John Wiley, 31–48.

Jenman B & Kitchen C (1998) A comparison of management and rehabilitation of two wet grassland nature reserves: The Nene Washes and Pevensey Levels, England. In: CB Joyce & PM Wade (eds) *European wet grasslands: biodiversity, management and restoration.* Chichester, John Wiley, 229–45.

Joyce CB & Wade PM (1998) Wet grasslands: A European perspective. In: CB Joyce &

PM Wade (eds) *European wet grasslands: biodiversity, management and restoration.* Chichester, John Wiley, 1–12.

Junk WJ, Bayley PB & Sparks RE (1989) The flood pulse concept in river–floodplain systems. In: DP Dodge (ed.) *Proceedings of the International Large River Symposium.* Canadian Special Publication of Fisheries and Aquatic Science **106**, 110–27.

Kingsford RT (1999) Managing the water of the Border Rivers in Australia: irrigation, Government and the wetland environment. *Wetlands Ecology and Management* **7**, 25–35.

Lemly AD, Kingsford RT & Thompson JR (2000) Irrigated agriculture and wildlife conservation: Conflict on a global scale. *Environmental Management* **25**(5), 485–512.

Lewis WM (ed.) (1995) *Wetlands: Characteristics and boundaries.* National Research Council (US) and Committee on Characterization of Wetlands. National Academy Press, Washington, DC.

Maltby E (1986) *Waterlogged wealth: why waste the world's wet places?.* London, Earthscan.

Maltby E (1991) Wetland management goals: Wise use and conservation. *Landscape and Urban Planning* **20**, 9–18.

McComb AJ & Davis JA (eds) (1998) *Wetlands for the future.* Adelaide, Gleneagles Publishing.

MAFF (1989) *Environmentally Sensitive Areas.* London, HMSO.

MAFF (1991) *Conservation guidelines for drainage authorities.* London, MAFF.

MAFF (1995) *The North Kent Marshes Environmentally Sensitive Area: guidelines for farmers.* London, MAFF.

Mitsch WJ (1988) Productivity–hydrology–nutrient models of forested wetlands. In: WJ Mitsch, M Straškraba & SE Jørgensen (eds) *Wetland modelling.* Amsterdam, Elsevier, 115–32.

Mitsch WJ (ed.) (1994) *Global wetlands: old world and new.* Amsterdam, Elsevier.

Mitsch WJ & Gosselink JG (1993) *Wetlands* (2nd edn). New York, Van Nostrand Reinhold.

Mitsch WJ, Mitsch RH & Turner RE (1994) Wetlands of the Old and New Worlds: Ecology and management. In: WJ Mitsch (ed.) *Global wetlands: old world and new.* Amsterdam, Elsevier, 3–56.

Montes C & Bifani P (1989) *An ecological and economic analysis of current status of Spanish wetlands.* Report to OECD, Autonomous University of Madrid.

Newbold C & Mountford JO (1997) *Water level requirements of wetland plants and animals.* English Nature Freshwater Series No. 5, Peterborough, English Nature.

Niering WA (1988) Hydrology, disturbance and vegetation change. In: JA Kusler & G Brooks (eds) *Wetland hydrology.* Proceedings of a national wetland symposium, Chicago, USA, 16–18 September 1987. Association of State Wetland Managers, New York, 50–4.

Proctor MCF (1995) The ombrotrophic bog environment. In: BD Wheeler, SC Shaw, WJ Fojt, & RA Robertson (eds) *Restoration of temperate wetlands.* Chichester, John Wiley, 287–303.

Psilovikos A (1992) Changes in Greek wetlands during the twentieth century: The Cases of the Macedonian inland waters and of the river deltas of the Aegean and Ionian coasts. In: PE Gerakis (ed.) *Conservation of Greek wetlands.* Proceedings of a Greek wetlands workshop, Thessaloniki, Greece, 17–21 April 1989. Gland, IUCN, 175–96.

Ramsar Convention Bureau (1990) *Directory of wetlands of international importance.* Gland, Ramsar Convention Bureau.

Rheinhardt RD & Hershner C (1992) The relationship of below-ground hydrology to canopy composition in five tidal freshwater swamps. *Wetlands* **12**(3), 208–16.

Rodwell JS (ed.) (1991) *British plant communities, Vol. 2. Mires and heaths.* Cambridge, Cambridge University Press.

Ross SM (1995) Overview of the hydrochemistry and solute processes in British wetlands. In: JMR Hughes & AL Heathwaite (eds) *Hydrology and hydrochemistry of British wetlands.* Chichester, John Wiley, 133–81.

Royal Society for the Protection of Birds, English Nature and Institute of Terrestrial Ecology (1997) *The wet grassland guide: managing floodplain and coastal wet grasslands for wildlife.* Sandy, Royal Society for the Protection of Birds.

Semeniuk CA (1987) Wetlands of the Darling system—A geomorphic approach to habitat classification. *Journal of the Royal Society of Western Australia* **69**, 95–112.

Semeniuk CA & Semeniuk V (1995) Geomorphic approach to classifying wetlands in tropical north Australia. In: CM Finlayson (ed.) *Wetlands research in the wet–dry tropics of Australia.* Supervising Scientist Report **101**, Supervising Scientist Group, Environment Australia, Canberra, 123–28.

Semeniuk V & Semeniuk CA (1997) A geomorphic approach to global classification for natural wetlands and rationalization of the system used by the Ramsar Convention—A discussion. *Wetlands Ecology and Management* **5**, 145–58.

Stevenson N & Frazier S (1999) Review of wetland inventory information in Western Europe. In: CM Finlayson & AG Spiers (eds) *Global review of wetland resources and priorities for wetland inventory.* CD-ROM version. Supervising Scientist Group, Environment Australia, Jabiru.

Storrs MJ & Finlayson CM (1997) *Overview of the conservation status of wetlands of the Northern Territory.* Supervising Scientist Report **116**, Supervising Scientist Group, Environment Australia, Canberra.

Stromberg JC, Fry J & Patten DT (1997) Marsh development after large floods in an alluvial arid-land river. *Wetlands* **17**(2), 292–300.

Sutcliffe JV & Parks YP (1989) Comparative water balances of selected African wetlands. *Hydrological Sciences Journal* **34**(1), 49–62.

Thompson JR & Hollis GE (1995) Hydrological modelling and the sustainable development of the Hadejia–Nguru Wetlands, Nigeria. *Hydrological Sciences Journal* **40**(1), 97–116.

Thompson JR & Hollis GE (1997) Wetlands and integrated river basin management: Experiences in Asia and the Pacific: Chapters 1–4 and Annex I. In: Anonymous (ed.) *Wetlands and integrated river basin management: experiences in Asia and the Pacific.* United Nations Environment Programme, Nairobi, and Wetlands International, Kuala Lumpur.

Thornton D & Kite DJ (1990) *Changes in the extent of the Thames Estuary grazing marshes.* Peterborough, Nature Conservancy Council.

Tidy M (1995) ESA as a wetland rehabilitation scheme: Principles, policy and practice in the Somerset Levels and Moors. In: M Cox, V Straker & D Taylor (eds) *Wetlands: archaeology and nature conservation.* London, HMSO, 207–17.

Tiner RW (1984) *Wetlands of the United States: current status and recent trends.* United States Fish and Wildlife Service, Washington, DC.

Vardavas IM (1989) A water budget model for the tropical Magela flood plain. *Ecological Modelling* **46**, 165–94.

Weller MW (1994) *Freshwater marshes: ecology and wildlife management* (3rd edn). Minneapolis, University of Minnesota Press.

Wheeler BD (1995) Introduction: Restoration and wetlands. In: BD Wheeler, SC Shaw, WJ Fojt & RA Robertson (eds) *Restoration of temperate wetlands.* Chichester, John Wiley, 1–18.

Wheeler BD & Shaw, SC (1995) Plants as hydrologists? An assessment of the value of plants as indicators of water conditions in fens. In: JMR Hughes & AL Heathwaite (eds) *Hydrology and hydrochemistry of British wetlands*. Chichester, John Wiley, 63–82.

Williams M (1990) Understanding wetlands. In: M Williams (ed.) *Wetlands: a threatened landscape*. Oxford, Blackwell, 1–41.

Williams WD (1998) Dryland wetlands. In: AJ McComb & JA Davis (eds) *Wetlands for the future*. Adelaide, Gleneagles Publishing, 33–47.

Winter TC & Rosenberry DO (1995) The interaction of ground water with prairie pothole wetlands in the cottonwood lake area, east-central North Dakota, 1979–90. *Wetlands* **15**, 193–211.

8 Physical Contexts for Saltmarsh Conservation

J. R. FRENCH[1] AND D. J. REED[2]
[1]*University College London, UK*
[2]*University of New Orleans, USA*

INTRODUCTION

After centuries of loss through reclamation, hydrological alteration and destructive industrial use (Teal and Teal 1969; Josselyn 1983; Chabreck 1988; Doody 1992), saltmarshes are now recognised as essential components of coastal and estuarine ecosystems. In addition to their function in the maintenance of estuarine food chains (Mitsch and Gosselink 1993) the conservation value of saltmarsh habitats is reflected in the protection of a large proportion of remaining saltmarsh under recent environmental legislation (Burd 1989; Vairin 1997). In comparison with other major habitats, saltmarshes are typically highly fragmented, yet they offer much more valuable 'ecosystem services' than their limited spatial extent might suggest (Costanza et al. 1997).

Within the estuaries and coastal plains of the developed world, economic pressures in favour of large-scale agricultural land claim have eased substantially (Brooke 1992; Wolff 1992). However, although engineered conversion of intertidal to dryland habitat is diminishing, the unplanned reversion of saltmarsh to open water is now of major concern in subsiding coastal plains. A large literature concerns the more general problem of land loss in modern delta plain settings, most notably in coastal Louisiana (Britsch and Dunbar 1993; Turner 1997), but also in the Rhône (Ibanez, Day and Pont 1999) and Nile deltas (Stanley 1988) and in the Venice Lagoon (Day et al. 1998). Particular emphasis has been placed on the elevational stability of marshes subject to high rates of subsidence-driven relative sea-level rise and managerial interventions which adversely impact their sedimentary and hydrographic environment (Boyer, Harris and Turner 1997; Reed, DeLuca and Foote 1997; Kuhn, Mendelssohn and Reed 1999). Delta plain environments provide at least a partial analogue for the impacts of the accelerated sea-level rise that is expected elsewhere (Day and Templet 1989). In estuarine settings, net loss of marsh area is occurring due to lateral erosion, especially

Habitat Conservation: Managing the Physical Environment. Edited by A. Warren and J. R. French.
© 2001 John Wiley & Sons Ltd.

where geomorphological adjustments and natural ecological transitions are checked by fixed sea defences (Titus 1991; Turner and Dagley 1993; Burd 1995). Unless these constraints can be removed or bypassed in some way, an acceleration in the rate of sea-level rise will inevitably lead to significant areal losses through erosion in addition to the *in-situ* habitat transitions which might be expected due to increased frequency and duration of inundation (Boorman, Goss-Custard and McGrorty 1989; Titus *et al.* 1991; Nicholls, Hoozemans and Marchand 1999).

In developing strategies for the conservation of saltmarsh resources, it is important to embed broader goals—such as the retention and restoration of key ecosystem functions, habitat extent and type and biodiversity (National Research Council 1987)—within the wider context of coastal and estuarine management. As Clifford (2001) observes in relation to river restoration, conservation and management are becoming increasingly difficult to disentangle. This is equally true of coastal and estuarine wetland environments. Together with coastal dunes (Arens, Jungerius and van der Meulen 2001), saltmarshes are increasingly appreciated as natural dissipators of wave and tidal energy and, as such, form an integral part of strategic thinking in relation to sustainable approaches to flood defence management and coastal protection (Titus 1991; Agriculture Select Committee 1998). This is significant for two reasons. First, it assigns additional economic significance to previously understated natural environmental functions. Second, it provides an impetus for the engineered creation of new areas of saltmarsh in order to replace functions lost through erosion or development. Taken together with the broadening scope of environmental legislation, these developments provide tangible opportunities to incorporate conservation priorities into coastal and estuarine management thinking.

Utilisation of these opportunities requires a multidisciplinary perspective in order that biological conservation requirements can be effectively meshed with broader management agendas driven by economic and political priorities. The goals of saltmarsh conservation include the prevention of further habitat loss; the maintenance of function within systems threatened by degradation; and the creation or restoration of new habitat in compensation for past or future loss (Zedler 1996). The extent to which these goals are achievable depends, in turn, upon the appreciation of the physical processes responsible for the realisation of ecological functions and values. Within this context, this chapter aims to:

- Outline the key physical processes responsible for the emergence of ecological function and habitat character in 'natural' saltmarshes
- Identify those aspects of 'natural' morphodynamic behaviour which must be engineered in managed or restored saltmarshes
- Evaluate recent experiences with saltmarsh restoration

It is argued that saltmarsh conservation management must be founded upon a sound appreciation of natural morphodynamic processes coupled with a realistic assessment of the extent to which these can be engineered. Further, the conceptual framework within which conservation goals, management strategies and ecological engineering techniques are developed must necessarily be context-specific — a point which is sometimes missed when conservationists are forced to respond to oversimplified 'issue-led' representations of the environment, such as sea-level rise and coastal erosion.

FIRST-ORDER CONTROLS ON SALTMARSH MORPHODYNAMIC BEHAVIOUR

GEOLOGICAL, CONFIGURATIONAL AND BIOGEOGRAPHIC FACTORS

Saltmarshes are not all alike. Indeed, the interaction of physical and biological factors give rise to a remarkable diversity in form and function within what are often considered to be rather uniform, featureless, environments. Globally, major differences result from the interaction between the character and diversity of flora with climatic, edaphic and hydrographic influences (Frey and Basan 1985). These are mediated at a regional scale by the nature and abundance of fine sediments, and by the range of depositional settings afforded by geological (including tectonic) context. Geographical variation in vegetation zonation has been used to infer differences in succession, and provides an obvious basis for the identification of distinctive marsh types (see, for example, Beeftink 1966; Chapman 1974). More recent work has employed numerical techniques to elucidate subtle differences in plant habitat (for example, Cantero *et al.* 1998; Visser *et al.* 1998), and to relate these to patterns of invertebrate and bird usage (for example, Norris *et al.* 1997; Levin, Talley and Hewitt 1998) and to both natural and anthropogenic influences (Adam 1978, 1990). This kind of research has traditionally formed the basis for saltmarsh resource inventories and evaluations of conservation value.

From a conservation *management* perspective, it is also important to emphasise the physical determinants of habitat distribution and type, and the manner in which these govern the morphodynamic behaviour of both 'natural' and restored systems (Goodwin and Williams 1992; Haltiner *et al.* 1997; Zeff 1999). Interestingly, recent work on the morphodynamics of tidally dominated marshes (see, for example, Allen and Pye 1992) has tended to assign a secondary, largely opportunistic, role to the colonisation of intertidal surfaces by halophytic vegetation. Vegetation–substrate–fauna interactions obviously contribute to habitat characteristics within these systems, but specific ecological outcomes are contingent upon the provision of viable substrates for initial vegetation colonisation. Regional variations in four main sets of physical factors — fine sediment regime; tidal conditions; coastal configuration; and relative sea-level history — define a *physical context* within which the objectives and practicalities of conservation management must be established.

Fine Sediment Regime

Stratigraphic and geomorphological studies have reinforced the conception of saltmarshes as efficient sediment-trapping systems, which have accreted vertically and laterally in response to post-glacial sea-level rise (Redfield 1972; McCaffery and Thompson 1980; Pethick 1981; Shaw and Ceman 1999). However, the nature of this sedimentary function varies markedly between *allochthonous* systems, characterised by the deposition of externally derived inorganic sediments, and *autochthonous* systems, dominated by the accumulation of internally produced organic material (Reed 1995). The relative importance of inorganic and organic matter accumulation determines the nature of saltmarsh morphodynamic development as well as the ability of both physical and ecological components of the system to adjust to changes in environmental boundary conditions (notably tidal range and mean sea-level).

Tidal Regime

The conventional depiction of saltmarshes as low-energy depositional environments is somewhat misleading, since some of the best-developed examples occur at high tidal energies (for example, Gordon, Cranford and Despanque 1985). Tidal range defines a zone within which vertical and lateral sedimentary infilling can potentially occur (Allen 1990). A number of studies have also indicated a correlation between tidal range and the potential for sediment transport, as indicated by sedimentation rate (Harrison and Bloom 1977; Stevenson, Ward and Kearney 1986) or tidal current velocities (Friedrichs 1995). These simple relationships are complicated by interactions with other factors. For example, the relative importance of non-tidal inundation and sedimentation events (including storm surges) is greater in micro-tidal settings (Stumpf 1983; French and Spencer 1993; Cahoon, Reed and Day 1995), although regional trends in storm surge variance may still contribute significantly to differences in sedimentation within meso- and macro-tidal marshes (see following section). Furthermore, the tidal control on sediment transport which is apparent in predominantly allochthonous systems is much less evident in marshes dominated by *in-situ* accumulation of organic matter: this finding applies to both micro-tidal and meso-tidal settings (see, respectively, Callaway, DeLaune and Patrick 1997; Kelley *et al.* 1988)

Local tidal characteristics also give rise to a series of 'reference levels' which have, over time, acquired a multitude of physical, ecological and legal connotations. For example, the levels associated with Mean High Water (MHW) or Mean High Water Spring Tides (MHWST) have been variously associated with approximate equilibrium marsh elevations (Kestner 1975; Krone 1987; Pethick 1981), and transitions between key habitat types (Collins, Collins and Leopold 1987; Coats *et al.* 1995), as well as being used to define property ownership. Ecologically, any systematic relationship between community structure and tidal levels is potentially very useful as a basis for understanding and predicting saltmarsh response to a change in inundation (Nydick *et al.* 1995) or in the creation of artificial habitats (Coats *et al.* 1995). As van der Molen (1996) has argued, however, the distortion of deepwater tides as they propagate into shallow wetland areas can introduce significant errors into reconstructions of past sea-level change based on floral and faunal relationships with average tidal levels. By implication, such errors may be similarly important when tidal reference levels are used in the design of artificial saltmarshes located in extensive shallow water embayments or estuaries.

Even where good tidal data exist, supposed associations with marsh ecology frequently break down under detailed scrutiny. McKee and Patrick (1988) showed the growth limits of *Spartina alterniflora* to usually lie between Mean Low Water (MLW) and MHW within a very broad sample of US Atlantic and Gulf Coast saltmarshes. However, systematic differences in actual growth limits were apparent, and these appeared to be correlated with mean tidal range, with the likelihood of additional variation due to, *inter alia*, salinity, drainage, soil chemistry and interspecific competition (for a UK perspective, see Gray 1992). Significant within-marsh variability occurs in relation to tidal creeks, with *S. alterniflora* tending to be taller and more productive along channel margins. This observation has been variously related to the supply of nutrients (especially nitrogen; Valiela and Teal 1974) and to spatial contrasts in interstitial salinity (Nestler 1977). At local to regional scales, other researchers have associated productivity with tidal range (for example, Steever, Warren and Niering 1976). This is consistent with Odum's (1974) concept of tidal 'energy subsidy', whereby regular inundation facilitates mineral cycling, nutrient transport and waste removal. The interaction of all these factors generates spatial variation in productivity and community structure which confounds simple interpretation of habitat type in terms of average reference elevations.

Coastal Configuration

At a continental scale, structural controls on the configuration and extent of coastal margins exert a first-order control on both sediment supply and hydrography. The tendency of the world's major river systems to drain onto passive continental margins means that coasts on active margins are often sediment-poor (Milliman and Meade 1983). With reference to large-scale topography, Frey and Basan (1985) draw attention to physiographic contrasts in the space available for saltmarsh development between the Pacific and Atlantic coasts of North America. On the tectonically active Pacific coast, saltmarsh is highly fragmented and is restricted to relatively narrow fringes around protected embayments, estuaries and (in the north), intricate fijord networks. In comparison, Atlantic coastal plain marshes are much more extensive and are essentially continuous over large areas. At a regional scale, variation in the width of the continental shelf exerts a control on tidal range (Hayes 1979) and, therefore, on the potential for saltmarsh development within the intertidal zone. Wide shallow coastal shelves favour a flood-dominated hydrodynamics regime, leading to a net landward transport of sediments, and the retention of sediments with estuaries and embayments (Allen and Pye 1992).

More locally, wind–wave climate exerts an important influence on marsh extent, even in otherwise sheltered embayments and estuaries, where subtle geographical variations in fetch may give rise to significant differences in the character and energetics of the intertidal zone (Knutson and Inskeep 1982). Wave-induced stresses determine the viability of halophytic vegetation establishment (Eisma and Dijkema 1997), although the influence of waves on the stability of the underlying substrate seems to be of more importance than the mechanical strength of the plants (Redfield 1972; Pethick and Reed 1987; Allen 1989, 1997). Wave climate also determines the morphology of the saltmarsh–tidal flat transition (Allen 1989; Pethick 1992).

Sea-level History

Stratigraphic studies have shown the formation of saltmarshes to be intimately related to movements in mean sea-level and for this to be recorded in long-term plant community structure (Orson, Warren and Niering 1998). Some individual saltmarshes are known to have maintained their integrity through vertical and lateral accretion during long periods of sustained sea-level rise (Redfield 1972; McCaffery and Thompson 1980). Oscillations in sea-level tendency, both positive and negative, appear to have triggered expansion of saltmarsh within settings naturally conducive to fine sediment accumulation (Pethick 1980; Orson, Warren and Niering 1987; Shaw and Ceman 1999).

Sea-level rise thus provides a moving boundary condition under which the sediment-accumulating potential (both organic and inorganic) of the saltmarsh environment can be most fully realised. This 'depositional paradigm' (Stevenson, Ward and Kearney 1988) has been challenged by the discovery of apparent sedimentary deficits within subsiding deltaic marshes, some of which do not appear to be able to accumulate sufficient material to keep pace with higher rates of relative sea-level rise (Baumann, Day and Miller 1984; Day and Templet 1989). The likelihood of accelerated and more widespread sea-level rise has given rise to a 'submergence paradigm', under which it is envisaged that more frequent inundation will force a reversion of the vegetation succession (Kana *et al.* 1988; Boorman, Goss-Custard and McGrorty 1989; Huiskes 1990; Warren and Niering 1993), possibly culminating in physical degradation and reversion to intertidal flat or shallow water pond habitat. However, the concept of 'submergence' constitutes a gross oversimplification of the complex (and still poorly documented) linkages between sea-level, sedimentation accumulation, topography and biotic response.

PHYSICAL CONTEXTS FOR CONSERVATION MANAGEMENT

Considered together, the physical factors outlined above provide one means of making sense of the apparent complexity and diversity of saltmarsh characteristics. Significantly, the identification of distinctive marsh types based upon extrinsic forcing factors aids in the elucidation of basic functional differences among different systems (Stevenson, Ward and Kearney 1986), and permits the formulation of realistic conservation goals and management strategies for systems impacted by environmental change and anthropogenic activity. Mitigation of physical deterioration in habitat quality demands consideration of regionally-specific processes and the extent to which intervention to restore desired levels of complexity and function is technically feasible. For example, where deterioration under higher rates of sea-level rise is of primary concern, understanding of historical sea-level change and the balance between autochthonous and allochthonous sedimentation are extremely important: marshes already subject to high rates of subsidence may already lie close to the limits imposed by *in-situ* plant productivity and organic matter accumulation.

An appreciation of physical context also aids the identification of appropriate natural 'reference systems' and their intepretation in terms which can inform the ecological engineering of restored systems of comparable complexity and function. Key issues here concern the morphodynamic process through which these reference habitats have evolved (French 1996), and their inherent variability at scales commensurate with the implementation and monitoring of restoration schemes (Zedler 1999). The integration of broader conservation principles with a regional knowledge base is critical to the success of restoration schemes (see, for example, Race 1985).

The implication here is that specification of the overriding physical controls on marsh characteristics is as important for the management of their conservation function as the discernment of geographical variations in vegetation type and community structure. The most useful classificatory frameworks are thus those which distinguish between predominantly allochthonous and predominantly autochthonous sedimentary environments and between systems subject to strong tidal forcing and those in which inundation is more episodic and associated with low-frequency storm surges or river flood events (Frey and Basan 1985; Stevenson, Ward and Kearney 1986; Dijkema 1987).

GEOMORPHOLOGICAL FRAMEWORKS

SEDIMENT BALANCE MODELS

Vertical Infilling of the Tidal Frame

Vegetated surfaces constitute the most extensive morphological element of a saltmarsh and elevation, through its influence on sedimentation, inundation and drainage is a key determinant of habitat structure. Important conceptual insights into elevational adjustment under the joint influences of sedimentation and inundation have been obtained from relatively simple one-dimensional mass-balance models. The important linkages involved are depicted schematically in Figure 8.1(a). Following Reed (1990), the term 'hydroperiod' refers to the frequency and duration of inundation due to both regular tidal action and aperiodic meteorologically induced surges. Hydroperiod is a function of surface elevation, but itself influences the processes of autochthonous and (especially) allochthonous sedimentation which drive elevation change. Through its effect on hydroperiod, any change in elevation may force responsive transitions in the flora and fauna (see, for example, Niering and Warren 1980; Zedler, Josselyn and Onuf 1982). Numerical modelling studies (French 1993, 1994; Allen 1990) show that these feedbacks between form (elevation) and process (sedimentation) have a profound influence on the rate and manner of marsh growth within the tidal frame. By implication, these adjustments have major implications for the development and maintenance of habitat structure, since slight changes in elevation can result in large changes in frequency and duration of inundation (Barnby, Collins and Resh 1985; Reed and Cahoon 1992). Biotic responses to this kind of hydroperiod forcing result from short-term (interannual) variability in mean sea-level and storminess (see, for example, Morris, Kjerfve and Dean 1990) , as well as secular trends in sea-level.

Allochthonous Systems

In marshes dominated by inorganic sedimentation, an increase in hydroperiod (due to sea-level rise and/or an increase in tidal range) tends to favour an increase in the rate of sedimentation. Given an adequate sediment supply, additional sedimentation can potentially drive elevational change at a rate to match the rise in sea-level. This has several important implications. First, under stable sea-level, upward marsh growth occurs at an asymptotically decreasing

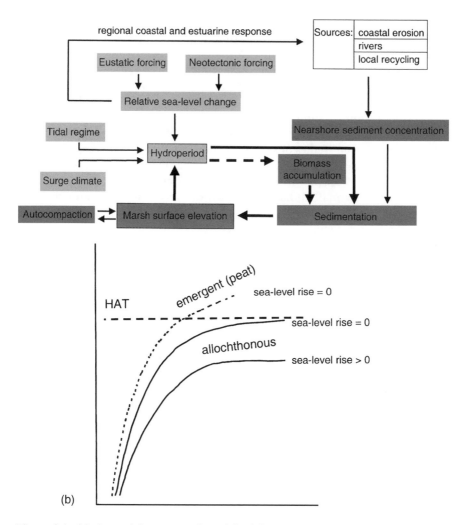

Figure 8.1. (a) A spatial conceptual model of form–process linkages controlling the adjustment of saltmarsh surfaces due to the accumulation of inorganic and organic sediments within a vertical tidal frame; (b) modes of vertical marsh adjustment within the tidal frame (modified from Allen 1990, with permission from Cambridge University Press). HAT = Highest Astronomical Tide.

rate such that, theoretically, the transition to upland habitat is never accomplished (Allen 1990; Figure 8.1(b)). Second, under rising sea-level, marsh elevation will evolve towards a state of equilibrium with the rate of sea-level rise (French 1993). The elevation at which this is achieved will be some distance below that of the highest tides: the magnitude of this difference will be greater under more rapid rates of sea-level forcing. Third, the rate at which equilibrium is approached is strongly dependent upon sediment supply. In moderately turbid settings, the 'equilibrium time' is on the order of, at most, a few centuries (French 1993; Allen 1990).

There is empirical support for these conclusions. The few examples of successional transition to dryland habitat can probably be attributed to significant accumulation of organic matter (no marsh is wholly allochthonous), to supra-tidal sedimentation during infrequent extreme events, or to a local fall in sea-level. More generally, the asymptotic form of elevation growth curves is well established from studies of spatial chronosequences of marshes. One of the best examples is Pethick's (1981) analysis of saltmarshes developed behind prograding barrier beaches in Norfolk, UK—later modelled numerically by French (1993). Allen (1990) has also shown this mode of marsh growth to be consistent with stratigraphic sequences within the highly inorganic marshes fringing the Severn Estuary, south-western UK. Analyses of marsh elevations relative to local tidal datums indicate some consistency in the degree of vertical infilling, suggesting that most established marshes lie close to their theoretical equilibrium elevation, rather than far from it. However, there is also evidence for geographical variations in the location of established marsh elevations within the tidal frame (Figure 8.2). These may reflect spatial gradients in sediment supply, storm surge variance and/or relative sea-level rise.

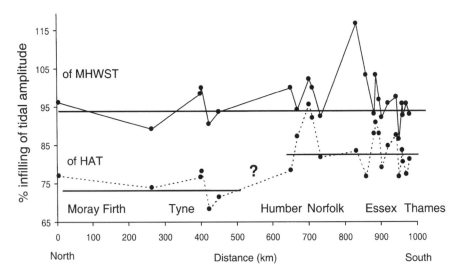

Figure 8.2. Geographical variation in saltmarsh infilling of the tidal frame along the UK east coast. Spatial variation in the degree of infilling may be due to either differences in sediment supply or non-tidal surge variance (both of which increase southwards), although it is possible that relative sea-level tendency is also a factor. Based on data in French (1994) and Pye and French (1993) and various unpublished surveys held by the Coastal and Estuarine Research Unit at University College London.

Autochthonous Systems

As shown in Figure 8.1(a), processes of organic sedimentation also operate within a feedback loop. Elevational changes arising from sedimentation influence hydroperiod which, in turn, affects the health and vigour of the marsh halophytes. In general terms, a decrease in hydroperiod (as would be expected during vertical accretion under stable sea-level) will exert a positive influence on plant productivity (Reed 1995; Nyman *et al.* 1995), organic matter accumulation, and elevation. The increase in elevation and changes in soil hydrology (such as improved drainage) will lead to the appearance of higher marsh plant species. In time, there is the possibility of emergence beyond the tidal frame (Allen 1990; Figure 8.1(b)), although this may be limited by the effects of autocompaction within highly organic marsh peats (see, for example, Kaye and Barghoorn 1964; Cahoon, Reed and Day 1995).

Conversely, any increase in hydroperiod may increase plant stress, with a corresponding decrease in productivity and the potential for organic matter accumulation. Experimental studies indicate significant (though variable) reductions in above-ground biomass and stem density within freshwater to brackish species subjected to increases in salinity and increased hydroperiod (Pezeshki, DeLaune and Patrick 1987; McKee and Mendelssohn 1989). The viability and productivity of plants subjected to reducing soil conditions depends not only on the capacity to maintain root metabolism and nutrient uptake, but also on the existence and effectiveness of mechanisms for detoxifying and adapting to soil phytotoxins (Koch and Mendelssohn 1989). Again adapatibility varies between species. In the genus *Spartina*, Gleason and Zieman (1981) found *S. patens* to be less effective than *S. alterniflora* at supplying oxygen to belowground tissues during simulated inundations.

Plant production thus partly defines a threshold submergence that coastal and estuarine marshes can accommodate. Changes in plant growth (and overall marsh health) due to sea-level rise or altered hydrological conditions are likely to be strongly mediated through soil chemistry changes associated with hydroperiod and substrate drainage (Reed 1995). In a South Carolina saltmarsh, Morris, Kjerfve and Dean (1990) correlated variations in the above-ground productivity of *S. alterniflora* with interannual sea-level anomalies. At a decadal timescale, positive sea-level anomalies were found to be associated with *enhanced* productivity, an effect attributed to reduction in soil salinity as a consequence of increased frequency of flooding and efficacy of tidal exchange. Longer-term responses to a rise in sea-level may be more negative. Nyman *et al.* (1995) demonstrate higher rates of organic matter burial in less saline marshes: this is attributed to both higher productivity and lower decomposition rates. Sea-level rise may thus lead to spatial gradients in marsh deterioration, as presently stable marshes become adversely impacted by saline intrusion and reduced rates of organic matter production and burial. The rate of change is clearly important, however, in determining biotic changes at a community level. In a coastal Louisiana marsh, Taylor, Day and Nuesaenger (1989) found that gradual changes in salinity over several decades led to the invasion of more salt-tolerant species, with the healthiest plant cover developing in areas of greatest change in salinity.

Accretionary Status

In areas of rapid coastal submergence, attention has been focused on the ability of saltmarshes to keep pace vertically with high rates of sea-level rise. Empirical comparisons of vertical accretion and sea-level rise (Baumann, Day and Miller 1984; Walker *et al.* 1987; Bricker-Urso *et al.* 1989) have given rise to the concept of 'accretionary balance'. Although some of the earliest studies were decidedly pessimistic (for example, National Research Council 1987), other work has indicated a wide range of marsh responses. Stevenson, Ward and Kearney (1986) computed the accretionary balance for a large sample of US Atlantic and Gulf Coast marshes (Figure 8.3). They found a correlation with tidal range, implying that tidal energy overrides local variations in sediment supply as the key physical control on the maintenance of elevation. Presently submerging marshes occur at low tidal ranges and depend for their integrity on high rates of organic matter burial. Although these sites (all in of which are in Louisiana) exhibit some of the highest rates of sediment accretion (often in excess of $1 \, cm \, yr^{-1}$; Penland and Ramsey 1990), this is insufficient to offset the effects of subsidence.

When comparable results from the more tidally dominated, highly allochthonous, European marshes are included, it becomes clear that the data are better described by an 'envelope' than by a straight line fit (French 1994; Figure 8.3). The upper boundary of this envelope presumably defines a process limit for sedimentation. This is likely to be determined by both the opportunity for sedimentation within the upper part of the intertidal zone and by the availability of sediment to drive it (Reed 1989). Tidally dominated systems, which tend also to be allochthonous in character, are bounded by a lower limit which represents a state of equilibrium with relative sea-level forcing (i.e. where the elevational increase due to continued sedimentation is just offset by sea-level rise; French 1994). Within this region of the data envelope, temporal elevation trajectories (cf. Figure 8.1(b)) are evident within data from neighbouring sites of different age. Prograding spits and barrier islands are one of the few settings where this can be readily observed. Thus the Norfolk data in Figure 8.3 refer to discrete backbarrier marshes and reveal a sequence of decreasing accretionary surplus with increasing age (French 1993). Similar variability in the rate of accretion and its relationship with local tidal range, was observed by Wood, Kelley and Belknap (1989) within backbarrier, estuarine, and bluff-toe settings along coast of Maine, USA.

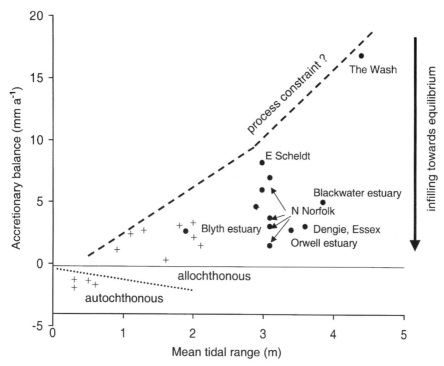

Figure 8.3. Plot of vertical 'accretionary balance' (defined as accretion − local relative sea-level rise) against mean tidal range for North American Atlantic and Gulf coast ('+'; data from Stevenson, Ward and Kearney 1986; other references) and European marshes ('•'; data from studies reviewed in French 1994). For interpretation, see text.

These studies emphasise the long-term sediment sink function of salt-marshes, and the ability of the associated processes to maintain marsh integrity as measured by a single attribute (overall elevation) in the face of a single forcing factor (sea-level rise and its influence on hydroperiod). In reality, saltmarshes are subject to a much wider range of processes and the nature of environmental forcing is more complex. The sediment budget will be determined not just by the accretionary processes noted above, but also by the hydrodynamic regime that develops as tidal action is modified by the evolving morphology, as well as faunal effects (i.e. biodeposition and bioturbation) which may either enhance or reduce sediment stability (Frey and Basan 1985; Stevenson, Ward and Kearney 1988; Hughes 1999). These hydrological pro-cesses, and their influence on the sediment budget, can be expected to vary as marsh morphology and the relative proportion and distribution of sub-habitats (including channel bed and banks; levee marsh; interior marsh; and salt pan) evolves.

External forcing due to sea-level rise also encompasses erosional adjustments (Orson, Warren and Niering 1987; French and Spencer 2001) which, in modern coastal settings operate within the spatial constraints imposed by historical land claim, and under a process environment changed through the effects of altered hydrodynamics (dredging, bridge and jetty construction, river training) and sediment supply (catchment land-use change, damming, dredging). Such adjustments are reflected in the erosional cliffs cut into the seaward margins of many saltmarshes (van Eerdt 1985; Allen 1989), and in the degraded, 'mud mound' topography (Greensmith and Tucker 1966) which characterises marshes exposed to high storm wave climates (Pethick 1992). High rates of historical retreat have been recorded for some marsh systems (Phillips 1986; Harmsworth and Long 1986; Pethick and Reed 1987; Pringle 1995). In some cases, the sediment released through erosion may drive high rates of vertical accretion on the remaining landward marsh (Reed 1988). Even if sediment volumes are conserved, however, topography may militate against the preservation of marsh area. The conceptual model presented by Phillips (1986; Figure 8.4(a)) envisages that the vertical accretion rate, a, needed to maintain the area of a marsh eroding at a rate, b, is related to upland slope, S, as follows:

$$a = bS \qquad\qquad (8.1)$$

Even where upland slopes are gentle, as in Pamlico Sound, North Carolina (Phillips 1986), the elevation change necessary to ensure that landward migration occurs without loss of marsh area generally exceeds measured vertical accretion rates (e.g. a sedimentation rate of about $32\,\mathrm{mm\,yr^{-1}}$ is needed to offset lateral erosion at $0.9\,\mathrm{m\,yr^{-1}}$ along a gradient of $2°$). Thus, although accretion may track sea-level rise, this will probably be at the expense of habitat transitions effected by changes in their elevational distribution. Kana et al. (1988) have quantified these ecological consequences for the coastal wetlands around Charleston, South Carolina (Figure 8.4(b)), and Boorman, Goss-Custard and McGrorty (1989) present similar calculations for seawall-backed saltmarshes in the UK. In essence, these studies show that while the various saltmarsh habitats (especially floristically diverse high marsh sub-habitats) are squeezed, the proportion of tidal flat habitat is expected to increase.

Implications for Conservation Management

The nature of the sedimentary balance has a number of implications for saltmarsh management. First, it is important to place the problem of saltmarsh maintenance, and the threat posed by external change, into an appropriate context. Some systems are inherently vulnerable to change, since they are

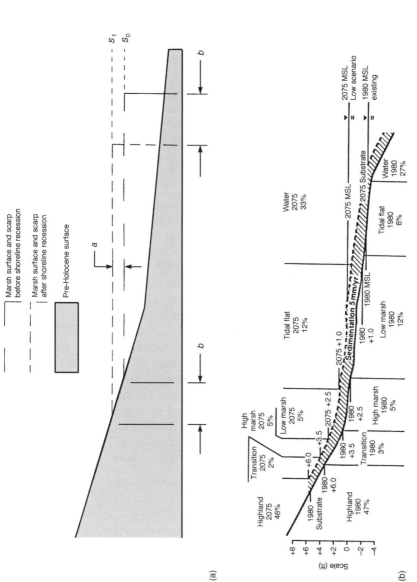

Figure 8.4. (a) Conceptual model showing accretion required to maintain marsh area against given rate of lateral erosion within constraint imposed by upland topography (after Phillips 1986; with permission of Coastal Education & Research Foundation); (b) consequences of vertical and lateral adjustment to sea-level rise on the distribution of coastal wetland habitat around Charleston, South Carolina (reproduced from Kana *et al.* 1988).

already operating close to thresholds of submergence which define important transitions in habitat. Other systems are potentially more resilient but still depend for their future viability on the availability of replacement locations (whether engineered or otherwise) to which existing ecological functions can be transferred. The key physical variables here are sediment regime and tidal range, although it is recognised that gradation between allochthonous and autochthonous marshes is also mediated by biogeographical factors.

Irrespective of marsh type, elevation and hydroperiod are key variables determining habitat character — owing to its strong association with intertidal zonation of both flora and fauna, and its sensitivity to tidal, hydrological and sedimentary forcing. These relationships are especially relevant to marshes which are spatially coherent over large areas. Many saltmarshes are highly fragmented, however (Hollis, Thomas and Heard 1989). Thus it is misleading to rely on elevation adjustment as the primary indicator of marsh sustainability, where marginal erosion is leading to net reduction in habitat area. Sediment budgets provide an alternative basis for the assessment future habitat viability (in terms of areal extent and elevational distribution), although the data requirements are high given the need to integrate spatially complex patterns of erosion and sedimentation (French 1994; see below). Where topographic constraints lead inevitably to habitat loss (*sensu* Phillips 1986), some form of erosion control may be desirable (Pethick and Reed 1987), although this must be planned with a view to its effect on the wider adjustment of the whole intertidal zone (Pethick 1994).

Insights obtained at this level offer few pointers to the design and creation of artificial marsh habitat, aside from the obvious need to achieve the minimum elevations required by particular plant species, and the importance of an adequate sediment supply to maintain elevation in the face of sea-level rise. Rather, ecological engineering must be informed by a knowledge of within-marsh processes, especially those which concern the interaction between major morphological elements (surfaces, channels) and physical processes (hydro-dynamics, sedimentation), and which ultimately govern the longer-term sustainability of a created habitat.

SPATIALLY DISTRIBUTED FORM–PROCESS FEEDBACKS

Spatial Variation in Sedimentation

Saltmarshes exhibit distinctive hydrodynamic regimes, which lead to particular morphological outcomes and which are in turn modified by them. Associated with these are marked spatial contrasts in sedimentation (Stoddart *et al.* 1989; French *et al.* 1995; Leonard, Hine and Luther 1995), and a temporal evolution in marsh morphology and habitat strucuture which is not related to average elevation alone. Of particular importance is the hydraulic interaction between channels and adjacent vegetated surfaces (French and Stoddart 1992). Understanding this interaction is crucial if artificial marshes are to recreate the physical and ecological characteristics of the corresponding natural systems (French 1996; Zeff 1999; Reed *et al.* 1999).

Studies of backbarrier settings which have experienced minimal human impact have provided valuable insights into the sedimentary dynamics of 'natural' saltmarshes. The aerial view of Scolt Head Island, Norfolk, UK (Figure 8.5(a)) thus provides a useful context within which the factors influencing contemporary sedimentation can be identified. These are:

- Within-marsh location (in relation to inherited larger-scale topography)
- Elevation (as a determinant of hydroperiod)
- Proximity to channels (associated with lateral hydraulic gradients)

At the largest scale, the topography of the pre-existing intertidal flats defines the potential for sedimentation. At Scolt Head Island, sandy and muddy flats are formed in the lee of prograding gravel and dune barriers. Tidal flat gradients are significantly higher than those of the established saltmarshes: sedimentary infilling within the tidal frame is thus a levelling process (Figure 8.5(b)), an observation which accounts for the near-horizontal surfaces of the oldest saltmarshes. This reflects the elevational control on the equilibrium relationship between hydroperiod and sedimentation (Figure 8.5(c)). The narrower range of elevations on the modern marsh surface still experience very different hydroperiods, which depend strongly on local tidal and surge characteristics. The essential form-process feedback depicted in Figure 8.1(a) is, therefore, spatially distributed according to both inherited and acquired topography. A third control results from the interaction of channel and surface hydrodynamics and is responsible for significant local variation in sedimentation. For the marsh depicted in Figure 8.5(a), French and Spencer (1993) recorded as much variability in vertical accretion within 25 m of a major channel as over the whole of the saltmarsh (Figure 8.5(d)). Essentially similar findings have been reported for a wide range of tidally-influenced marshes elsewhere (Stumpf 1983; Leonard, Hine and Luther 1995).

These same general controls on physical sedimentary processes can be observed in anthropogenically impacted marshes. In the Dutch Wadden Sea, for example, extensive saltmarshes were formed through artificial accretion works (Kamps 1962) with the aim of reclamation for agriculture. This aim is no longer an economic priority and large areas of essentially artificial marsh have been either abandoned to nature or subject to periodic clearance of the drainage systems with the aim of maintaining the marshes for grazing. Many of these marshes have subsequently been incorporated into nature reserves

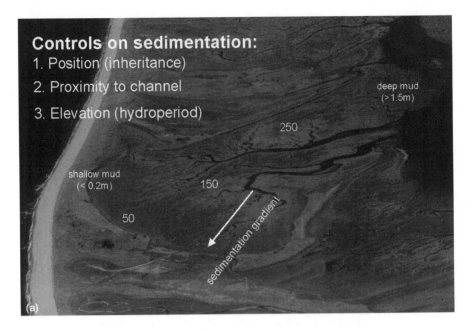

Figure 8.5. (a) Physical controls on sedimentation, illustrated within a backbarrier saltmarsh, Scolt Head Island Norfolk (aerial photograph reproduced with permission of Cambridge University Collection of Air Photographs, copyright reserved); (b) sedimentation 1983–96 as a function of underlying tidal flat elevation; (c) sedimentation as a function of marsh surface elevation; (d) sedimentation as a function of proximity to channel system.

(Ovensen 1990), and there is some uncertainty over the most appropriate management for the maintenance of conservation interest.

Patterns of sedimentation in the abandoned marshes of the Ems Dollard estuary have been studied by Esselink *et al.* (1998). Elevation changes surveyed between 1984 and 1992 were found to correlate with present elevation, proximity to both major and minor drainage channels within the abandoned drainage system (Figure 8.6(a)) and distance from the seawall (presumably related to the original tidal flat profile). Elevation change was also found to be influenced by variations in vegetation species composition, density and height associated with different grazing practices. Levee development along the margins of the drainage system leads to the creation of badly drained depressions. The vegetation in these (*Puccinellia maritima*, with an increasing annual growth over the study period of *Salicornia europea* and *Suaeda maritima*) suffers damage from cattle trampling, which may further impede sedimentation (Figure 8.6(b)). Ungrazed marsh interior areas are dominated by *Spartina anglica*, *Aster tripolium* and *Phragmites australis*. Owing to the

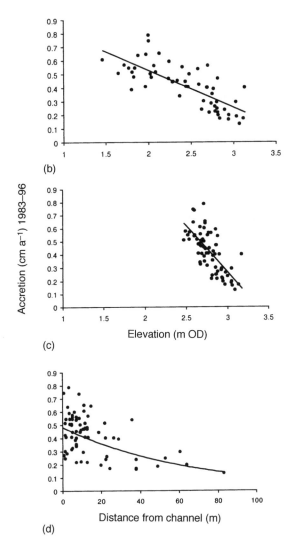

Figure 8.5 (b), (c) and (d)

brackish estuarine conditions, abandonment of grazing here would probably lead to an overall dominance of *Phragmites* (Esselink *et al.* 1998). Continuation of grazing is thus desirable from the point of view of maintaining diversity, but this implies some additional management to avoid the deterioration of the vegetation within the interior depressions. This requires an understanding of the processes associated with the channel system and its role in the tidal exchange of water and sediment.

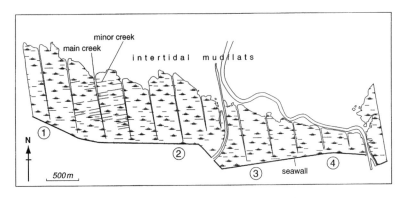

(a)

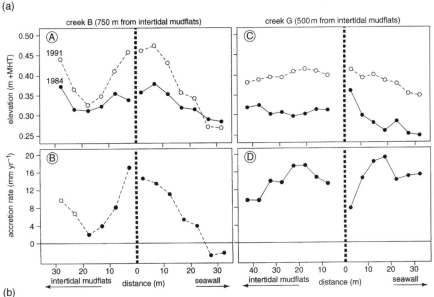

(b)

Figure 8.6. (a) Elevational changes within abandoned artificial saltmarshes in the Ems Dollard estuary, Netherlands, in relation to tidal creeks; (b) interrelationships between physical sedimentary processes and grazing effects on vegetation structure (both adapted from Esselink *et al.* 1998). Reproduced by permission of Coastal Education & Research Foundation.

CHANNEL–SURFACE INTERACTIONS

Well-defined channel systems (also variously termed 'creeks' or 'sloughs') are a near-ubiquitous morphological feature of natural saltmarshes. They are especially well-developed in marshes which occur at the upper end of the tidal energy spectrum, where they often form intricate branching or anastamosing networks (Chapman 1977; Ashley and Zeff 1988; French and Stoddart

1992). Channel systems impart additional diversity to marsh habitats shaped primarily through the first-order geological, geomorphological and biogeographical controls on marsh type. Important sub-habitats include subtidal and intertidal bottoms; cliffed and sloping banks; and channel-side levee. Tidal channels are utilised by juvenile fish species (Roza, McIvor and Odum 1988; Rountree and Able 1992; Williams and Zedler 1999; Desmond, Zedler and Williams 2000). Tidal channel banks, despite their limited area, provide important habitats for burrowing invertebrates (McIvor and Odum 1988), with additional biodiversity arising from variations in sediment type (i.e. sand or mud), slope (related to inundation frequency) and stability (i.e. eroding or depositing). Channel margins are also associated with distinctive plant habitats and patterns of bird usage (Bradley and Morris 1990; Zedler et al. 1999).

From a geomorphological perspective, the near-ubiquity of channels within tide-dominated settings implies that they are an intrinsic feature of saltmarsh morphodynamic development (Frey and Basan 1985; French and Stoddart 1992). There is evidence for inheritance of major channel alignments and the subsequent evolution of higher channel densities as marsh elevations are built up through accretion (Redfield 1972; Shi, Lamb and Collin 1995). Following the establishment of a continuous vegetation cover, and the subsequent build-up of surface elevation by sedimentation, tidal exchange changes from being primarily via sheet flow (tidal flat stage) to being substantially channelised (saltmarsh stage). The longer-term adjustment of channel morphology is only partially understood, although some generalisations are possible. First, it is clear that saltmarsh channels are far less mobile than their tidal flat counterparts (Garofalo 1980; Ashley and Zeff 1988). Although narrow deep sections, with cliffed margins, may give the impression of active lateral migration, actual rates of bank retreat are generally quite low (Gabet 1998). Second, vertical and lateral marsh growth is accompanied by an expansion of the network. This has been documented for sites where sequential aerial photography spans the early stages of marsh formation (see, for example, Shi, Lamb and Collin 1995), with erosional extension of channel headwaters being the primary mechanism of network growth.

Increasing surface elevations clearly force adjustments in channel gradient, which are accommodated through spatial extension of the network. Allen (1997) has also emphasised the role of tidal prism changes which are accommodated by rationalisation of the network and cross-sectional enlargement/infilling of individual sections. In most marshes, the total tidal prism at high spring tides is dominated by the shallow water overlying the marsh surface, rather than the comparatively small volume contributed by the deeper, but much less areally extensive channels. However, since much of this flow will be routed via the major channels, the tidal prism provides a crude surrogate for the discharge that they must convey. For this reason, marsh area versus prism and prism versus creek size relationships have found extensive application as a basis for designing tidal channels in restored marshes (Coats *et al.* 1995; see below). Exactly which tidal prism to use (i.e. mean diurnal or semi-diurnal; spring tide; highest astronomical tide; or extreme surge tide) is not immediately clear, however.

In established marshes, channelisation of a major proportion of the tidal exchange leads to spatial contrasts in flow intensity, such that suspended sediment is more readily transported within the channels but readily deposited once it is transported over-bank and across the marsh surface (Stumpf 1983; French, Clifford and Spencer 1993). This explains why channel-related processes appear to cause much of the spatial variation in sedimentation on the marsh surface (Reed *et al.* 1999). Furthermore, the interaction of shallow water tides with marshes composed of deep channels and shallow surfaces generates highly distinctive flow regimes. In many marshes, well-defined flow peaks are observed during both the flood and ebb, around the time at which inundation or drying of the marsh surface occurs (Bayliss-Smith *et al.* 1979; Dankers *et al.* 1984; Reed 1988). The mechanics of these transients have been investigated by French and Stoddart (1992), who argue that the marsh surface acts as a topographic threshold separating two distinctive flow regimes: a flood-dominated, 'below-marsh' regime, and an ebb-dominated, 'over-marsh' regime (Figure 8.7). The former is associated with neap tides and sediment deposition within the channel system: the latter is associated with spring tides, scouring of the channel system and the introduction of suspended sediments onto the marsh surface. The flood transient occurs just above channel bankfull, due to the sudden increase in tidal prism at a time when overall tidal exchange is still largely routed via the channel system. High turbulent stresses generated at this time (French and Clifford 1992) appear to be important in maintaining the suspension of sediment within the upper part of the water column, such that a proportion of this material can be advected laterally over the adjacent marsh surfaces, where it tends to deposit within a very short distance (French, Clifford and Spencer 1993; Leonard, Hine and Luther 1995; Reed *et al.* 1999). On the ebb, maximum channel velocities and stresses occur just below bankfull, associated with steep water surface slopes generated as water retained on the vegetated marsh provides a delayed inflow to the channels. These ebb-flow velocities are often (though not always) greater than the flood maxima, and are clearly important in scouring the channel system, and in extending it through erosion in headwater regions.

Importantly, even ebb-dominated saltmarshes can be highly efficient sediment sinks, since a large proportion of the sediment introduced on the flood is retained on the surface (Stumpf 1983; Dankers *et al.* 1984; French and Stoddart 1992; Wang, Lu and Sikora 1993). Ebb-dominance within a marsh does not, therefore, carry the same connotation as it does in a wider estuarine or embayment context, where it may be associated with a tendency for sediment export. Local flood-tidal dominance has been reported for some

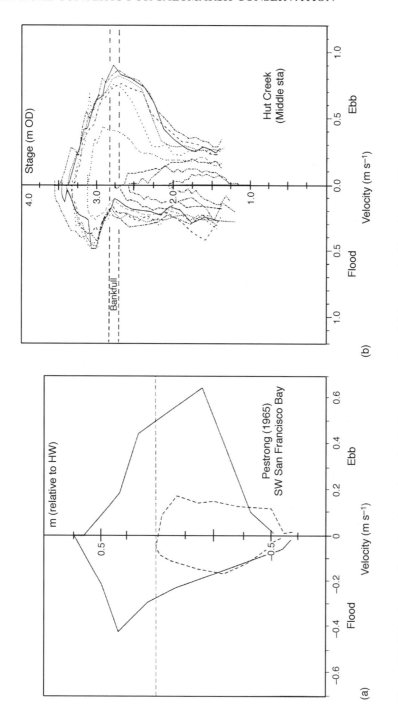

Figure 8.7. Velocity transients in saltmarsh creeks during overmarsh and belowmarsh tides: (a) Norfolk, UK; and (b) southern San Francisco Bay. Both redrawn from French and Stoddart (1992); (b) based on data in Pestrong (1965).

locations (Leonard, Hine and Luther 1995), where it is possibly a result of lower tidal range, or diurnal rather than semi-diurnal tides (making the momentum contrasts between channel and surface less marked), or because strongly flood-dominant characteristics of the broader coastal setting override any internal form-process interaction (see, for example, Friedrichs and Aubrey 1988).

The hydrodynamic and sedimentary consequences of channel–surface interactions in saltmarshes are only *partly* dependent upon vegetation characteristics. Extremely steep sedimentation gradients relative to channel margins have been observed within uniform vegetation (French and Spencer, 1993; see also above), implying that channel–surface hydraulic gradients are more important than local variation in plant-induced roughness. Also, gross hydrodynamic characteristics (i.e. flood- and ebb-velocity and stress transients, and the resulting tidal asymmetry), whilst modified by the effects of vegetation-induced roughness are not necessarily dependent upon them (French and Stoddart 1992; see also Dronkers 1986; Friedrichs 1995). However, in the rather different setting of a north Australian tidal mangrove swamp, Wolanski, Jones and Bunt (1980) used a numerical flow model to demonstrate that the maintenance of deep self-scouring channels was dependent upon the high surface friction generated by the dense vegetation. Here, extreme vegetation roughness is a critical factor determining both the size and geometry of the channel system.

DISSIPATIVE FUNCTIONS PERFORMED BY NATURAL SALTMARSHES

Wave Attenuation

Although the formation of saltmarshes is restricted to the lower end of the wind–wave energy range, established marshes are extremely effective at attenuating incident wave energy. This energy dissipation function contributes, in turn, to the stability of the marsh sediments (Pestrong 1972; Gray 1974) and favours further sedimentation under the processes outlined previously (Dodd and Webb 1975; Frey and Basan 1985; Leonard, Hine and Luther 1995).

In North America, the effectiveness of saltmarsh grasses (chiefly *Spartina alterniflora*, but also *S. foliosa*, *Carex lyngbyei* and *Deschampsia caespitosa*) in attenuating wave energy and encouraging sedimentation has been documented through both field and laboratory studies (Wayne 1976; Knutson *et al.* 1982). Both wave attentuation and the potential for sediment retention have been shown to be related to vegetation density and marsh width (Dean 1978; Gleason *et al.* 1979; Knutson *et al.* 1982). The joint effect of these variables has been modelled by Knutson (1988) by approximating the relatively smooth stems of *S. alterniflora* by cylindrical roughness elements. The ratio of incident wave height, H_i, to marsh wave height, H_m, observed at distance x is given by

$$\frac{H_i}{H_m} = \frac{1}{1 + AH_ix} \tag{8.2}$$

where

$$A = \frac{C_dC_p}{3\Pi s^2 d} \tag{8.3}$$

and where C_d and C_p = surface and vegetation drag coefficients respectively (C_d being around 1 and C_p being about 5 for *S. alterniflora*), s = mean stem spacing and d = stem diameter. An inference from this model is that most of the dissipation occurs within a short distance of the marsh edge: in Knutson's study of Chesapeake Bay marshes, 50% of the energy was lost with 2.5 m of the edge, although incident wave heights were very low (< 0.2 m). These dissipative and stabilising functions mean that the creation of saltmarsh has been quite widely adopted as a non-structural technique for erosion control (Knutson *et al.* 1981; Knutson and Inskeep 1982) and the stabilisation of dredge material dumps (Seneca 1974), with habitat creation being an important secondary objective (Seneca, Woodhouse and Broome 1975; Landin, Webb and Knutson 1989).

In Europe, physical scale-model experiments (Hydraulics Research, 1980) and field observations (Erchinger 1995; see also Figure 8.8(a)) have documented the effectiveness of even a narrow fringe of saltmarsh in reducing wave energy in front of sea defence structures. Reduction in wave runup and in the probability of overtopping (Brampton 1992) translate into lower construction cost for an equivalent standard of protection (Turner and Dagley 1993; Leggett and Dixon 1994). In Essex, UK, some 330 of the 440 km of seawall currently maintained by the Environment Agency rely on fronting saltmarsh as a first line of defence (Dixon, Leggett and Weight 1998). Recent work by Möller *et al.* (1996, 1999) has quantified and modelled this dissipative function for the meso- to macro-tidal marshes which are typical of the UK east coast. Field measurements along an open coast sand-flat to saltmarsh transect in Norfolk indicate that incident wave energy is reduced by an average of 82% over the saltmarsh, compared to only 29% over an equivalent length of tidal flat. Application of a one-dimensional model (Möller *et al.* 1999) confirms that the additional dissipation over the saltmarsh is primarily due to the additional friction associated with the vegetation. These findings provide additional empirical justification for protection and/or creation of saltmarsh as part of a more sustainable approach to flood defence management and also aid the specification of the necessary structural elements of particular schemes (Dixon, Leggett and Weight 1998).

The stabilising effect of vegetation also has some consequences for marsh

Figure 8.8. (a) Estuarine fringing saltmarsh, Roach estuary, Essex, UK. Note installation of concrete facing on seawall where saltmarsh has been lost through erosion; (b) restored tidal wetland at Tollesbury Fleet managed realignment site, Blackwater estuary, Essex, UK. This view taken 15 months after breaching: *Salicornia* spp. and *Suaeda maritima* have since become established immediately in front of the seawall; (c) trial sediment recharge using cohesive maintenance dredgings pumped behind a gravel-retaining bank, Orwell estuary, Suffolk, UK. Depth of recharge mud is about 1 m along the inner edge of the gravel bank. This scheme does not provide a direct replacement for saltmarsh but helps to protect eroded remnants of high saltmarsh and dissipates wave energy in front of a deteriorating seawall.

morphodynamics. Surface erosion is rare (the frequently cited observation by Pethick (1992) of surface lowering and post-storm recovery in Essex marshes applied only to a narrow zone at the seaward edge) and the interior of most marshes can be considered an effective sink for inorganic sediment. However, erosion does occur in response to extreme wave conditions or as part of adjustment to a trend in sea-level and/or tidal regime. A cliffed seaward margin is a common feature in many marshes (Redfield 1972; Jacobsen 1980; van Eerdt 1985; Allen 1989), and often exhibits high rates of erosional retreat (for example, Rosen 1980; Harmsworth and Long 1986; Phillips 1986; de Jong *et al.* 1994). Whether this represents a natural morphodynamic adjustment triggered by the steepening of the intertidal profile through vertical marsh growth (possibly being followed by the development of secondary marsh at a lower level), or a response to some external forcing (be it sea-level rise, or a

Figure 8.8. (b) and (c)

geomorphological or ecological change on the fronting tidal flats) is not always clear (see, for example, Gray 1972; Harmsworth and Long 1986; Pringle 1995).

Tidal Energy Dissipation

Whereas the dissipation of wind–wave energy is of primary importance in open coastal settings tidal energy provides the dominant physical forcing within the restricted fetch environment of estuaries and embayments. At this scale, intertidal morphology also exerts a considerable influence on the tidal hydro-dynamics (Dronkers 1986), including the transition between flood- and ebb-tidal dominance (Friedrichs, Lynch and Aubrey 1992) and the spatial variation in tidal range and local hydroperiod (Pethick 1994; van der Molen 1996).

Within-marsh reduction in tidal energy occurs through the extraction of momentum by vegetation 'roughness elements' (Shi, Pethick and Pye 1995). The diversity of marsh plant geometries and important differences in their mechanical properties (such as rigidity and buoyancy) make quantitative generalisation difficult. A few field studies (for example, Stumpf 1983; Wang, Lu and Sikora 1993 and Leonard, Hine and Luther 1995) have documented the reduction in horizontal tidal current velocity within marsh plant canopies. The associated reduction in bottom stress means that overmarsh tidal flows alone (i.e. in the absence of extreme waves) are only rarely competent to resuspend deposited sediment that has consolidated over more than a few tidal cycles. Similar conclusions have been reached from laboratory experiments (Shi, Pethick and Pye 1995).

Tidal energy dissipation has also been identified as a primary control on marsh morphodynamics. Pethick (1992) has argued that both the cross-sectional and planform morphology of channel systems evolve towards equilibrium between the stresses imposed by tidal flows and the resistance of the sediments as stabilised by vegetation, a response which is also conditioned in the longer term by changes in tidal prism through vertical accretion or lateral erosion (Kelly, Gehrels and Belknap 1995; Allen 1997). More explicitly, Pethick (1992, p. 54) asserts that 'creeks are not a morphological response to drainage of the ebbing tide from the marsh . . .' but instead represent '. . . a morphological device to dissipate tidal energy'. Saltmarsh channel systems certainly appear to exhibit many of the dissipative characteristics of larger estuaries developed in unconstrained coastal plain settings, such as an exponential decrease in width and cross-sectional area with distance landwards from the mouth (Wright, Coleman and Thom 1973). Pethick (1992) also makes two further observations. First, that the sum of all the channel widths decreases exponentially with distance from the mouth; and second, that the total length of the channels system is dependent upon the width of the channel mouth. A consequence of this is that some channel networks (for example, in some of the lower marshes in Norfolk, UK) end abruptly, some distance from the landward margin of the marsh. That is, a substantial part of the upper marsh appears undrained by channels — which appears inconsistent with their formation as a drainage feature under ebb-dominated flows.

As French (1996) has observed, however, there is some confusion in the literature between the various physical *functions* performed by saltmarsh channel systems (most obviously sediment supply, drainage and the dissipation of tidal energy) and the processes by which those channel systems are actually *formed*. It is hard to envisage how 'excess flood tidal energy' (*sensu* Pethick 1992) can be the prime agent of geomorphic work in systems characterised by relatively resistant cohesive sediments. An alternative explanation for truncated channel systems lies in the control exerted by marsh height on overall channel gradient. As marsh height increases, long-profile adjustment can be expected to occur in response to excessive stream power in the upper reaches: this is consistent with the network extension by headward erosion observed at decadal scales in rapidly developing saltmarshes, and with the lack of channels in the landward regions of young marshes. Despite the greater efficacy of ebb-tide processes in channel formation (French and Stoddart 1992), the resulting morphological outcomes may still lie close to the optimum for tidal energy dissipation (French 1996).

Implications for Conservation Management

Morphodynamic behaviour resulting from within-marsh variability in physical processes is clearly important in determining the distinctive characteristics of marshes developed within the broader environmental settings considered previously. If marshes are to be managed, or new systems created, it is important that the local operation of these processes be understood. For authochthonous systems, it seems clear that both hydrological and sedimentary processes are implicated in the formation and maintenance of spatial habitat diversity (i.e. streamside–interior–backmarsh transitions). Where tidal action is limited, management of hydrological conditions is especially important.

In tide-dominated allochthonous marshes, coastal configuration (planform and topography) defines a spatial envelope for marsh formation. Within this envelope, sedimentation is strongly controlled by proximity to the channel system connecting the marsh to the adjacent estuarine or coastal water body, and by the abundance of sediment within adjacent systems. The presence and nature of this channel system is partly inherited but also evolves under a process regime conditioned by feedbacks between the channels and the marsh surface (in the longer term through changes in tidal prism; in the short term, through hydrodynamic contrasts between channel and surface flows).

Although these controls are not yet fully understood, it is clear that attempts to engineer saltmarsh formation must consider not only elevation (as a control on sedimentation potential) but also the provision of a functional channel system (to realise sedimentation potential in a way that is consistent with the development of viable vegetation and invertebrate habitats). The nature of these linkages is time-dependent and the characteristics of established marshes represent the outcome of a long interplay between the various system components. It follows that, in attempting to create new marshes, engineers must achieve something that will evolve naturally along a similar trajectory, rather than one that takes the initial conditions rapidly away from the design goals (Haltiner *et al.* 1995).

The basis for adapting this accumulated scientific understanding to the needs of conservation managers and ecological engineers is not immediately clear. There are still too few truly comparative studies to permit the generation of empirical 'rules of thumb' that have anything more than local applicability. At best, the established knowledge has been used to infer modes of marsh formation in nearby 'reference' systems and to use these (usually qualitatively) to inform attempts at restoration. A more robust approach would be to define more precisely the important physical functions performed by the corresponding natural system, and to generalise from these a set of guidelines for the optimal engineering of desired characteristics. Such an approach requires further basic research.

SALTMARSH RESTORATION

INTRODUCTION

The accumulated technical experience of saltmarsh restoration is now considerable, and encompasses approaches ranging from *in-situ* management to the engineered creation of new habitat. In the United States, there is quite a long history of saltmarsh restoration within the context of broader wetland mitigation policies. Many of these schemes have been undertaken with conservation as a primary goal, often involving 'habitat exchange' whereby developers are permitted to utilise natural wetland sites in return for the creation of artificial habitats elsewhere. The results of these projects have been mixed (Race 1985; Zedler and Callaway 1999). This can be attributed in part to the tendency to view saltmarsh habitat restoration as an essentially ecological problem, whereas from the preceding discussion it is clear that an understanding of physical sedimentary and hydraulic processes is vital if erosion is to be mitigated or if new resources are to be created. Thus, as Williams (1994) has observed, lack of success with wetland restoration has most often resulted from a failure to consider the geomorphic and hydraulic functioning of the corresponding natural system and integrate this into the ecological design of the new one. For example, morphological features, such as channel networks, cannot be considered simply as a 'cosmetic extra' with which to endow landscape aesthetic qualities or additional biodiversity: they are an integral component of the saltmarsh morphodynamic system and are crucial to the generation and maintenance of the associated physical and biological functions.

This observation is also pertinent to saltmarsh restoration as part of sustainable flood-protection strategies. In the United Kingdom, particular emphasis is now being placed on the accommodation of sea-level rise and the erosion of the intertidal zone through managed realignment of existing lines of defence (or 'managed retreat'). Such adaptive strategies (Agriculture Select Committee 1998) rely on the creation of new intertidal habitat by allowing presently reclaimed (primarily agricultural) land to revert to tidal action (Burd 1995). Experience to date is restricted to a limited number of experimental and small-scale operational 'setbacks'. Crucially, these schemes are designed and financed primarily with flood defence in mind (Dixon, Leggett and Weight 1998), although it is recognised that important secondary ecological benefits accrue (including the establishment of marine invertebrate communities and nesting, roosting and feeding sites for wading birds and wildfowl). From both engineering and conservation perspectives, it is clear that many lessons remain to be learned. Small-scale realignments undertaken to date appear qualitatively successful, although their long-term effectiveness remains to be judged. Crucially, experience with unplanned realignments (due to past seawall failures) indicates that the ecological success of large-scale realignments — as measured by the establishment of a patchwork of vegetated marsh, channel and tidal flat habitats — cannot be taken for granted (Brooke 1992; French, French and Watson 1999).

CASE STUDIES IN SALTMARSH HABITAT RESTORATION

US West Coast: San Francisco Bay

Massive changes in the extent and character of the intertidal zone in San Francisco Bay in the last century (Josselyn 1983), coupled with increasing concern regarding the conservation of saltmarsh-dependent fauna, have produced a new emphasis on habitat restoration in the area. Two human disturbances in the region have influenced the saltmarshes of the Bay in different ways. In the north Bay, extensive areas of saltmarsh were diked and drained for agriculture, including grazing and some row crops. In the south Bay, where less freshwater was available for irrigation and to flush salt from the drained soils, salt ponds became very common in the mid-1800s. In addition, in the more brackish Suisin Bay, over 55 000 acres (22 250 ha) of land that was historically tidal marsh are now hydrologically managed with levees and water-control structures to support migratory waterfowl. This loss of natural saltmarsh habitat to other uses came in the same era that hydraulic mining for gold in the Sierra Nevada (1855–84) increased sediment delivery to the Bay and resulted in the progradation of some remaining saltmarsh areas. Jaffe, Smith and Zink (1999) note that this pulse of sediment associated with mining debris has now moved through the subtidal portions of San Francisco Bay and that sedimentation rates are now decreasing.

Within this context, one of the main concerns facing the design of saltmarsh restoration projects is the need to rebuild intertidal elevations within previously drained areas to levels appropriate for vegetation establishment. Some diked and drained former marsh areas have subsided more than 2 m, to well below mean tide level. However, the goal of these saltmarsh conservation efforts is not merely the provision of additional areas covered with halophytic vegetation. Particular interest attaches to the light-footed clapper rail, which forages within tidal creeks for crabs, snails and other foods. The birds, now under severe threat of extinction largely because of habitat loss, are poor in flight and so remain close to their feeding grounds. The birds nest on the ground but elevate their nests and build using hollow stems of the native cordgrass, *Spartina foliosa*, so that the nests float during extreme high tides. The birds also weave the cordgrass over their nests to form a canopy protecting them from predation by raptors. Thus the provision of both tidal creeks for food, and the *S. foliosa* which grows along creek banks in Bay salt marshes, are critical measures of restoration success in these systems.

Critically, such expectations require that the restoration of diked marshes encompasses a clear understanding of tidal creek dynamics and their relationship to the surrounding marsh plain. The restoration of tidal action to these diked and subsided former marsh areas by dike-breaching is only one step towards habitat conservation in the Bay. Natural sedimentation within these reopened areas may ultimately build elevations suitable for vegetation colonisation but both public and regulatory expectations of restoration usually require that changes are recognisable in the landscape within several years, rather than after several decades of restoration decision making. Thus, early efforts for saltmarsh restoration in the bay in the 1970s included the addition of dredged material to increase elevation before dike-breaching. Florsheim and Williams (1994) document how that approach ultimately provided vegetated areas but that natural channels developed in only limited areas apparently because of the limited ability of tidal action to 'carve' a channel system into the dredged material.

As a result, a 'second-generation' approach to saltmarsh conservation in the Bay area has evolved, based on mimicking, but accelerating, the natural physical processes which build tidal marshes. Natural sedimentation rates in the vicinity of the Sonoma Baylands restoration site in northern San Francisco Bay meant that a levee breach would result in saltmarsh habitat after about 35 years (Florsheim and Williams 1994). To reduce this time-frame, a design team adopted a two-faceted approach to increasing elevation while allowing intertidal gradients to facilitate creek development. Dredged material was used to infill the diked area to an elevation optimal for tidal channel development (rather than the optimal elevation for complete vegetation cover). In addition, dredged material was used to form wave barriers within the restoration area to limit the development of small waves and thus enhance the potential for deposition of suspended sediment and the building of a natural marsh topography on the dredged material platform.

Working with natural processes and allowing engineering techniques to enhance or modify the physical dynamics of the intertidal areas has proven fruitful in conservation efforts in the San Francisco Bay area. It has also shown how understanding the interplay between physical development and ecological function is a vital element of successful conservation planning in saltmarshes.

US Gulf Coast: Opportunities for Enhancing Sediment Supply in the Mississippi Delta Plain

One approach to the problem of mitigating saltmarsh loss is to intervene in sediment supply through the diversion of existing transport pathways or by artificially introducing material derived from navigational, or even dedicated dredging. The aim here is either to reduce regional sediment deficits (for example, those resulting from estuarine ebb-tidal dominance or rapid deltaic subsidence) or build up local elevations in order to effect habitat transitions.

In Louisiana, sediment diversions have been undertaken in order to fill shallow open-water bodies adjacent to the Mississippi River. Wetland loss in the Mississippi Delta Plain as a whole is attributable to alterations in flooding patterns and sediment supply at a number of scales. Contributing factors include the placement of flood-control levees along the river itself, isolating the delta plain from its natural source of sediment and freshwater, as well as the smaller-scale but very common construction of spoil banks along the extensive network of canals dredged for oil and gas exploration. Opportunities for addressing the results of these changes vary across the delta plain. In the Delta National Wildlife Refuge, an area within the 'birdsfoot delta' below the flood-control levees, 24 artificial 'crevasses' (cuts in the natural river levee) have been created by the US Fish and Wildlife Service. These have resulted in emergent marsh formation within about two years of breaching, with an average land gain of nearly 5 ha yr^{-1} (Boyer, Harris and Turner 1997). The cost of each crevasse averaged around US$20 000. Thus these schemes prove cost-effective relative to larger and more complex projects higher on the river (Turner and Boyer 1997) where costs must include the relocation of infrastructure such as roads and railways as well as the engineering challenge of making a controlled breach in major flood-control levee. While cost-effective, the artificial crevasses represent a technique which has limited application geographically and that creates new habitat in areas remote from hurricane-protection levees and human infrastructure where the need for conservation is perhaps greatest.

For sites remote from river distributaries with readily breachable levees, managed dredge material disposal offers an alternative means of increasing wetland elevations. Sediment recharges need to be carefully managed if they are to have the desired ecological effects. In particular, the thickness of material placement controls the mechanism of vegetative recolonisation. Wilber (1993, cited in Ford, Cahoon and Lynch 1999) suggests that thin deposits (< 0.15 m) can be recolonised by *in-situ* vegetation, while thicker deposits necessitate invasion by new plants. One means of ensuring recolonisation within existing marshes subject to rapid submergence is thin-layer deposition using a spray-dredging technique (Cahoon and Cowan 1988). Ford, Cahoon and Lynch (1999) have evaluated the performance of a spray disposal project undertaken along a dredged canal near Venice, Louisiana, also in the 'birdsfoot' delta. With application of a thin layer of material averaging 23 mm, stems of *Spartina alterniflora* were initially flattened but soon recovered such that overall cover actually increased threefold after a year. Marsh elevation was significantly increased (a total of 62 mm of vertical accretion was recorded between 1996 and 1998 at the sprayed site). The technique also proved successful in raising the elevation of adjacent shallow pond environments (caused by deterioration of saltmarsh to open water). Here 129 mm of dredge material was applied and this raised bottom elevations to the extent that lateral colonisation by *S. alterniflora* occurred by rhizome growth from adjacent marsh. Mendelssohn and Kuhn (1999) examined the application of dredged material of varying depths to a *S. alterniflora* salt marsh and found positive relationships between plant production and depth of sediment, up to thicknesses of over 30 cm. They also note that this production increase appears to be mediated both by an increase in elevation as well as in the nutrients supplied with the added sediments. It is likely that the stimulus provided by the nutrients will be exhausted after a few growing seasons, but the increase in elevation may provide some years of sustainability to otherwise rapidly subsiding marshes.

Thus within the Mississippi delta plain opportunities exist for saltmarsh conservation concurrent with continuing industrial and commercial development. While industrial development within the coastal zone caused many problems for coastal saltmarshes, increased understanding of saltmarsh value and threats to its sustainability mean that dredging and development have not been halted — rather that dredged material is now used to improve and sustain otherwise degrading habitats.

South-east England: Managed Realignment and Sediment Recharge

Embanking and reclamation since medieval times has dramatically altered the planform and cross-sectional morphology of numerous small estuaries in south-east England, at the expense of more than 90% of the natural saltmarsh. The remaining saltmarsh fringe is now eroding — Burd (1993) estimated the loss of Essex saltmarshes at around 2% per annum between 1973 and 1983 — posing a problem for flood-defence engineers and conservation managers alike. Saltmarsh loss is usually attributed to wave action, as the morphology of the estuarine intertidal responds to relative sea-level rise (Pethick 1994; RSPB 1997), although Hughes (1999) has argued that herbivory and bioturbation by sediment in fauna may cause vegetation loss and marsh degradation in areas where wave energy is limited. Whatever the mechanism, it is clear that saltmarsh response is constrained by the legacy of flood-defence infrastructure such that natural upland-to-saltmarsh transitions are rare (Burd 1993).

Within this context, the creation of new intertidal habitats through the managed realignment of sea defences is currently favoured as a long-term adaptive policy for sustainable flood protection and estuary use (Agriculture Select Committee 1998). Managed realignment involves setting back defences to a new line some distance inland (preferably utilising natural contours), and at the same time promoting the establishment of new saltmarsh and tidal flat between the old and new defences (Burd, 1995). It is important to note that saltmarsh creation is not the sole objective here (Dixon, Leggett and Weight 1998), although the engineering and conservation significance of saltmarsh is emphasised. In Essex alone, there are about 440 km of seawall, of which 330 km rely on a strip of fronting saltmarsh as a first line of protection (Leggett and Dixon 1994). These fragments of saltmarsh also retain considerable value as habitats for wildfowl and waders and for important invertebrate species.

Experience with managed realignment in south-east England has been restricted to a small number of experimental and operational schemes undertaken on a fairly modest scale. The largest of experimental trial has been carried out at Tollesbury Fleet, in the Blackwater estuary, Essex, under the auspices of the Ministry of Agriculture, Fisheries and Food (MAFF). Here, 21 ha of former agricultural land were returned to tidal action in the summer of 1995, following the construction of a new landward seawall and breaching of the original outer wall (Figure 8.9). The land level within the site is significantly lower than that of the outlying saltmarshes, and the post-project monitoring includes measurement of both sediment accretion and net elevation change. Early results presented by Cahoon et al. (2000) show build-up occurring at 9–40 mm yr^{-1} within the lowest parts of the site, close to the breach (sites 3 and 5 in Figure 8.9(b)), with more modest increases of around 4–5 mm yr^{-1} within the higher marginal zone (4 and 6 in Figure 8.9(b)). The outer marshes here (sites 2 and 1 in Figure 8.9(b)) are building vertically at between 3 and 7 mm yr^{-1} against a mean annual sea-level rise of between 2 and 3 mm yr^{-1}. Five years after breaching, vegetation establishment has been restricted to the appearance of a 30–50 m wide zone of pioneer marsh species (*Salicornia* spp. and *Suaeda maritima*) along the upper edges of the site (especially around sites

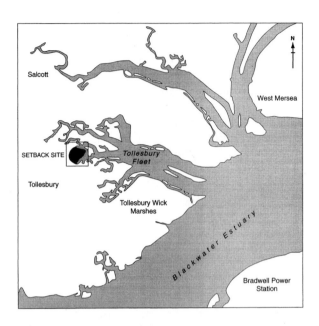

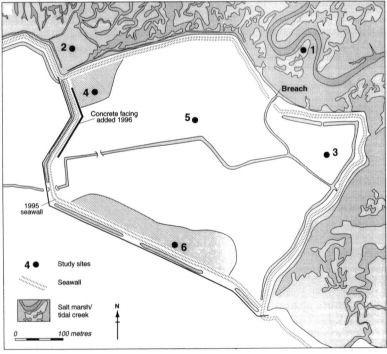

4 and 6 in Figure 8.9(b)). Much of the site remains too low to support marsh plants and drainage may also be impeded by the lack of any engineered channels to supplement the relic field drain (Figure 8.9(b); French 1996). The need to minimise wave action was one factor behind the decision to leave the old seawall *in-situ* either side of the 70 m wide breach (another factor being the cost of removal). However, wave action within the site is sufficient to erode the newly constructed seawall in places, and concrete facing has had to be installed where it faces the breach.

In the outer Blackwater estuary, an operational realignment has been carried out by the Environment Agency at Orplands (adjacent to Bradwell power station in Figure 8.9(a)). Erosion of the fronting saltmarsh since the 1950s led to extensive damage to the 2 km seawall which protected 38 ha of grazing marsh (with SSSI status) and arable production. By 1993, the cost of repairing the seawall was estimated at £350 000, with a projected £250 000 of additional expenditure required over the next 20 years (Dixon and Weight 1995) — a protection cost of nearly £16 000 ha^{-1} relative to a market value of < £4000 ha^{-1}. Accordingly, a realignment option was chosen which involved two breaches in the seawall, minor enhancement of an existing counter wall to separate the site into two management units, and the excavation of a initial creek system. From a conservation perspective, key features of this scheme include the provision of additional gravel habitat for wading birds, the excavation of a shallow scrape for wader feeding at low water, and the utilisation of existing topographic contours as the boundary. Post-project monitoring and evaluation has been very encouraging, with earlier than expected colonisation by marsh plants, rapid development of marine invertebrates, and extensive use by waders and waterfowl (Dixon, Leggett and Weight 1998).

The longer-term potential for saltmarsh establishment within managed realignment sites is difficult to judge given that experience with such schemes remains limited. Studies of natural breaches following seawall failure provide useful, though sometimes discouraging, insights into the longer-term viability of managed realignment. In the Blyth estuary, Suffolk, over 250 ha of reclaimed land have reverted to tidal action at various times since the 1920s. Saltmarsh formation on the resulting mudflats has been restricted to colonisation by *Spartina anglica* along the southern margin of the estuary, where terrain provides shelter from the predominant southwesterly winds. Locally generated wind waves have been shown to be effective in reducing the stability of the intertidal sediments, limiting net elevation increase and causing

Figure 8.9. *(opposite)* (a) Tollesbury Fleet managed realignment site, Blackwater estuary, Essex, UK; (b) site numbers refer to elevation change measurement locations of Cahoon *et al.* (2000). Shaded area represents extent of *Suaeda maritima* and *Saliconia* spp. colonisation in 1999, 3 years after breaching. For further explanation, see text.

the near-complete erosion of the former earth river walls (French *et al.* 1999, 2000). Hughes (1999) has also questioned whether managed realignment will inevitably lead to saltmarsh development, on the basis of observation of natural seawall breaches in the Crouch estuary, Essex. Here, although pioneer saltmarsh plants were able to colonise the mud, saltmarsh growth has been prevented by bioturbation and herbivory by the invertebrate infauna (chiefly *Nereis diversicolor* and *Corophium volutator*).

These studies indicate the importance of providing an environment which is conducive to sedimentation if managed realignment sites are to achieve any significant re-establishment of saltmarsh. Dixon, Leggett and Weight (1998) describe an innovative solution to the problem of low site elevation, which allows tidal sedimentation and the progressive re-establishment of saltmarsh prior to seawall breaching. At Abbots Hall (Blackwater estuary), 20 ha of arable land have been allowed to revert to saline habitat, with tidal exchange being effected through two sluices through the intact seawall. The land surface here lies up to 2 m below the outlying saltmarshes, as a consequence of past levelling and infilling of former marsh creeks, underdrainage and compaction by livestock and agricultural machinery. To encourage sediment and water exchange, a network of meandering creeks was created, and the excavated material used to raise the adjacent surface. This Environment Agency scheme was completed in 1996 and since then has resulted in colonisation by marsh halophytes, extensive colonisation by marine invertebrates, and nesting by birds including shelduck (*Tadorna tadorna*), redshank (*Trina totanus*) and oystercatcher (*Haematopus ostralegus*). Experimental schemes such as this may point the way towards a more enlightened approach to managed realignment as an environmentally-sensitive approach to long-term flood-defence management and planning.

While long-term adaptive strategies involving managed realignment are formulated (and the political difficulties tackled), there is an immediate need to mitigate existing damage to defences. This need is urgent where sites of high environmental capital value are at risk of saline inundation and 'unplanned realignment' through seawall failure. One approach involves the beneficial use of dredge material to mitigate estuarine sediment deficits, and to restore the engineering and ecological function of intertidal flats and saltmarshes. Several variations of this idea are under evaluation in south-east England. These include (1) recharge of reclaimed marshes to raise elevations prior to managed realignment; (2) sub-tidal placement of sediment to reduce the tendency for erosion of adjacent intertidal margins; and (3) foreshore placement in order to increase the dissipation of wave energy, reduce erosion and/or trickle-feed sediment back into the estuarine system.

Foreshore recharge has been most the widely tried of these techqniues, and schemes using $10–100 \times 10^3 \, m^3$ of sands and gravels have performed well in terms of habitat creation as well as their effectiveness in reducing erosion

(Carpenter and Brampton 1997). However, maintenance dredgings in southeast England are mainly cohesive muds. These constitute a resource that might be used to mitigate sediment deficits, but the fate of fine material under wave and tidal action is harder to predict and placement is problematic owing to the need to limit resuspension while consolidation and de-watering occurs. However, a trial $22\,000\,m^3$ recharge of an eroding foreshore in the Orwell estuary, Suffolk, has shown the practicality of pumping cohesive dredgings within a retaining bank of coarse gravel. This scheme, undertaken in late 1997 by Harwich Haven Authority (with Environment Agency support), extends over approximately 450 m of foreshore (Figure 8.8(c)). Two years after placement, minimal sediment loss due to resuspension has occurred, although a trickle-feed of mud into the estuary is occurring as compacted mud outcrops exposed by onshore movement of the retaining gravel are slowly eroded as the modified foreshore adjusts to prevailing hydrodynamic conditions (French et al. 2000). The elevation of the intertidal flat has been increased by up to 1 m, leading to a significant reduction in wave energy at the seawall toe. Encouragingly, there has also been a rapid establishment of a diverse and abundant invertebrate fauna and an increase in bird usage as wading species are able to feed later into the tidal cycle. Such a scheme does not directly replace the much higher saltmarsh fringe, and halophyte establishment has been very limited. It is possible, however, that one or more 'top-up' recharges might be used to achieve elevations capable of regenerating a saltmarsh vegetation cover.

CONCLUSIONS

Although coastal and estuarine saltmarshes continue to be lost due to development and through geomorphological and hydological responses to environmental change, the recognition of the diverse physical and ecological functions that they provide, together with the increasing scope of environmental legislation, provides tangible opportunities for their conservation. These opportunities arise directly through policies explicitly aimed at compensating damage to wetlands and endangered species habitats, as well as indirectly, through the emergence of more sustainable and environmentally sensitive approaches to erosion control and flood defence. In consequence, it is becoming increasingly difficult to isolate the requirements of conservation from wider concerns in coastal and estuarine management. It is especially important, therefore, that conservation goals are formulated with reference to the best possible scientific understanding, not just of the ecological characteristics of natural saltmarshes and their vulnerability to change, but also of the processes by which important functions and values can be realised in restored systems.

If it is to be effective, saltmarsh conservation management must be under-pinned by an appreciation of physical environmental processes operating at two distinct scales. First, it is important to understand the distinctiveness of saltmarshes formed in particular regional geomorphological and hydrological contexts, and the extent to which these have been anthropogenically influenced. The identification of marsh types based upon physical forcing factors provides a basis for assessments of their vulnerability to anthropogenic activity and environmental change, and enables the identification of management con-straints. Second, the morphodynamic processes by which saltmarshes have naturally evolved their ecological and physical habitat functions must be elucidated at a more local scale. This is not simply a matter of relating habitat type and pattern to physical parameters. Rather, it is necessary to know how particular ecological outcomes can arise from a given starting point. Recent studies of saltmarsh morphodynamics are of great value in this respect, and provide a conceptual framework within which ecological engineering tech-niques can be used to produce systems which evolve towards, rather than away from, desired ecological outcomes. Such studies also aid in the understanding of inherent system variability, and in the formulation of realistic criteria with which to evaluate performance against these conservation targets.

The ecological success of saltmarsh remediation and restoration schemes has thus far been mixed. This is partly a consequence of insufficient emphasis being placed on the engineering of appropriate geomorphological and hydrological functions, but also results from a limited understanding of how subtle differ-ences in habitat type, pattern and usage arise in nature. Important advances are being made in both these areas, and scientific research and accumulating engineering and management experience is laying the foundations for future success.

REFERENCES

Adam P (1978) Geographical variation in British saltmarsh vegetation. *Journal of Ecology* **66**, 339–66.
Adam P (1990) *Saltmarsh ecology*. Cambridge, Cambridge University Press.
Agriculture Select Committee, House of Commons (1998) *Sixth Report: Flood and coastal defence, Volume 1*. London, HMSO.
Allen JRL (1989) Evolution of salt-marsh cliffs in muddy and sandy systems: a qualitative comparison of British west-coast estuaries. *Earth Surface Processes and Landforms* **14**, 85–92.
Allen JRL (1990) Salt-marsh growth and stratification: a numerical model with special reference to the Severn Estuary, southwest Britain. *Marine Geology* **95**, 77–96.
Allen JRL (1997) Simulation models of salt-marsh morphodynamics: some implications for high-intertidal sediment couplets related to sea-level change. *Sedimentary Geology* **113**, 211–23.
Allen JRL & Pye K (1992) Coastal saltmarshes: their nature and importance. In: JRL

Allen & K Pye (eds) *Saltmarshes: morphodynamics, conservation and engineering significance*. Cambridge, Cambridge University Press, 1–18.

Arens SM, Jungerius PD & van der Meulen F (2001) Coastal dunes. In: A Warren & JR French (eds) *Habitat conservation: managing the physical environment*. Chichester, John Wiley, 229–72.

Ashley GM & Zeff ML (1988) Tidal channel classification for a low-mesotidal salt marsh. *Marine Geology* **82**, 17–32.

Barnby MA, Collins JN & Resh VH (1985) Aquatic macroinvertebrate communities of natural and ditched potholes in a San Francicsco Bay saltmarsh. *Estuarine Coastal and Shelf Science* **20**, 331–47.

Baumann RH, Day JW & Miller CA (1984) Mississippi deltaic wetland survival: sedimentation versus submergence. *Science* **224**, 1093–5.

Bayliss-Smith TP, Healey MG, Lailey R, Spencer T & Stoddart DR (1979) Tidal flows in salt marsh creeks. *Estuarine Coastal and Marine Science* **9**, 235–55.

Beeftink WG (1966) Vegetation and habitat of the salt marshes and beach plains in the south-western part of the Netherlands. *Wentia* **15**, 83–108.

Boorman LA, Goss-Custard JD & McGrorty S (1989) *Climate change, rising sea level and the British coast*. Research Publication No. 1, Institute of Terrestrial Ecology, London, HMSO.

Boyer ME, Harris JO & Turner RE (1997) Constructed crevasses and land gain in the Mississippi River. *Restoration Ecology* **5**, 85–92.

Bradley PM & Morris JT (1990) Influence of oxygen and sulfide concentration on nitrogen uptake kinetics in *Spartina alterniflora*. *Ecology* **71**, 282–7.

Brampton AH (1992) Engineering significance of British saltmarshes. In: JRL Allen & K Pye (eds) *Saltmarshes: morphodynamics, conservation and engineering significance*. Cambridge, Cambridge University Press, 115–22.

Bricker-Urso S, Nixon S, Cochran JK, Hirschberg DJ & Hunt C (1989) Accretion rates and sediment accumulation in Rhode Island saltmarshes. *Estuaries* **12**, 300–17.

Britsch LD & Dunbar JB (1993) Land loss rates: Louisiana coastal plain. *Journal of Coastal Research* **9**, 324–38.

Brooke JS (1992) Coastal defence: the retreat option. *Journal Institute of Water and Environmental Management* **6**, 151–6.

Burd F (1989) *The saltmarsh survey of Great Britain: an inventory of British saltmarshes*. Peterborough, Nature Conservancy Council. Research and Survey in Nature Conservation Series 17.

Burd F (1993) *Erosion and vegetation change in the saltmarshes of Essex and North Kent (1973 to 1983)*. Peterborough, English Nature, Research and Survey Publication 42.

Burd F (1995) *Managed retreat: a practical guide*. Peterborough, English Nature.

Cahoon DR & Cowan JH Jr (1988) Environmental impacts and regulatory policy impplications of spray disposal of dredged material in Louisiana wetlands. *Coastal Management* **16**, 341–62.

Cahoon DR, French JR, Spencer T, Reed DJ & Möller I (2000) Vertical accretion versus elevational adjustment in UK saltmarshes: an evaluation of alternative methodologies. In: K Pye & JRL Allen (eds) *Coastal and estuarine environments: sedimentology, geomorphology and geoarchaeology*. Geological Society of London Special Publication 175, 223–38.

Cahoon DR, Reed DJ & Day JW (1995) Estimating shallow subsidence in microtidal saltmarshes of the southeastern United States: Kaye and Barghoorn revisited. *Marine Geology* **128**, 1–9.

Callaway JC, DeLeaune RD & Patrick WH (1997) Sediment accretion rates from four coastal wetlands along the Gulf of Mexico. *Journal of Coastal Research* **13**, 181–91.

Cantero JJ, Leon R, Cisneros JM & Cantero A (1998) Habitat structure and vegetation relationships in central Argentina salt marsh landscapes. *Plant Ecology* **137**, 79–100.

Carpenter KE & Brampton AH (1997) *Maintenance and enhancement of saltmarshes.* Bristol, National Rivers Authority R&D Note 567/1/SW.

Chabreck RA (1988) *Coastal marshes: ecology and wildlife management.* Minneapolis, University of Minneapolis Press.

Chapman VJ (1974) *Saltmarshes and salt deserts.* Lehre, J. Cramer.

Chapman VJ (ed.) (1977) *Wet coastal ecosystems.* Amsterdam, Elsevier.

Clifford NJ (2001) Conservation and the river channel environment. In: A Warren & JR French (eds) *Habitat conservation: managing the physical environment.* Chichester, John Wiley, 67–104.

Coats RN, Williams PB, Cuffe CB, Zedler JB, Reed DJ, Waltry SM & Noller JS (1995) *Design guidelines for tidal channels in coastal wetlands.* Report for US Army Corps of Engineers, Waterways Experiment Station. San Francisco, Phillip Williams and Associates Ltd.

Collins LM, Collins JN & Leopold LB (1987) Geomorphic processes of an estuarine marsh: preliminary results and hypotheses. In: V Gardiner (ed.) *International Geomorphology 1986, Part 1.* Chichester, John Wiley, 1049–72.

Costanza R, dArge R, deGroot R, Farber S, Grasso M, Hannon B, Limburg K, Naeem S, ONeill RV, Paruelo J, Raskin RG, Sutton P & vandenBelt M (1997) The value of the world's ecosystem services and natural capital. *Nature* **387**, 253–60.

Dankers N, Binsberger M, Zegers K, Laane R & van de Loeff MR (1984) Transportation of water, particulate and dissolved organic and inorganic matter between a salt marsh and the Ems–Dollard estuary, the Netherlands. *Estuarine, Coastal and Shelf Science* **19**, 143–65.

Day JW, Scarton F & Rismondo A (1998) Rapid deterioration of a saltmarsh in Venice Lagoon, Italy. *Journal of Coastal Research* **14**, 583–90.

Day JW & Templet PH (1989) Consequences of sea level rise: implications from the Mississippi Delta. *Coastal Management* **17**, 241–57.

Dean RG (1978) Effects of vegetation on shoreline erosional processes. *Proceedings National Symposium on Wetlands.* American Water Resources Association, 415–26.

Desmond JS, Zedler JB & Williams GD (2000) Fish use of tidal creek habitats in two southern Californian salt marshes. *Ecological Engineering* **14**, 233–52.

Dijkema KS (1987) The geography of salt marshes in Europe. *Zeitschrift für Geomorphologie* NF **31**, 489–99.

Dixon AM & Weight RC (1995) *Managing coastal re-alignment: case study at Orplands sea wall, River Blackwater, Essex.* Ipswich, National Rivers Authority, Anglian Region.

Dixon AM, Leggett DJ & Weight RC (1998) Habitat creation opportunities for landward coastal re-alignment: Essex case studies. *Journal Institution of Water and Environmental Management* **12**, 107–12.

Dodd JD & Webb JW (1975) *Establishment of vegetation for shoreline stabilisation in Galveston Bay.* Fort Belvoir, US Army Corps of Engineers, CERC Report MP 6-75.

Doody P (1992) Saltmarsh conservation. In: JRL Allen & K Pye (eds) *Saltmarshes: morphodynamics, conservation and engineering significance.* Cambridge, Cambridge University Press, 80–114.

Dronkers J (1986) Tidal asymmetry and estuarine morphology. *Netherlands Journal of Sea Research* **20**, 117–31.

Eisma D & Dijekema KS (1997) The influence of salt marsh vegetation on sedimentation. In: D Eisma (ed.) *Intertidal deposits*. Boca Raton, CRC Press, 403–14.

Erchinger HF (1995) Intaktes Deichvorland für Kustenschutz unverzichtbar. *Wasser & Boden* **47**, 48–53.

Esselink DE, Dijkema KS, Reents S & Hageman G (1998) Vertical accretion and profile changes in abandoned man-made tidal marshes in the Dollard estuary, the Netherlands. *Journal of Coastal Research* **14**, 570–82.

Florsheim J & Williams PB (1994) *Physical criteria for defining success: a review of the physical performance of tidal marshes constructed with dredged materials in San Francisco Bay, California.* Report 867 for San Francisco District, Corps of Engineers. San Francisco, Phillip Williams and Associates.

Ford MA, Cahoon DR & Lynch JC (1999) Restoring marsh elevation in a rapidly subsiding salt marsh by thin-layer deposition of dredged material. *Ecological Engineering* **12**, 189–205.

French CE, French JR, Clifford NJ & CJ Watson (2000) Sedimentation–erosion dynamics of abandoned reclamations: the role of waves and tides. *Continental Shelf Research* **20**, 1711–33.

French CE, French JR & Watson C (1999) Abandoned reclamations as analogues for sea defence re-alignment. In: NC Kraus & WG McDougal (eds) *Coastal Sediments 99: Proceedings of the 4th International Symposium on Coastal Engineering and Science of Coastal Sediment Processes.* New York, American Society of Civil Engineers. Volume 3, 1912–26.

French JR (1993) Numerical modelling of vertical marsh growth and response to rising sea-level, Norfolk, UK. *Earth Surface Processes and Landforms* **18**, 63–81.

French JR (1994) Tide-dominated coastal wetlands and accelerated sea level rise: a Northwestern European perspective. *Journal of Coastal Research Special Issue* **12**, 91–101.

French JR (1996) Function and design of tidal channel networks in restored saltmarshes. In: P Gardiner (ed.) *Proceedings Tidal 96 Interactive Symposium for Practising Engineers.* University of Brighton, Department of Civil Engineering, 128–37.

French JR & Clifford NJ (1992) Characteristics and 'event-structure' of near-bed turbulence in a macro-tidal saltmarsh channel. *Estuarine and Coastal Shelf Science* **31**, 49–69.

French JR & Spencer T (2001) Sea-level rise. In: A Warren & JR French (eds) *Habitat conservation: managing the physical environment.* Chichester, John Wiley, 305–47.

French JR & Spencer T (1993) Dynamics of sedimentation in a tide-dominated backbarrier salt marsh, Norfolk, UK. *Marine Geology* **110**, 315–31.

French JR & Stoddart DR (1992) Hydrodynamics of salt marsh creek systems: implications for marsh morphological development and material exchange. *Earth Surface Processes and Landforms* **17**, 235–52.

French JR, Clifford NJ & Spencer T (1993) High frequency flow and suspended sediment measurements in a tidal wetland channel. In: NJ Clifford, JR French & J Hardisty (eds) *Turbulence: perspectives on flow and sediment transport.* Chichester, John Wiley, 249–78.

French JR, Spencer T, Murray AL & Arnold NA (1995) Geostatistical analysis of sediment deposition in two small tidal wetlands, Norfolk, UK. *Journal of Coastal Research* **10**, 308–21.

French JR, Watson CJ, Möller I, Spencer T, Dixon M & Allen R (2000) Beneficial use of cohesive dredgings for foreshore recharge. *35th MAFF Conference of river and coastal engineers — Innovation Forum*, Keele, 11.10.1–4.

Frey RW & Basan PB (1985) Coastal salt marshes. In: RA Davis (ed.) *Coastal sedimentary environments*. New York, Springer-Verlag, 225–301.

Friedrichs CT (1995) Stability shear stress and equilibrium cross-sectional geometry of sheltered tidal channels. *Journal of Coastal Research* **11**, 1062–74.

Friedrichs CT & Aubrey DG (1988) Non-linear tidal distortion in shallow well-mixed estuaries: a synthesis. *Estuarine Coastal and Shelf Science* **27**, 521–45.

Friedrichs CT, Lynch DR & Aubrey DG (1992) Velocity asymmetries in frictionally-dominated tidal embayments: longitudinal and lateral variability. In: D Prandle (ed.) *Dynamics and exchanges in estuaries and the coastal zone*. Washington DC, American Geophysical Union, Coastal Studies **40**, 277–312.

Gabet EJ (1998) Lateral migration and bank erosion in a saltmarsh tidal channel in San Francisco Bay, California. *Estuaries* **21**, 745–53.

Garofalo D (1980) The influence of wetland vegetation on tidal stream migration and morphology. *Estuaries* **3**, 258–70.

Gleason ML & Zieman JC (1981) Influence of tidal inundation on internal oxygen supply of *Spartina alterniflora* and *Spartina patens*. *Estuarine Coastal and Shelf Science* **13**, 47–57.

Gleason MC, Elmer DA, Pien NC & Fisher JS (1979) Effects of stem density upon sediment retention by salt marsh cordgrass, *Spartina alterniflora* Louisel. *Estuaries* **2**, 271–3.

Goodwin P & Williams PB (1992) Restoring coastal wetlands: the California experience. *Journal Institute of Water and Environmental Management* **6**, 709–719.

Gordon DC, Cranford P & Despanque C (1985) Observations on the ecological importance of salt marshes in the Cumberland Basin, a macrotidal estuary in the Bay of Fundy. *Estuarine Coastal and Shelf Science* **20**, 205–27.

Gray AJ (1972) The ecology of Morecambe Bay V: the saltmarshes. *Journal Applied Ecology* **9**, 207–20.

Gray AJ (1992) Saltmarsh ecology: zonation and succession revisited. In: JRL Allen & K Pye (eds) *Saltmarshes: morphodynamics, conservation and engineering significance*. Cambridge, Cambridge University Press, 63–79.

Gray DM (1974) Reinforcement and stabilisation of soil by vegetation. *ASCE Journal of the Geotechnical Engineering Division* **100**, 696–9.

Greensmith JT & Tucker EV (1966) Morphology and evolution of inshore shell ridges and mud mounds on modern intertidal flats, near Bradwell, Essex. *Proceedings of the Geologists Association* **77**, 329–46.

Haltiner J, Zedler KE, Boyer ME, Williams GD & Callaway JC (1997) Influence of physical processes on the design, functioning and evolution of restored tidal wetlands in California. *Wetlands Ecology and Management* **4**, 73–91.

Harmsworth GC & Long SP (1986) An assessment of saltmarsh erosion in Essex, England, with reference to the Dengie Peninsula. *Biological Conservation* **35**, 377–87.

Harrison EZ & Bloom AL (1977) Sedimentation rates on tidal salt marshes in Connecticut. *Journal of Sedimentary Petrology* **47**, 1484–90.

Hayes MO (1979) Barrier island morphology as a function of tidal and wave regime. In: SP Leatherman (ed.) *Barrier islands*. Orlando, Academic Press, 1–27.

Hollis GEH, Thomas D & Heard S (1989) *The effects of sea-level rise on sites of conservation value in Britain and northwest Europe*. London, University College London. Report to WWF, Project 120/88.

Hughes RG (1999) Saltmarsh erosion and management of saltmarsh restoration: the effects of infaunal invertebrates. *Aquatic Conservation: Marine and Freshwater Ecosystems* **9**, 83–95.

Huiskes AHL (1990) Possible effects of sea level changes on salt marsh vegetation. In: JJ Beukama (ed.) *Expected effects of climate change on marine coastal ecosystems.* Dordrecht, Kluwer Academic, 167–72.

Hydraulics Research Ltd (1980) *Design of sea walls allowing for wave overtopping.* Wallingford, Hydraulics Research Ltd, Report EX924.

Ibanez C, Day JW & Pont D (1999) Primary production and decomposition of wetlands of the Rhône delta, France: interactive impacts of human modifications and relative sea level rise. *Journal of Coastal Research* **15**, 717–31.

Jacobsen NK (1980) Form elements of the Wadden Sea area. In: KS Dijkema (ed.) *Geomorphology of the Wadden Sea Area.* Rotterdam, Balkema, 50–71.

Jaffe B, Smith R & Zink L (1998) *Sedimentation and bathymetry changes in San Pablo Bay: 1856–1983.* United States Geological Survey Open File Report 98-759.

Josselyn M (1983) *The ecology of San Francisco Bay tidal marshes: a community profile.* Slidell, US Fish and Wildlife Service Report FWS/OBS-83/23.

Kamps LF (1962) Mud distribution and land reclamation in the eastern Wadden shallows. *Rijkwaterstaat Communications* **4**, 1–73.

Kana TW, Eiser WC, Baca BJ & Williams ML (1988) Charleston case study. In: JG Titus (ed.) *Greenhouse effect, sea level rise and coastal wetlands.* Washington DC, US EPA, 37–59.

Kaye CA & Barghoorn ES (1964) Late Quaternary sea-level change and crustal rise at Boston, Massachusetts, with notes on the autocompaction of peat. *Geological Society of America Bulletin* **75**, 63–80.

Kaznowska S & Waller C (1993) Buying time for Benacre. *Enact* **1**, 10–11.

Kelley JT, Gehrels WR & Belknap DF (1995) Late Holocene relative sea-level rise and the geological development of tidal marshes at Wells, Maine, USA. *Journal of Coastal Research* **11**, 136–53.

Kelley JT, Belknap DF, Jacobson GL & Jacobsen HA (1988) The origin and evolution of salt marshes along the glaciated coastline of Maine, USA. *Journal of Coastal Research* **4**, 649–65.

Kestner JT (1975) The loose boundary regime of The Wash. *Geographical Journal* **141**, 389–414.

Koch MS & Mendelssohn I (1989) Sulphide as a soil phytotoxin: differential responses in two marsh species. *Journal of Ecology* **77**, 565–78.

Knutson PL (1988) Role of coastal marshes in energy dissipation and shore protection. In: DD Hook (ed. with 12 others) *The ecology and management of wetlands. Volume 1: Ecology of wetlands.* London, Croom Helm, 161–75.

Knutson PL & Inskeep MR (1982) *Shore erosion control with salt marsh vegetation.* Fort Belvoir, US Army Corps of Engineers CERC Technical Aid 82-3.

Knutson PL, Brochu RA, Seelig WN & Inskeep (1982) Wave damping in *Spartina alterniflora* marshes. *Wetlands* **2**, 87–104.

Knutson PL, Ford JC, Inskeep MR & Oyler J (1981) National survey of plated salt marshes (vegetative stabilization and wave stress). Wetlands **1**, 129–57.

Krone RB (1987) A method for simulating historic marsh elevations. In: NC Kraus (ed.) *Proceedings Coast Sediments '87.* New York, American Society of Civil Engineers, 316–23.

Kuhn NL, Mendelssohn IA & Reed DJ (1999) Altered hydrology effects on Louisiana salt marsh function. *Wetlands* **19**, 617–26.

Landin MC, Webb JW & Knutson PL (1989) *Long-term monitoring of eleven Corps of Engineers habitat development field sites built of dredged material, 1974–1987.* US Army Corps of Engineers, WES Technical Report D-89/1.

Leggett DJ & Dixon M (1994) Management of the Essex saltmarshes for flood defence. In: R Falconer & P Goodwin (eds) *Proceedings, International Conference on Wetland Management*. London, ICE, 232–45.

Leonard LA, Hine AC & Luther ME (1995) Surficial sediment transport and deposition processes in a *Juncus-Roemerianus* marsh, west-central Florida. *Journal of Coastal Research* 11, 322–36.

Levin LA, Talley TS & Hewitt J (1998) Macrobenthos of *Spartina foliosa* (Pacific cordgrass) salt marshes in southern California: Community structure and comparison to a Pacific mudflat and a *Spartina alterniflora* (Atlantic smooth cordgrass) marsh. *Estuaries* 21, 129–44.

McCaffery RJ & Thompson J (1980) A record of the accumulation of sediment and trace metals in a Connecticut salt marsh. *Advances in Geophysics* 22, 165–236.

McIvor CC & Odum WE (1988) Food, predation, risk and microhabitat selection in a marsh fish assemblage. *Ecology* 69, 1341–51.

McKee K & Mendelssohn IA (1989) Response of a freshwater plant community to increased salinity and increased water level. *Aquatic Botany* 34, 301–16.

McKee K & Patrick WH (1988) The relationship of Smooth Cordgrass (*Spartina alterniflora*) to tidal datums: a review. *Estuaries* 11, 143–51.

Mendelssohn IA & Kuhn NL (1999) The effects of sediment addition on salt marsh vegetation and soil physico-chemistry. In: LP Rozas, JA Nyman, CE Proffitt, NN Rabalais, DJ Reed & RE Turner (eds) *Recent research in coastal Louisiana: natural system function and response to human influences*. Baton Rouge, LA, Louisiana Sea Grant College Program, 55–61.

Milliman JD & Meade RH (1983) Worldwide delivery of river sediment to the oceans. *Journal of Geology* 91, 1–21.

Mitsch WJ & Gosselink JG (1993) *Wetlands*. New York, Van Nostrand Reinhold.

Möller I, Spencer T & French JR (1996) Wind wave attenuation over saltmarsh surfaces: preliminary results from Norfolk, England. *Journal of Coastal Research* 12, 1009–16.

Möller I, Spencer T, French JR, Dixon M & Leggett DJ (1999) Wave transformation over salt marshes: a field and modelling study from North Norfolk, England. *Estuarine Coastal and Shelf Science* 49, 411–26.

Morris JT, Kjerfve B & Dean JM (1990) Dependence of estuarine productivity on anomalies in mean sea level. *Limnology and Oceanography* 35, 926–30.

National Research Council (1987) *Responding to changes in sea level*. Washington DC, National Academy Press.

Nestler J (1977) Interstitial salinity as a cause of ecophenic variation in *Spartina alternifora*. *Estuarine Coastal and Marine Science* 5, 707–14.

Nicholls RJ, Hoozemans FMJ & Marchand M (1999) Increasing flood risk and wetland loss due to global sea-level rise: regional and global analyses. *Global Environmental Change — Human and Policy Dimensions* 9, 69–87.

Niering WA & Warren RS (1980) Vegetation patterns and processes in New England salt marshes. *Bioscience* 30, 301–7.

Norris K, Cook T, O'Dowd B & Durdin C (1997) The density of redshank *Tringa totanus* breeding on the salt-marshes of the Wash in relation to habitat and its grazing management. *Journal of Applied Ecology* 34, 999–1013.

Nydick KR, Bidwell AB, Thomas E & Varenkamp JC (1995) A sea-level rise curve from Guildford, Connecticut, USA. *Marine Geology* 124, 137–59.

Nyman JA, DeLaune RD, Pezeshki SR & Patrick WH (1995) Organic-matter fluxes and marsh stability in a rapidly submerging estuarine marsh. *Estuaries* 18, 207–18.

Odum EP (1974) Halophytes, energetics and ecosystems. In: RJ Reimold & WH Queen (eds) *Ecology of halophytes*. New York, Academic Press, 599–602.

Orson RA, Warren RS & Niering WA (1987) Development of a tidal marsh in a New England River valley. *Estuaries* **10**, 20–27.

Orson RA, Warren RS & Niering WA (1998) Interpreting sea level rise and rates of vertical marsh accretion in a southern New England tidal salt marsh. *Estuarine Coastal and Shelf Science* **47**, 419–29.

Ovensen CH (ed.) (1990) Salt marsh management in the Wadden Sea region. Proceedings 2nd Trilateral Working Conference, Rømø, Demark. Copenhagen, National Forest and Nature Agency.

Penland S & Ramsey KE (1990) Relative sea-level rise in Louisiana and the Gulf Coast of Mexico: 1908–1988. *Journal of Coastal Research* **6**, 323–42.

Pestrong R (1965) The development of drainage patterns on tidal marshes. *Stanford University Publications in Geology* **10**, 1–81.

Pestrong R (1972) The shear strength of tidal marsh sediments. *Journal of Sedimentary Petrology* **39**, 322–6.

Pethick JS (1980) Salt marsh initiation during the Holocene transgression: the example of the north Norfolk marshes. *Journal of Biogeography* **7**, 1–9.

Pethick JS (1981) Long-term accretion rates on tidal marshes. *Journal of Sedimentary Petrology* **61** 571–7.

Pethick JS (1992) Saltmarsh geomorphology. In: JRL Allen & K Pye (eds) *Saltmarshes: morphodynamics, conservation and engineering significance*. Cambridge, Cambridge University Press, 41–62.

Pethick JS (1994) Estuarine wetlands — function and form. In: RA Falconer & P Goodwin (eds) *Wetland management*. London, Thomas Telford, 75–87.

Pethick J & Reed D (1987) Coastal protection in an area of salt marsh erosion. In: NC Kraus (ed.) *Proceedings Coast Sediments '87*. New York, American Society of Civil Engineers, 1094–1104.

Pezeshki SR, DeLaune RD & Patrick WH (1987) Response of the freshwater marsh species, *Panicum hemitomon* Schult., to increased salinity. *Freshwater Biology* **17**, 195–200.

Phillips JD (1986) Coastal submergence and marsh fringe erosion. *Journal of Coastal Research* **2**, 427–36.

Pringle AW (1995) Erosion of a cyclic salt-marsh in Morecambe Bay, northwest England. *Earth Surface Processes and Landforms* **20**, 387–405.

Pye K & French PW (1993) *Erosion and accretion processes on British saltmarshes. Volume 2: Database of British saltmarshes*. MAFF Research Report ES19B(2). Cambridge, Cambridge Environmental Consultants Ltd.

Race MS (1985) Critique of present wetland mitigation policies in the United States based on an analysis of past restoration projects in San Francisco Bay. *Environmental Management* **9**, 71–82.

Redfield AC (1972) Development of a New England salt marsh. *Ecological Monographs* **41**, 201–37.

Reed DJ (1988) Sediment dynamics and deposition in a retreating coastal salt marsh. *Estuarine Coastal and Shelf Science* **26**, 67–79.

Reed DJ (1989) Patterns of sediment deposition in subsiding coastal marshes, Terrebonne Bay, Louisiana: the role of winter storms. *Estuaries* **12**, 222–7.

Reed DJ (1990) The impact of sea-level rise on coastal salt marshes. *Progress in Physical Geography* **124**, 24–40.

Reed DJ (1995) The response of coastal marshes to sea-level rise: survival or submergence? *Earth Surface Processes and Landforms* **20**, 39–45.

Reed DJ & Cahoon DR (1992) The relationship between marsh surface topography and vegetation parameters in a deteriorating Louisiana *Spartina alterniflora* salt marsh. *Journal of Coastal Research* **8**, 77–87.

Reed DJ, DeLuca N & Foote AL (1997) Effect of hydrologic management on marsh surface sediment deposition in coastal Louisiana. *Estuaries* **20**, 301–11.

Reed DJ, Spencer T, Murray A, French JR & Leonard L (1999) Marsh surface sediment deposition and the role of tidal creeks: implications for created and managed coastal marshes. *Journal of Coastal Conservation* **5**, 81–90.

Rosen PS (1980) Erosion susceptibility of the Chesapeake Bay shoreline. *Marine Geology* **34**, 45–59.

Rountree RA & Able KW (1992) Fauna of polyhaline subtidal marsh creeks in southern New Jersey: composition, abundance and biomass. *Estuaries* **15**, 171–85.

Rozas LP, McIvor CC & Odum WE (1988) Intertidal rivulets and creekbanks: corridors between tidal creeks and marshes. *Marine Ecology Progress Series* **47**, 303–7.

RSPB, National Trust, Norfolk Wildlife Trust, Suffolk Wildlife Trust & English Nature (1997) *Coasts in crisis: world famous wetlands at risk in Norfolk and Suffolk*. Norwich, RSPB.

Seneca ED (1974) Stability of coastal dredge spoil with Spartina alterniflora. In: RJ Reimold & WH Queen (eds) *Ecology of halophytes*. New York, Academic press, 525–30.

Seneca ED, Woodhouse WW & Broome SW (1975) Salt-water marsh creation. In: LE Cronin (ed.) *Estuarine research, Volume 2*. New York, Academic Press, 427–38.

Shaw J & Ceman J (1999) Salt-marsh aggradation in response to late-Holocene sea-level rise at Amherst Point, Nova Scotia, Canada. *The Holocene* **9**, 439–51.

Shi Z, Lamb HF & Collin RL (1995) Geomorphic change of saltmarsh tidal creek networks in the Dyfi Estuary, Wales. *Marine Geology* **128**, 73–83.

Shi Z, Pethick JS & Pye K (1995) Flow structure in and above the various heights of a saltmarsh canopy: a laboratory flume study. *Journal of Coastal Research* **11**, 1204–9.

Stanley DJ (1988) Subsidence in northeastern Nile Delta: rapid rates, possible causes, and consequences. *Science* **240**, 497–500.

Steever ZE, Warren RS & Niering WA (1976) Tidal energy subsidy and standing crop production of *Spartina alterniflora*. *Estuarine Coastal and Marine Science* **4**, 473–8.

Stevenson JC, Ward LG & Kearney MS (1986) Vertical accretion in marshes with varying rates of sea-level rise. In: DA Wolfe (ed.) *Estuarine variability*. Orlando, Academic Press, 241–60.

Stevenson JC, Ward LG & Kearney MS (1988) Sediment transport and trapping in marsh systems: implications of tidal flux studies. *Marine Geology* **80**, 37–59.

Stoddart DR, Reed DJ & French JR (1989) Understanding saltmarsh accretion, Scolt Head Island, Norfolk, England. *Estuaries* **12**, 228–36.

Stumpf RP (1983) The processes of sedimentation on the surface of a saltmarsh. *Estuarine Coastal and Shelf Science* **17**, 495–508.

Taylor NC, Day JW & Nuesaenger GE (1989) Ecological characterization of Jean Lafitte National Historical Park, Louisiana: basis for a management plan. In: WG Duffy & D Clark (eds) *Marsh management in coastal Louisiana: effects and issue — Proceedings of a symposium*. US Fish and Wildlife Service Biological Report 89/22, 247–77.

Teal JM & Teal M (1969) *The life and death of a saltmarsh*. Boston, Little Brown & Co.

Titus JG (1991) Greenhouse effect and coastal wetland policy: How Americans could abandon an area the size of Massachusetts at minimum cost. *Environmental Management* **15**, 39–58.

Titus JG, Park RA, Leatherman SP, Weggel JR, Greene MS, Mausel PW, Brown S,

Gaunt C, Trehan M & Yohe G (1991) Greenhouse-effect and sea-level rise: the cost of holding back the sea. *Coastal Management* **19**, 171–204.

Turner K & Dagley J (1993) What price seawalls? *Enact, Managing Land for Wildlife* **1**, 8–9.

Turner RE (1997) Wetland loss in the northern Gulf of Mexico: Multiple working hypotheses. *Estuaries* **20**, 1–13.

Turner RE & Boyer ME (1997) Mississippi River diversions, coastal wetland restoration/creation and an economy of scale. *Ecological Engineering* **8**, 117–28.

Vairin BA (1997) *Caring for coastal wetlands: the Coastal Wetlands Planning, Protection and Restoration Act*. Lafayette, United States Geological Survey National Wetlands Research Center.

Valiela I & Teal JM (1974) Nutrient limitation in salt-marsh vegetation. In: RJ Reimold & WH Queen (eds) *Ecology of halophytes*. New York, Academic Press, 547–63.

van der Molen J (1996) Tidal distortion and spatial differences in surface flooding characterstics in a salt marsh: implications for sea-level reconstruction. *Estuarine Coastal and Shelf Science* **45**, 221–33.

van Eerdt MM (1985) Salt marsh cliff stability in the Oosterschelde. *Earth Surface Processes and Landforms* **10**, 95–106.

Visser JM, Sasser CE, Chabreck RH & Linscombe RG (1998) Marsh vegetation types of the Mississippi River Deltaic Plain. *Estuaries* **21**, 818–28.

Walker HJ, Coleman JM, Roberts HH & Tye RS (1987) Wetland loss in Louisiana. *Geografiska Annaler* **69A**, 189–200.

Wang FC, Lu TS, & Sikora WB (1993) Intertidal marsh suspended sediment transport processes, Terrebonne Bay, Louisiana. *Journal of Coastal Research* **9**, 209–20.

Warren RS & Niering WA (1993) Vegetation change on a northeast tidal marsh: interactions of sea-level rise and marsh accretion. *Ecology* **74**, 96–103.

Wayne CJ (1976) The effects of sea and marsh grass on wave energy. *Coastal Research Notes* **4**, 6–8.

Williams GD & Zedler JB (1999) Fish assemblage composition in constructed and natural marshes of San Diego Bay: relative influence of channel morphology and restoration history. *Estuaries* **22**, 702–16.

Williams P (1994) From reclamation to restoration: changing perspectives in wetland management. In: RA Falconer & P Goodwin (eds) *Proceedings international conference on wetland management*. London, Institution of Civil Engineers, 1–6.

Williams SJ, Dodd K & Gohn KK (1991) *Coasts in crisis*. United States Geological Survey, Circular 1075.

Wolanski E, Jones M & Bunt JS (1980) Hydrodynamics of a tidal creek–mangrove system. *Australian Journal of Marine and Freshwater Science* **31**, 431–50.

Wolff WJ (1992) The end of a tradition: 1000 years of embankment and reclamation of wetlands in the Netherlands. *Ambio* **21**, 287–91.

Wood ME, Kelley JT & Belknap DF (1989) Patterns of sediment accumulation in the tidal marshes of Maine. *Estuaries* **12**, 237–46.

Wright LD, Coleman JM & Thom BG (1973) Processes of channel development in a high-tide range environment, Gulf-Ord River Delta, Western Australia. *Journal of Geology* **81**, 15–41.

Zedler JB (1996) Ecological isses in wetland mitigation: an introduction to the forum. *Ecological Applications* **6**, 33–7.

Zedler JB & Callaway JC (1999) Tracking wetland restoration: do mitigation sites follow desired trajectories? *Restoration Ecology* **7**, 69–73.

Zedler JB, Callaway JC, Desomn JS, Smith VG, Williams GD, Sullivan G, Brewster AE

& Bradshaw BK (1999) Californian salt-marsh vegetation: an improved model of spatial pattern. *Ecosystems* **2**, 19–35.

Zedler JB, Josselyn M & Onuf C (1982) Restoration techniques, research and monitoring: Vegetation. In: M Josselyn (ed.) *Wetland restoration and enhancement in California*. La Jolla, California Sea Grant Report T-CSGCP/007, 63–72.

Zeff ML (1999) Salt marsh tidal channel morphometry: applications for wetland creation and restoration. *Restoration Ecology* **7**, 205–11.

9 Coastal Dunes

S. M. ARENS,[1,2] P. D. JUNGERIUS[1,3] AND F. VAN DER MEULEN[4]

[1]Netherlands Centre for Geo-Ecological Research ICG, University of Amsterdam, The Netherlands
[2]Bureau for Beach and Dune Research, The Netherlands
[3]Bureau G & L, The Netherlands
[4]Ministry of Transport, Public Works and Water Management, National Institute for Marine and Coastal Management, The Netherlands

COASTAL DUNES AS A DYNAMIC SYSTEM

INTRODUCTION

In essence, the coastal dune environment can be seen as a system: a working system of elements, biotic and abiotic by nature and interacting with each other and their environment. For the purpose of this book it is necessary to define the steering processes of that system. Once knowledge has been obtained about the steering processes, the manager can choose to take that knowledge into account. Management activities can then be applied in such a way that the manager is working with nature rather than against it.

Knowledge of the processes in dunes and the application of that knowledge for their management has become increasingly customary. This chapter discusses recent studies of coastal dune processes and, using examples from various landscapes, illustrates the problems faced and the lessons learned. We emphasise the Dutch situation, but at the same time also try to extrapolate to the European situation.

THE COASTAL DUNE SYSTEM

The physical setting is essential in every natural system. It also determines its main abiotic processes. For example, is the system mainly aquatic, marine or terrestrial? Are the processes predominantly geomorphological or biological in character (in other words, are processes with wind and water as agents determinant) or is the production of biomass and vegetation structures predominant? In the former case, the system is likely to be very dynamic; in the latter case it is probably a more stable system where plants and animals build up complex and long-living communities.

Habitat Conservation: Managing the Physical Environment. Edited by A. Warren and J. R. French.
© 2001 John Wiley & Sons Ltd.

Figure 9.1 presents a simplified model of the coastal dune system as seen from a sedimentary perspective. Note that, from a systems approach, dunes do not stop at the beach, nor do they stop at the inner dune ridge. The offshore zone, beach and dune should be seen as one interconnected landscape entity. Ideally this entity should be managed as such. In practice, the entity is often divided into many parts and managed as many (often even separate) parts.

The coastal dune system consists of three main subsystems: an offshore zone, a transit zone and a resting zone. In Figure 9.1 these are seen from left to right. The coastline itself, the transition from land to sea, lies in the middle from top to bottom.

The Offshore Zone

This is the marine part of the dune system. The offshore zone is the sediment bank. When there is a positive sediment budget, more sand is transported

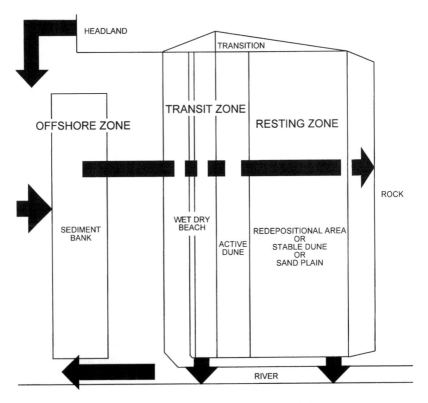

Figure 9.1. The coastal dune system (from Van der Meulen 1993).

towards the right of Figure 9.1 (i.e. from the offshore zone to the beach and subsequently to the dunes beyond the beach). An accreting coast can be the result. In the case of a negative sediment budget, more sand is transported from the beach, and perhaps also the dunes, to the offshore zone. This means erosion. So far, we have only mentioned movements along a transect perpendicular to the coast (on/offshore). But most coasts also have longshore currents such that sediment is transported from one part of the sediment bank to the next. Such transport may aggravate the transport movements that have just been outlined. Apart from the sources mentioned, sediments in longshore transport can enter the sediment bank, for example by erosion of a headland further down the coast (top of Figure 9.1) or by discharge of a river entering the sea (bottom of Figure 9.1) or from the deeper sea bottom (extreme left of Figure 9.1). Examples of coastal erosion caused by sediment starvation in rivers that have been dammed upstream are well known. This illustrates the fact that integrated coastal management is often equivalent to catchment management. We will not elaborate on this point here. Long Point peninsula in Lake Erie, Canada, is an example of a sandy coast where erosion of a headland occurs, causing growth of the coast further along the shore; but this co-acting erosion–accretion system moves along the coast, one process following another at any particular spot.

The Transitional Zone

This is the wet-dry beach and the active dune or, as we call it here, foredune. Sandy coasts with steep beaches and foreshores and low tidal differences have a narrow transitional zone. The opposite is found on gently sloping shores with great tidal differences. Once the sand is deposited on the beach by tides and waves, it can be transported by the wind. Depending on whether the prevailing winds are on- or off-shore, sand reaches the foredunes or not. The foredune is the first place where sand is effectively trapped by vegetation and where real dune formation can begin (Arens 1996). The evolution, ecological processes, aerodynamics and morphology of new foredunes on the upper beach has been reviewed by Hesp (1989). A boulevard (or worse, one with buildings) along the beach can cut off the active dune from sediment supply and thus remove its conditions for life. Depending on the intensity of development, foredunes have been manipulated in many ways, particularly for recreation and coastal defence. This kind of interference is drastic. For example, the Dutch foredunes have often been transformed into a high, straight sand dyke where wind activity and the formation of blow-outs is not allowed. Beyond the influence of water and waves, the transitional zone is dominated by aeolian processes. Natural vegetation, which is essential for trapping the sand, usually does not develop much beyond the pioneer stage.

The Resting Zone

This is the zone of inland dunes behind the foredunes. The resting zone can be up to several kilometres wide, as is the case on the Dutch mainland coast and along the French coast of Aquitaine (Paskoff 1994). In most cases, this zone is a depositional area inherited from the past. Wind activity is reduced to occasional blow-outs. Plants have stabilised the surface and various kinds of vegetation have developed. The vegetation succession in dunes with higher calcium carbonate contents and more nutrients leads to a mosaic of grasslands, shrublands and woodlands. Stabilisation measures have been carried out over many centuries. In the twentieth century, large-scale afforestation, mainly with pines, helped to stabilise this zone. Vegetation development is associated with soil development. When the soils have developed humic horizons the dune surface is more resistant to wind erosion. Since the influence of the wind has been reduced, we can characterise the resting zone as a subsystem where biological processes have become dominant over geomorphological processes. This theme is elaborated in the next section. On many parts of the European coast, for example northern France and Great Britain, the sand dunes meet solid deposits further inland, on which they often partly rest. Such deposits usually rise in height above the dunes, going inland. On low-lying delta coasts, like the Dutch mainland coast, which is essentially a coastal barrier system, dunes border former lagoonal areas, later filled in with clay and peat, and now turned into polders. The polders are lower than the sea and the dunes protect them from flooding.

THE STEERING PROCESSES

Seen at a general systems scale and looking at steering processes, the coastal system consists of a marine and a terrestrial part. The marine part is dominated by hydraulic processes. The terrestrial part is the visual dune landscape which is our main concern. This part is characterised by land forms and vegetation cover. The seaward zone is dominated by geomorphological processes. Their influence diminishes inland as biological processes become dominant. This means that every part of the dune landscape is determined by the interaction of geomorphological and biological processes, as shown in the model of Figure 9.2. The degree to which each of these processes dominates, at each location, is reflected in the soil profile.

The geomorphological processes operating in dunes are perhaps more dynamic than anywhere else. They comprise the action of wind and water. Essentially they lower the surface by erosion, or raise the surface by accumulation, thereby changing relief. There is little or no soil profile development as long as these processes keep the surface unstable. Biological processes result in

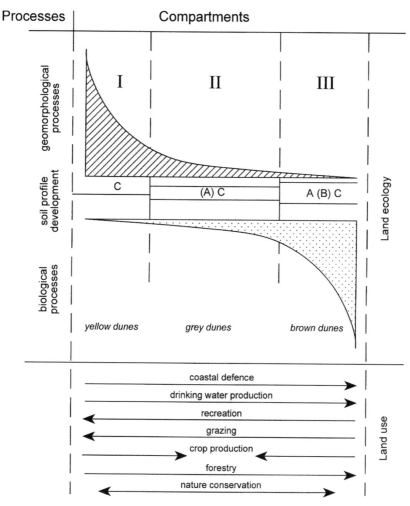

Figure 9.2. Schematic representation of the dune ecosystem and the effect of land use (after Jungerius and Van der Meulen 1988; with permission from Elsevier Science).

the establishment and development of vegetation. By biological processes we mean the production of biomass, the building up of vegetation structures, the change in species composition and increase in biodiversity. Underground biological processes include the formation of organic horizons leading to soil profile development. The role of animals is indirect. Rabbits and other grazers reduce the vegetation cover which apparently induces instability.

The first compartment of Figure 9.2 represents the beach and the foredunes where new landforms are created. Some plants such as marram grass

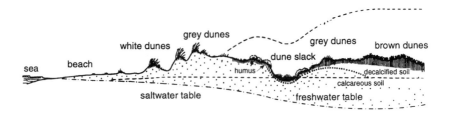

Figure 9.3. Cross-section of a coastal dune area along the North Sea (from Ellenberg 1978:493; with permission from Verlag Eugen Ulmer).

(*Ammophila arenaria*) are adapted to the geomorphological dynamics of this zone, but biological processes are generally of minor importance. No soil profile is found here. These are the white or yellow dunes, *Weissdünen* (Figure 9.3) or *dunes blanches*. The second compartment comprises the inner dune landscape where various combinations of geomorphological and biological processes prevail. This results in great landscape diversity. The soil is coloured grey by organic matter and develops an AC profile (grey dunes, *Graudünen* or *dunes grises*). The third compartment corresponds to inner dunes that are completely covered with vegetation (*Braundünen*).

Besides these, another group of processes which cannot be readily inferred from Figure 9.1 is of vital importance. These are hydrological processes (Bakker 1990). Once dunes have been formed, a geohydrological equilibrium causes the formation of a bell-shaped freshwater body within the dunes. The top is the phreatic level; the base is the brackish interface between fresh water and salt water in the deeper underground. The width of the terrestrial part of the coastal system and the average sea level are the main determinants for the shape of this fresh water body (see Carter 1991). The response of ground-water to long-term shoreline erosion and sea level rise is shown in Figure 9.4. In the terrestrial part of the dune, especially the resting zone subsystem, blow-outs can form dune slacks: when wind excavation of sand reaches the phreatic groundwater level, aeolian processes cease. From then on, the development of the slack is largely dominated by hydrological processes, for example the seasonal fluctuations of the groundwater level. Dune slacks have become the prime subject of conservation management because of their botanical values.

The steering processes give coastal environments a highly dynamic character that should be seen both in a temporal and a spatial context. Understanding this context is essential to management for conservation or for any other function. Restoring these processes means the (re)introduction of the natural dynamics into the coastal environment (see, for example, Van der Meulen and Jungerius 1989).

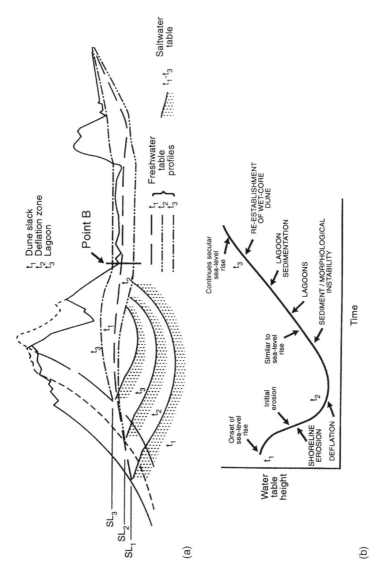

Figure 9.4. (a) Response of groundwater in a dune system to long-term shoreline erosion; (b) graphic representation of water table height over time through a phase of sea-level change (from Carter, 1991; with permission from Kluwer Academic).

THE SCOPE OF DUNE COAST CONSERVATION

HISTORICAL CHANGES IN THE CONSERVATION ISSUES

The need for conservation comes with the human use of coastal dunes. The conservation tradition is, therefore, longest in the **inner dunes**. From recent historical records it is clear that the resting zone has not always been so stable as it is today. People have used sand dunes for purposes of hunting and various forms of agriculture since prehistoric times (Higgins 1933) and along many European coasts there were mobile dunes, often threatening the local population. For at least a thousand years in most of Europe, this zone has received some form of protection. Various techniques of sand fixation were applied and legislation was developed, for example in Denmark, where the first royal decree was as early as 1539 (Skarregaard 1989). Westhoff (1989) believes that in The Netherlands 'the dune area was rather like a desert' in the past, but evidence to support his point is conflicting. Paintings from the end of the sixteenth century and photographs from the end of the nineteenth century show that the landscape of much of the inner dunes was, in fact, little different from that of today apart from the forests that have since been planted. The large fossilised blow-outs that have been observed from the Pyrenees to Jutland must have been spectacular spots of deflation in the past, but their very existence is also proof that the surface in between the blow-outs was stable.

Agriculture was never very successful and gradually the use of the dune landscape became more diversified. Each type of land use left its imprint on the landscape and determined future management. An interesting case is presented by De Raeve (1989). From Cap Blanc Nez to the Wadden islands, the Flemish coastal plains and the Dutch lowlands are part of a common geological setting and are both entirely bordered by coastal dunes. These dunes belong to the same system as far as climatic and floristic conditions are concerned. This holds true also for the properties of the sand (Depuydt 1972). Contrasting with the natural similarities are the strong differences in human interference which have developed over the last 100 years. The Flemish plain almost everywhere rises 2–4 m above sea level, so the dunes were not needed to protect the hinterland. Moreover, the hydrogeological conditions of inner Belgium render the dunes redundant for drinking-water supply. Flemish coastal dunes therefore kept their former status as waste land. Their only use were for sand exploitation, afforestation, and house and hotel building for the growing tourist industry. The chaotic developments linked with the '*laissez-faire*' mentality in the Belgian landscape caused the loss of 70% of the Flemish dune area but, on the other hand, left practically intact the major natural processes in the few unexploited areas, including active parabolic systems and large masses of moving sand (De Raeve 1989).

In contrast, in The Netherlands where coastal defence and the extraction of water were essential functions of the dune landscape, dune management was equivalent to dune conservation. The large-scale measures to stabilise blow-outs (which reached lengths of 50 m and more) came into effect from the middle of the nineteenth century and are in places still applied today. The effect of the stabilisation measures has been that the young phases of the vegetation succession have disappeared, the areas covered by mature stages have increased and there is consequently a loss of natural variation. This development results in a landscape with low ecological value.

This approach in which 'the total mastery of aeolian dynamics' is the main aim of dune management persists in some countries (Barrère 1992). But dunes under natural conditions are the product of accumulation, not erosion, and they disappear only on a receding coastline. In this case the sea is to blame, not the wind.

In The Netherlands, the character of coastal dune conservation has changed considerably under the influence of a growing appreciation of natural processes. Legal obstacles against tolerating wind activity are being removed (Anon. 1994). The policy of non-intervention was first introduced in the inner dunes after it was shown by many years of monitoring that blow-outs are stabilised by a number of natural processes (Van der Meulen and Jungerius 1989). The change in conservation attitude has been drastic. The very acts which were condemned in the past — rabbit burrowing, heavy grazing pressure by cattle and horses and recreational use — are now stimulated to destabilise the vegetated surface of the dune. However, it takes many years before the dunes are restored to their natural state. Experiments with artificial reactivation of blow-outs have been carried out to accelerate the process (see below).

The historical development of the management of the **foredunes** followed a different path. Protecting the coastline against a landward shift, including the conservation of the foredunes, traditionally posed different management problems from conserving the landscape of the inner dunes. The processes that create problems are mainly marine, and the conservation measures to control unwanted developments are usually carried out by engineers. They include the construction of groynes, beach nourishment and sand trapping to increase the volume of the foredunes. Traditionally, the approach adopted by engineers does little to enhance the ecological quality of the coast, but subsequent to the change in attitude towards a more natural approach to the conservation of the inner dunes there has recently been a marked switch in the appreciation of ecological values by coastal management organisations. A policy analysis study for coastal defence management in The Netherlands carried out by Rijkswaterstaat (1990) resulted in a more nature-oriented policy for coastal defence management. Part of this was the decision by the government to maintain the 1990 coastline by 'dynamic preservation' (Hillen and Roelse

1995). This was achieved with regular beach nourishment which is the most natural conservation method (Carter 1988).

The effect of the shift in attitude on the stabilisation of sand was comparable with the development in the inner dunes: suspension of the erection of fences or planting marram grass, allowing blow-outs to develop in the foredunes, and recently even excavating trenches through the foredune ridge to stimulate the formation of parabolic dunes and 'slufters' which are low-lying area through which the sea has access to the area behind the foredunes (see below).

The **offshore** zone is the realm of marine processes such as longshore currents, waves and tides. All these processes result in the movement of sand masses. In their interaction they produce a specific foreshore profile perpendicular to the coast and a specific configuration of the coastline. The control of these processes needs insight into the hydrodynamics involved, and engineering methods to remedy any unwanted developments. There have as yet been few efforts to manipulate the processes of the foreshore along dune coasts in order to conserve the coastline, apart from sand nourishment, not on the beach, but on the foreshore (De Ruig 1997). However, it is considered feasible to make use of geological and geomorphological principles to expand the Dutch coastline (see below). This is an important break from the traditional method of making new land by dumping sand and rubble on the sea bottom.

STEERING THE PROCESSES

The landscape is the result of the interaction of a number of processes. The dune manager has to possess a certain knowledge of the way these processes operate, because his decisions usually affect the characteristics of the dune landscape which means he interferes with the processes involved. He can steer the geomorphological and biological processes in many ways. Basically, it comes down to increasing or decreasing the effectiveness of these processes. But there are many ways to achieve this goal. The interference with processes will result in a shift along the stabilisation–destabilisation axis. The direction of development for a number of land use types is indicated in Figure 9.2. For example, coastal defence benefits from a stable system: geomorphological processes are suppressed, whereas recreation tends to destabilise the system because of physical destruction of plants. Even measures taken for nature conservation interfere with natural development.

A number of well-known measures are:

- Groynes, breakwaters, dykes, dune-toe protection and banquets to counteract wave erosion
- Beach and dune nourishments to affect sediment transfer
- Planting, fences, creating open spaces to affect aeolian processes
- Grazing, sod removal, burning to influence biological processes.

Measures have to be taken with caution, because there are often side effects. For example, stabilisation of the surface results not only in a decrease of geo-morphological activity but also in a change in species through vegetation succession. Beach nourishment reduces the erosional characteristics of an area. Placement of fences or planting of marram grass in the foredunes reduces the potential inland transport of sand, which affects dune morphology and vegetation. Hard structures make a system static and consequently enforce repeated interference and often increase coastal erosion elsewhere along the coast.

THE NEED FOR SPACE

Conservation strategies are strongly dependent on the size of the management unit. Figure 9.5 is a translation of Figure 9.1 in terms of management. Figure

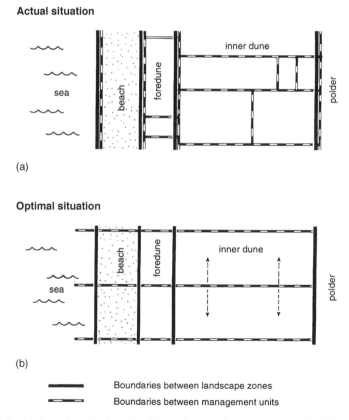

Figure 9.5. (a) Actual and (b) optimal boundaries of management units (from Van der Meulen and Van der Maarel 1989).

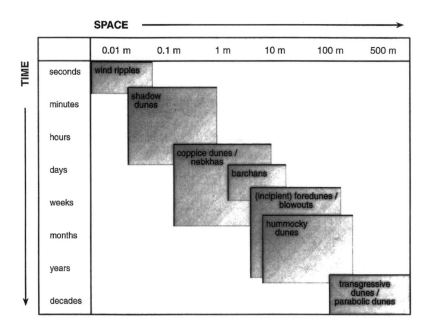

Figure 9.6. Requirements in space and time for the formation of aeolian features.

9.5(a) describes the situation which has often developed historically and which still prevails in many dune terrains: a pattern of small management units each with its own, short-term aims and package of measures. The situation of Figure 9.5(b) is preferable from the viewpoint of management: it has large multi-functional landscape units for which an integrated conservation plan can be devised. The larger the area, the more room there is for undisturbed natural processes (Wanders 1989). Figure 9.6 shows the space required for the formation or restoration of specific aeolian features. Large units also allow spatial planning such as the installation of zones with different functions.

Management of large units can be achieved even if the unit is divided among several owners. Houston (1989) gives an example for the Sefton Coast, a 25 km stretch of dunes in north-west England. The approach is based on the Coast Management Scheme established in 1978. The fundamental principle of the Sefton Coast Management Scheme is that the dune system must be managed as a single unit, rather than the fragmentation in the early twentieth century which had led to an overall loss of land quality. The Scheme has been able to restore a cooperative approach to the management of dune areas with such different functions as nature conservation, forestry and recreation. The appointment of a project manager to promote inter-functional cooperation is central to the concept of coordinated land management (Houston 1989).

PRACTICAL ASPECTS OF DUNE MANAGEMENT

Van der Meulen and Van der Maarel (1989) published an impact matrix for more than ten land-use types in coastal areas, arranged according to the ecosystem component they are affecting. The aim of the management is to sustain all the selected functions, be they nature conservation, coast protection, forestry, water extraction, recreation or any other activity. Therefore, the aims of management first must be made clear. Maps and other inventories of relevant terrain features are prepared in the field, from aerial photographs and satellite images. Research must be carried out to identify the requirements for the various sectors of dune users. Conflicting functions have to be brought in harmony. The desired steps and activities are finally listed in a management plan, again with mapped documentation. Various GIS techniques are available to assist in this kind of spatial planning. Many of these aspects of dune management are elaborated in Van der Meulen, Jungerius and Visser (1989) and Carter, Hesp and Nordstrom (1990).

Planning is followed by the execution of conservation activities, such as organisation of the work, planting marram grass, placing fences, shooting animals, guiding visitors, carrying out research, preparing reports and other documentation, to name a few identified by Wanders (1989). Practical handbooks are available in some countries. The conservation handbook written by Brooks and Agate (1986) is a good example. An environment checklist such as proposed by Davies, Williams and Curr (1995) can also be useful.

PREREQUISITES FOR DUNE CONSERVATION

Wanders (1989) lists a number of prerequisites for modern dune landscape conservation. The management organisation needs expertise that covers the entire range of functions. The financial means should be sufficient not only for the actual work in the dune area itself but also for research facilities and management planning. Biological research cannot stand alone. Research has to be extended to the physical environment, because insight into the physical processes is necessary to appreciate the dune systems' essential characteristics. Therefore, hydrology, geomorphology, soil science and related sciences of the physical environment should be integrated in management research. Politicians and all those who are responsible for utility functions must realise that dune management not only takes care of animals and plant species but is involved in all aspects of a dune landscape. Ecologists and geomorphologists see dunes as a dynamic system, while policy makers, planners and the general public see dunes as stable or static features. This creates the dilemma between maintaining processes and maintaining forms (Carter 1988).

Other requisites can be added, such as the availability of historical documentation about the area, and the provision of facilities for (public) education and information.

THE SUBSYSTEMS

The processes, types of land-use and management issues are different in the three subsystems. We will begin with the transitional and resting zones because they are traditionally the main concern of dune conservation. However, the technical possibilities of offshore management are rapidly increasing and should therefore be included in a chapter on dune management.

THE SUBSYSTEM OF BEACH AND FOREDUNES

Characterisation

General Characterisation

The transitional zone is characterised by a dynamic, geomorphologically active environment. Geomorphological processes operate at a short timescale, causing continuous change. The landscape comprises two linked geomorphological units, the beach and the active dune. In most systems, the active dune consists of distinctive dune ridges, called foredunes, backed up by less active dune forms. In this text, the foredune is defined as the first dune ridge, irrespective of the presence of specific plant species. The sharing of sediment, the active link in the sediment budget is an essential element of this zone. Its character is determined by the transgressive or regressive characteristics of the coastal zone and the magnitude of the processes in the beach and nearshore zone (Psuty 1989). In the development of foredunes, vegetation plays an important role.

Their constantly changing appearance makes foredunes different from inner dunes. Foredunes reflect present-day geomorphological processes, whereas the inner dunes (including fossilised foredunes) usually reflect geomorphological processes of the past. Geomorphological processes create a range of different depositional and erosional forms. The dunes themselves are depositional: smaller-scale embryonic and barchanoid dunes and sand sheets. Erosional forms are scarps, blow-outs, wind gullies and deflation surfaces (Carter, Hesp and Nordstrom 1990).

The impact of human action on foredunes is considerable along many coasts in temperate regions. Often, foredunes are stabilised, adapted in form or even artificially created (Nordstrom, McCluskey and Rosen 1986). This causes an important distinction between natural systems (undeveloped) and developed systems. Because of interference, foredunes are often considered as unnatural landscapes, resulting in limited scientific interest (Nordstrom, Psuty and Carter 1990).

Features

The predominant temperate coastal dune types are foredunes, blow-outs, parabolic dunes and transgressive dune sheets (including transverse, barchan and oblique dunes) (Hesp 1988). In most sandy coastal systems, some kind of foredune is present, but its extent depends on the specific local physical characteristics. Primarily, the action of the sea at the border of a large sand body is responsible for the creation of a linear feature. Second, foredune development depends on sand supply, degree and type of plant cover, rate of aeolian sand accretion or erosion, initial dune morphology, and magnitude and frequency of wave and wind forces (Hesp 1988). In general, extensive foredunes develop when there is a balance between the trapping ability of plants, the availability of sand and wind energy. Sand supply must not exceed the trapping capacity of the vegetation. Marram grass can outgrow deposition of up to $1\,\text{m yr}^{-1}$. If the vegetation cannot stabilise the sand, a precipitation ridge will develop. All sand is then trapped in a slipface which will move inland. The higher the dune, the slower the movement. There is no landward transport of individual grains, except in very high winds and through gaps.

In extreme cases, transgressive dunes may develop and the main form of dune development is inland and not by seaward extension (Hesp and Thom 1990). Other exceptions are 'climbing' dunes where sand may be blown some distance upslope behind a sandy beach, forming a veneer of sand on rock (Doody 1989). In the 'machair' of Scotland and Ireland, sand is trapped by short grasses (Angus and Elliott 1992).

The Processes

A thorough understanding of the landscape forms the basis for good management here as elsewhere. This is not always easy, because sediment exchanges in the beach–dune environment are governed by complex feedback mechanisms that may have important repercussions for the evolution of integrated beach-dune systems (Chapman 1989 in Sherman and Bauer 1993). For example, Arens and Wiersma (1994) in their classification of the Dutch foredunes make a major distinction between progressive, stable and regressive foredunes, which appear to have completely different characteristics with respect to management, features of aeolian activity and therefore conservation value.

The processes can be grouped under the headings shoreline dynamics, wind (energy) and vegetation.

Shoreline Dynamics

A number of marine processes contribute to the formation of beaches and foredunes (Figure 9.7). Longshore drift brings the products of coastal erosion

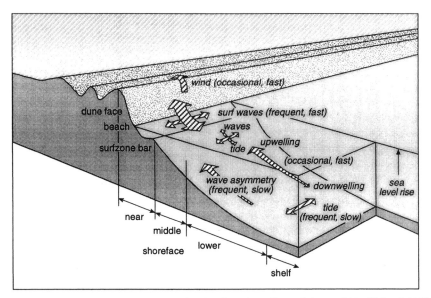

Figure 9.7. Hydrodynamic processes in the shoreface (from Stive and De Vriend 1995; with permission from Elsevier Science).

and material supplied by rivers to the offshore zone. From here the sediment is transported towards the beach by tidal currents and waves. On the beach it is taken up by the wind and deposited in a vegetated environment. Short and Hesp (1982) give a classification of dune coasts according to tidal regime, with transgressive dunes on high-energy coasts and foredunes, intermediate blowouts and parabolic dunes on low energy coasts. Constant, temporally asymmetric sediment exchange between beach and dune is an important natural process for maintaining both morphological stability and ecological diversity (Carter 1988).

Winds

Sand-transporting winds are necessary for dune formation, but onshore winds do not need to be dominant: the prevailing winds on the east coast of England are offshore, but there are dunes in many places. The direction of the most effective winds is often visible in the dune landscape. The more wind, the more sand can be transported, but the trapping capacity of plants may be a limit to dune building. The relationship between wind force and height of the foredunes is less clear: according to Depuydt (1972) higher foredunes are found where the beach and offshore profile are steep, whereas it is also known that high foredunes are associated with regressive coasts because of the large amounts of sand that are made available by the erosion processes.

Vegetation

Dune growth in temperate regions is caused by sand trapping in vegetation (so-called organogene Dünenbildung, Van Dieren 1934). If there is no vegetation, there will be no foredunes, but transgressive dune sheets as in South Africa (Short 1987). The species of trapping plants differs for different climatological conditions. A few large graminoid geophytes are the most effective (Westhoff 1989). A geophyte has (dormant) buds, corms or rhizomes buried well below the soil surface. In Europe, these species are: first the pioneer halophyte *Elymus farctus* or sand couch-grass, which is able to germinate in accumulations of organic tidal litter. The very first pioneers are some specialised halonitrophilous annuals, mainly sea rocket (*Cakile maritima*), but the dunes they form rarely survive winter storms. *Elymus farctus* is not an optimal sand binder; it is succeeded by the major sand accumulators *Ammophila arenaria* or marram grass and *Leymus arenarius* or lyme grass. The sterile hybrid between marram grass and wood small reed, named *Calammophila baltica* or *Ammocalamagrostis baltica* which can only propagate vegetatively but is often planted for dune stabilisation, is even more effective. Towards southern Europe *Elymus farctus* is replaced by *Leymus arenarius* (Westhoff 1989). In boreal climates *Elymus farctus* is replaced by *Euphorbia paralias*, sea spurge.

Continuous supply of fresh sand is needed for marram grass to evade the nematodes and fungi which affect its root system (Van der Putten 1989; Van der Putten, Van Dijk and Peters 1993). Inland, where salt spray diminishes, it is succeeded by other species including shrubs and trees (the latter mainly in calcareous dunes).

Human Interference

Three types of human interference can be distinguished. The first is intervention with hydrodynamic processes. This is an indirect influence on the dune system, since only the boundary conditions for dune development are affected. Examples of such activities include the construction of groynes and breakwaters, and shoreface nourishment. The second type is the direct intervention with dune-forming processes, thereby in some way stimulating dune development. Examples are the construction of sand fences to force and enhance sand deposition within a defined zone, or the planting of vegetation for the same reason. The third type of intervention is the adaptation of the dune form itself, for example the building of dunes to meet safety requirements. Examples are dune nourishment, or the creation of dunes by bulldozing beach sand. Other examples are the reshaping of foredunes by bulldozers to eliminate blow-outs, in order to prevent loss of sand by wind erosion.

Functions

Foredunes (the transition zone) are important for three completely different functions. Depending on the specific setting, there is an order in the importance of these functions.

Sea Defence

The main function of foredunes in many lowland coastal areas is sea defence. Dunes in general act as a buffer to extreme waves and wind because they are able to absorb wave attack (Carter 1988). In parts of Europe dunes form the only barrier to the sea. Foredunes here provide protection against flooding. In addition, foredunes may also safeguard specific coastal features against the dynamics of the transition zone, mainly against burial by sand. Examples are the protection of planted pine forests in Les Landes (Barrère, 1992; Favennec 1995), wet dune slacks in The Netherlands and tourist infrastructure, villages and settlements in many places.

Nature

Because of their characteristics, the landscape of foredunes form an ecological niche in which the plants are adapted to extreme conditions. These conditions involve (heavy) salt spray, intense activity of the wind and blown sand, an almost complete lack of nutrients and drought. Only a few plants are adapted to this environment. The gradients in the system and the intensity of the geomorphological processes make foredunes very important for nature. Tall pioneer stands with grasses having large rhyzomes (like marram grass, *Ammophila arenaria*) are common in exposed sites. In a natural system, the pioneer species disappear when sand accumulation decreases. More sheltered sites have low stands with more species, including mosses like *Tortula ruralis* and small herbs. In southern Europe, *Euphorbia panalis* and *Erynchyum maritimum* are characteristic. In the north, the geophytes of strictly coastal distribution are replaced by a group of tiny winter therophytes. The risk of severe drought in summer is a determining ecological factor, and further succession hardly goes beyond low shrubs of buckthorn (*Hippophae rhamnoides*). Conservation of this system is most effective when there is room enough or when the dunes are not developed for other purposes.

Recreation

The third important function is recreation. This function is mainly related to the beach, but the supporting infrastructure is mostly concentrated in the foredunes for safety. The foredunes are also much favoured for buildings, boulevards, hotels and houses because of the view they provide over the beach and the sea.

Implications for Management

In a completely natural landscape, species diversity and physical conditions are in balance. Physical processes do not threaten the natural function, they are part of it. This is often not understood by managers. Foredunes, especially along a prograding coast, require little or no management. In a natural system we can expect that sand taken from one area of the dune will be deposited in another, even in the case of increased erosion due to sea-level rise. The loss of dune areas in one place is compensated by a gain in another. However, since man likes to keep what he has got, he often upsets the balance between erosion and accretion of dune areas.

With respect to **sea defence**, the major threat is a loss of sedimentary volume due to dune erosion. Sand at the dune toe is removed by wave erosion. As a result the dune front collapses, a process known as scarping. Often a very steep, bare slope remains, vulnerable to wind erosion. A loss of volume in this way may result in the disappearance of the foredune and therefore in the degradation of the sea defence. Managers respond by trying to stabilise the foredune, to keep its volume intact. If the coastal sediment budget is slightly negative, stabilisation of foredunes may result in a loss of volume in the beach zone. This causes a gradual steepening of the coastal profile which may end up in a catastrophic loss of foredunes. It has also been shown that stabilisation by planting marram grass favours soil development which in turn stimulates soil pathogens. These, which include nematodes and fungi, affect the root system and may eventually destroy a naturally vital stand of marram grass (Van der Putten 1989).

The main threats for **nature** today are the loss of a very specific geo-ecological environment by erosion. This applies especially in the many situations where the landward boundary of the dune system is fixed. Other threats include pollution from beach littering and refuse from ships, deposited in the dunes during storm tides. The problem of littering by plastics is described in a number of publications (see Williams and Simmons 1996). Beach nourishment may have important effects on the landscape. For example, sand blown from a nourished beach often differs from the local sand in size, nutrient content or colour. Little is known about the effects on vegetation of this process.

Most of the physical processes impose a threat to the **recreational** function of foredunes. Facilities can be lost either by burial or by erosion. Blowing sand is a nuisance to visitors. Stabilisation of the foredune is not always the answer; it may result in a decreased beach width, which means a loss of recreational area.

The main conservation issue is how to reconcile the conflicting interests of the three functions. Excessive recreational use of beaches and foredunes may disturb natural processes, as may fixation of the beach and dune system for maximum security. Sea defence and recreation require stability, but these conflict with the interests of nature conservation (Van der Meulen and Van der Maarel 1989). Natural processes involve instability which conflicts with sea defence and recreation functions, but new insights are being developed (see next section).

Often, there is no conflict: sustainable development, natural beaches and foredunes are as attractive for recreation as for nature conservation. Nature education and guided tourism in nature areas nurture care for vulnerable areas. Beach nourishment as a new means for sea defence offers possibilities for nature by suspending the need for excessive stabilisation measures and for recreation (by providing wider beaches).

Case Studies: Dynamic Preservation of the Coastline

Scientific Research for Management

One of the fears of dune managers is a landward loss of sand by the wind. Research has found that a closed foredune acts as a very effective sand trap. Arens (1996) showed that the decrease in landward transport from beach to foredune is dramatic. Transport on the foredune, during a moderate storm event (average wind speed on the beach $15\,ms^{-1}$ at $5\,m$ height), is less than 0.1% of the transport on the beach. During lower wind speeds, transport into the foredunes declines to zero. Figure 9.8 shows two examples for some Dutch foredunes, one near Groote Keeten, in the northern part of North Holland, the second on the Wadden island of Schiermonnikoog. The figure shows the decrease in landward transport during onshore winds. Measurements indicated that due to the presence of vegetation, transport capacity decreased to zero at a small distance from the vegetation border. Both vegetation and lee-side effects prevent the sand from leaving the foredune system. The research indicated that the fixation of sand at the seaward side of foredunes is not necessary to prevent substantial landward losses.

Figure 9.9 shows the mechanisms of sand transport over vegetated and bare foredunes. In the case of a vegetated seaward slope, landward transport is negligible, because all sand is trapped in the vegetation. If the seaward slope is bare, some landward transport may occur. Depending on the steepness of the slope, sand will be transported in either saltation or suspension. In the last case, the sand will be transported further landward, but still, most of the sand will be arrested within about 100 m from the foredune.

Decreasing Management Efforts for Nature Development/restoration

A large part of the Dutch mainland coast consists of rigorously managed foredunes, resembling static sand dykes rather than dynamic geomorphological

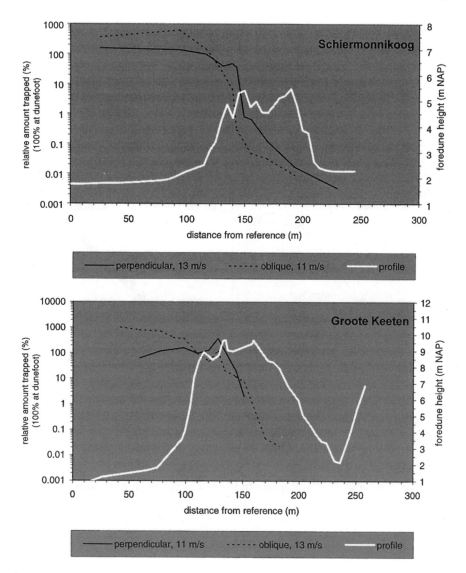

Figure 9.8. Sand transport over foredunes (after Arens 1996).

systems. Traditional management has involved immediate recovery of storm damage by the smoothing of cliffs and the planting of bare spots with marram grass, thereby reducing potential wind erosion. In 1990, the authority in charge of coastal safety decided to suspend the current management along a 1 km wide section of the shoreline, in line with a strategy of 'Dynamic Preservation'

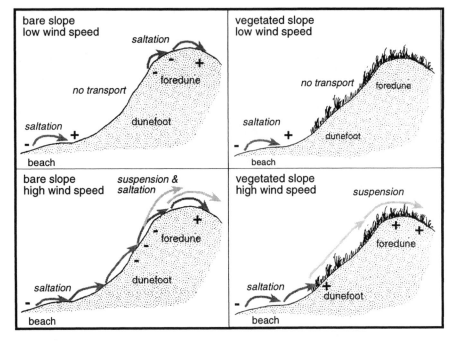

Figure 9.9. Conceptual scheme of landward transport over foredunes (from Arens 1996; with permission from Elsevier Science).

adopted by the Dutch government. The loss of sediment from the beach-foredune system by either wind or wave erosion was prevented by periodic beach nourishment. In 1999, the foredune exhibited a steep cliff-like slope which was mostly being eroded by the wind. Small blow-outs had been formed. The foredune slope was slowly adapting to the new situation, but it was not clear yet what the final form would be. The presence of a steep and bare cliff induces suspension transport (see Figure 9.9), causing some of the sand to be blown over the foredune, to be deposited at a few tens of metres in the lee. This gave some problems for the maintenance of a road behind the foredune, because of sand deposition on the road accompanied by an increase in verge height due to the increased growth of grass. The shape of the foredune contrasted with the static sand dyke feature it used to be, which is illustrated by Figure 9.10.

Active Interference for Nature Development including Habitat Restoration

Near Schoorl, in The Netherlands, the foredunes are backed by a dune area several kilometres wide. In the past, these foredunes were maintained as that part of the dune zone which ensured safety and on which safety regulations

Figure 9.10. The foredunes near Parnassia (mainland coast, The Netherlands).

were applied. The foredunes were managed by the Hoogheemraadschap Uitwaterende Sluizen, whereas the inner dune area is managed by the State Forestry Department. Because of grass encroachment and centuries of stabilising activities (Kooijman and Van der Meulen 1996; Veer 1998) dynamic geomorphological processes in the dune area have declined. Currently, only some small spots are active, mainly due to the digging activities of rabbits: most geomorphological activity is limited to the foredune zone. In the past, occasional major storms created gaps in the foredunes, which were closed immediately by the foredune managers. In 1995, a project proposal was presented to restore the natural dynamics (Stichting Duinbehoud 1995). The project involved cooperation between all the authorities in the region with responsibilities in coastal management. The main idea was artificially to create a gap in the foredunes, through which the sea would occasionally inundate the swale behind (Figure 9.11). To accomplish this, the idea of the foredune as a primary sea defence had to be abandoned. Because of the width of the dune zone behind, a new sea-defence zone could be defined, landward of the swale. In the swale, vegetation and soil were removed, to initiate aeolian activity. These activities should restore several gradients, such as fresh–salt, dry–wet, carbonate-rich–carbonate poor.

In 1997 the plan was executed. The sea has now entered and inundated the dune swale several times. Small parts at the border of the swale have been buried

Figure 9.11. Artificially created gap in the foredunes between Bergen and Schoorl.

by aeolian deposits. The project has drawn much attention in the media, and this has attracted large quantities of visitors, who further interfere with the intended 'natural' development. The initiators of the project have already expressed doubts that the gap will not be sealed by the sea and the wind within a few years.

One of the curious aspects of this kind of project is the eagerness for immediate results. One could ask whether, in a case like this, it would not have been preferable to let nature do the work in its own good time, and learn from the results. The answer to this question is not clear. After some years of non-management, natural blow-outs might have developed in the foredunes, which might gradually have evolved into natural gaps. However, the outcome of the strategy is very uncertain: there is no guarantee of success, and the desired result might take years to develop. Besides, it is very unlikely that aeolian activity in the completely vegetated dune swale could ever be created without human intervention.

THE SUBSYSTEM OF THE INNER DUNES

Characterisation

In the model presented in Figure 9.1, the subsystem of the inner dunes is shown as 'the resting zone'. Inner dunes are found in many countries between the active foredunes and the hinterland. They have also been called 'secondary dunes' (Klijn 1990). Ecologically, this zone covers the *Graudünen* (grey dunes) and the *Braundünen* (brown dunes) in Figure 9.3. There is a variety in dune forms including parabolic dunes, comb dunes, precipitation ridges, blow-outs and secondary dune slacks. Most of these dunes are remnants of the past (Klijn 1990). Secondary dune formation is triggered by the destruction of plant cover and subsequent wind erosion. Here, active processes create forms super-imposed on a fossil aeolian landscape.

The Processes

A number of processes are relevant in the conservation of the inner dunes such as erosion by wind and water, soil formation, biomass production, salt spray and the fluctuations of the groundwater level. These are grouped as geo-morphological and biological processes in Figure 9.2. In this figure, the inner dunes are represented by compartments II and III. The geomorphological and pedological processes are touched upon below.

Wind

The wind as a geomorphologic agent is much less important in the inner dunes than in the foredunes: it has been calculated that the wind on the beach can, at force 7, move more sand in two minutes than that produced by a sizeable blow-out in the inner dunes in one whole year, in terms of flux across a line of equal width (Jungerius 1989). The aeolian processes in the inner dunes are described by Carter, Hesp and Nordstrom (1990) and Hesp and Thom (1990). The effect of wind erosion depends on the relationship between the erosivity of the wind and the erodibility of the site. The erosivity of the wind is largest in exposed areas such as dune ridges and summits. The erodibility of the site is determined by vegetation and soil properties. Higher plants break the force of the wind, but a continuous cover of vegetation as low as moss and algae is also sufficient to protect the soil against erosion. Whether a surface without vegetation cover is affected by wind depends on the characteristics of the sand. The presence of loam, coarse material and even a thin coating of organic matter on sand grains will effectively reduce erodibility. Prevalent among the aeolian forms in the inner dunes is the blow-out (Carter, Hesp and Nordstrom 1990), a shallow depression of mostly elliptic outline which is free of vegetation.

Water

The traces of water erosion in the inner dunes are much less conspicuous than those of the wind and have generally not been subject to conservation measures. Yet much material is shifted downslope during rain, especially in summer when the sand of the A horizon is very dry and water-repellent (Rutin 1983). The main types of erosion by water are splash and surface wash. Their effect is the gradual flattening of relief. Colluvium is deposited at the base of the slope, with the seeds entrained from the surface soil in the upper slope positions. Water erosion often paves the way for wind erosion, by removing the humic surface soil on upper slopes.

Soil Formation

The development of soils on sand dunes has been described by Wilson (1990, 1992) and Jungerius (1989, 1990). Although soils are not generally studied in a dune terrain, their colour is often used to classify the dunes (Ellenberg 1978). Soil formation as part of the landscape is shown in Figure 9.2. In compartment I, where geomorphological processes prevail, there is no soil formation. These are the 'raw sands' of Wilson (1992) and consist of little altered mineral material.

Compartment II is characterised by great variability in the balance between geomorphological and biological processes. As a result there is also great variability in the development of the soil profile. On stable sites the formation of a B horizon is no exception, but shallow and truncated soils prevail where erosion is active, especially on the upper parts of steep slopes. Profiles lower downslope reflect the balance between colluvation and soil formation. Where soil formation can keep pace with the deposition of colluvium, abnormally thick A horizons are formed. Sites with a sequence of buried A horizons are an indication to the manager that periods of stability and plant growth have alternated with periods of instability and deposition of colluvium or aeolian sand. Wilson (1992) classifies the A–C profiles in this compartment as sand-pararendzinas and the A–B–C profiles which occur where acidification and decalcification can proceed for some time, as brown calcareous sands. In compartment III where soil development continues unhindered under a closed vegetation cover, podzols are eventually formed (Wilson, 1992). In The Netherlands this stage has nowhere been reached although there are podzolic trends in the northern districts where the calcium carbonate content of the dunes sand is low.

Functions

Nature Conservation

The outstanding natural and ecological value of the dune landscape has been praised in so many publications that there is no need to elaborate on this theme here. The European Union of Coastal Conservation was expressly founded in the 1980s for the purpose of safeguarding the 'golden fringe of Europe'. The proceedings of the congresses they organise contain contributions from all European countries where nature conservation of the inner dunes is an issue. Holistic nature conservation embraces all components of the landscape: climate, geology, landforms, soils, water, vegetation and fauna.

Recreation

The inner dunes combine outstanding scenic variation with a wealth of animal and plant life and an equally valuable abundance of geomorphologically interesting landforms. This makes them very attractive for human activities which at the same time constitute the main threat for conservation. Well-known sources of damage to vegetation and wildlife are trampling, the construction of roads and car parks, vandalism, camping, and the increased risk of fire (Brooks and Agate 1986). Trampling causes shifts in species composition and decreases in vegetation cover (see Williams and Randerson 1989).

Many dune managers are confronted with increasing use of the dunes for recreation. If the proper measures are taken, there does not need to be a problem. Management aspects are discussed by Carter (1988: Chapter 13).

Water Extraction

The groundwater stored in the dunes is an attractive source of drinking water and its extraction has become an important use of dunes in several countries. In The Netherlands, groundwater is the main reason that the dune landscape is protected from urban development and mass tourism. On the other hand, water extraction can cause ecological impoverishment and the construction of associated infrastructure can damage valuable dune forms. Lowering of the groundwater table is often followed by the disappearance of valuable dune slack ecotopes. In recent times these dune areas have been fed artificially with surface water from elsewhere.

Forestry

At present, forests cover extensive parts of the coasts of several European countries such as Denmark, Poland, the United Kingdom and southern France. Many of these forests are plantations, often of exotic pine species. The pines lower the groundwater table by transpiration. Moreover, the decaying needles have an acidifying effect on the soil. In countries where forestry is no longer a profitable enterprise, these plantations have been removed or converted into natural forests.

Implications for Management

The management issues depend on land use. Nature conservation management includes restoration, the maintenance of intrinsic landscape diversity, preserving the undisturbed relationship between biotic and abiotic processes, the regulation of groundwater level, etc. Obviously, for recreation the issues are quite different: the maintenance of carrying capacity, the development of an aesthetically pleasing landscape, and the creation of an adequate infrastructure with sufficient recreational amenities. For the extraction of drinking water it is also necessary to create a specific infrastructure and to control the hydrological properties of the terrain. Only the implications of management for nature conservation are elaborated in this section.

Management in this context is concerned with intervention in the landscape-forming processes. If these interventions cause unwanted disturbances for other functions, mitigating measures must be taken. Zoning the use of the dunes is one of the possibilities. But there is not always a need for conflict. As Westhoff (1989) puts it: 'Human impact on nature is both beneficial and deleterious. The main concern of environmental management is to promote the former and counteract the latter aspect'. There are a number of techniques to reach this goal.

Stabilisation–destabilisation

Stabilisation of moving sand is no longer seen as essential to the conservation of nature, unless the sand may bury rare plant species or valuable dune slacks. Where this applies, or infrastructure has to be protected, common stabilisation techniques can be applied (Brooks and Agate, 1986). For blow-outs, the common type of wind erosion in the inner dunes, this is generally not necessary. Often, a leeward dune develops which catches all the sand leaving the blow-out. Moreover, monitoring experiments have shown that there are a number of natural stabilisation processes and that most blow-outs have a limited lifespan if left alone (Van der Meulen and Jungerius 1989). Natural stabilisation mechanisms include the growth of algae which colonise deflational areas within the blow-outs, and the accumulation of sand within the blow-out when its ratios of length, width and depth are no longer aerodynamically appropriate, or the fact that the blow-out may become too deep for the sand to leave. If stabilisation measures have to be taken, it is recommended that they join forces with natural mechanisms. For example, to stabilise blow-outs it is best not to counteract the erosion of the inner parts, but to stimulate the growth of the leeward dune which will eventually prevent the sand from leaving the deflation area.

As we stated before, stabilisation is increasingly seen as a danger to nature because of its stifling effect on the landscape. People are not always to blame for the stabilisation. The decrease in the rabbit population in the 1950s also contributed to the increasing stabilisation of many European dunes. Human intervention may be needed to restore natural values. There are a number of measures to choose from: removal of vegetation, removal of sod or the introduction of grazers.

Grazing

Grass encroachment is one of the consequences of increased atmospheric deposition of acidifying and eutrophying components, although it can also be caused by the suppression of natural processes by continuous artificial stabilisation. Grass encroachment can be counteracted by grazing. The effect of grazing by rabbits was realised many years ago (Pickworth Farrow 1917). Where the number of rabbits was reduced, large grazers such as sheep, goats, cattle and horses were successfully introduced. Numerous studies have dealt with the effects of grazing on the vegetation of grassland communities (see Belsky 1992; Gibson and Brown 1992). Grazing results in a reduction of the standing crop and a more open vegetation structure, a change in humus form, and a shift in species composition in favour of smaller species, associated with conditions of more light near the soil surface. The specific effects of grazing are dependent on many factors including soil development, hydrology, management history, grazing density and type of grazer (Kooijman and Van der Meulen 1996). Until the beginning of the twentieth century, extensive grazing by domestic animals was common practice in dunes, so the acceptance of grazing as a management tool to restore biodiversity levels is in fact a reintroduction of this old practice (Kooijman and De Haan 1995).

Mowing (Anderson and Romeril 1992) and sod removal have also been applied to restore botanical diversity, but these methods may have deleterious effects on other components of the ecosystem: for example, both methods drastically reduced the number and depth of occurrence of soil fauna in the Dutch dunes (Jungerius et al. 1995).

Groundwater Control

Extraction of drinking water causes the lowering of the groundwater table and concomitant desiccation of the soil. Infiltration of water from outside usually remedies this problem but creates new ecological problems when this water is polluted or eutrophic. Van Dijk (1989) made a thorough analysis of the impact of drinking-water production on the dune landscape. Elaborate and costly measures are necessary to purify the water before it is suitable for infiltration. In The Netherlands, surface infiltration has gradually been substituted for artificial infiltration into deep aquifers.

Maintenance of wet dune slacks needs special care. For dune slacks to support a valuable vegetation it is necessary to keep the water table at a constant level. Restoration of dune slacks also depends on the quality and the flux of the groundwater in the surrounding area (Van der Meulen and Jungerius 1989).

Case Study: the Reactivation of Blow-outs as a Measure to Restore the Natural Dune Landscape

Active blow-outs are, at present, rare in the Dutch coastal dunes. Even in areas where stabilisation measures have been suspended and wind action is allowed, there has been a gradual decrease in the number of blow-outs. Reduced grazing pressure by rabbits due to repeated outbreaks of myxomatosis in the last few decades is part of the cause, but there are other factors such as the increased input of nutrients by atmospheric deposition. Eutrophication stimulates vegetation growth and is therefore thought to be responsible for the encroach-ment of certain algae (Pluis and Van Boxel 1993), grasses and shrubs. This may lead to the stabilisation of active blow-outs and prevent the formation of new ones.

Artificial reactivation of blow-outs can help to restore the natural dynamics of coastal dunes (Van Boxel et al. 1997). Around active blow-outs there is a range of deposition rates, depending on the size of the blow-out, from up to $50 \, cm \, yr^{-1}$ near the edge to a few $mm \, yr^{-1}$ at a distance of about 100 m from the blow-out. Calcareous and nutrient-poor sand deposited here can counteract the effects of acidification and eutrophication. Vegetation type and cover will react to the changes.

Within the framework of a programme sponsored by the Dutch government to test various measures for counteracting the deteriorating influence of air pollution (Van der Meulen et al. 1996), experiments were carried out to reactivate blow-outs. Two study areas were selected. The one discussed here is located in the inner dunes near Haarlem, at a distance of 1.5 km from the coast (Van Boxel et al. 1997). The relief of the study area is flat to undulating. The sand contains 7–9% calcium carbonate. The vegetation is open dune grassland, with increasing invasion of exotic moss (Campylopus introflexus) and sea buckthorn (Hippophae rhamnoides).

The blow-outs in this terrain were stabilised with branches in the 1970s. In order to reactivate them, the branches were removed, along with all the vegetation. However, the removal of vegetation was not sufficient to restore wind action. Organic matter in the sand acts as binder, so the whole of the A horizon had also to be removed. This meant an excavation to a depth of 20–30 cm even if the soil was still young, and more if the soil was old or if there were many roots of marram grass or other sand-binding plants in the soil. A mechanical shovel was used for this purpose, and special measures had to be taken to mitigate undue damage to the terrain. For the same reason, the sand could not be removed by trucks and had to be dumped close to the blow-out. Apart from the blow-outs, small bare patches were reactivated with spade and barrow. The effects of these measures were monitored with precision measuring techniques at yearly intervals, from 1991 to 1994.

The results (Van Boxel *et al.* 1997) showed that the small patches that were reactivated, stabilised spontaneously. This is not surprising, since most naturally formed blow-outs also stabilise in the first year after their formation and only a small percentage reach maturity (Jungerius and Van der Meulen 1989). The reactivation of the blow-outs was more successful. Their area and depth increased slowly. The deposition area is about four to six times the deflation area. Most of the sand was deposited within 30 m of the blow-out. This was as expected: it is rare to find sand more than 100 m away from even the largest blow-out in the inner dunes. *Campylopus introflexus* disappeared from the accumulation sites, but not the sea buckthorn. For this shrub to die off, the soil must be fully leached. This took at least 50–80 years on Texel, one of the Wadden islands, where the dune sand already has a low calcium carbonate content to begin with (Doing 1989).

Everyone would like deflation to continue until the groundwater level is reached, in which case a valuable dune slack vegetation might develop. However, experience has shown that the deflation stops when the capillary zone above the groundwater table is reached. This triggers plant growth in the blow-out. This not only stops further deflation but even promotes sand trapping. Where a dune slack is expected, a new dune develops! Even if a dune slack is created, geomorphic processes will effectively prevent plant growth if the banks are steeper than about three degrees.

THE OFFSHORE SUBSYSTEM

Characterisation

For our purpose, the subsystem of the offshore is defined as the shoreface. This is the realm of hydrodynamic processes. The shoreface (Figure 9.7) is that part of the sea-bottom lying between the dry–wet beach and the shelf. It has a slope of 1:100 to 1:1000. On the Dutch coast this more or less coincides with a narrow zone up to 20 m below average sea level.

The processes

The main processes of the shoreface are shown in Figure 9.7. The coast has been schematised and there are no inlets. This could well be the Dutch mainland coast. In cells of several kilometres in length, the offshore profile responds to a variety of hydrodynamic and sediment-transport processes. These differ in different compartments (near, middle and lower shoreface). As indicated by the thickness of the arrows, the main activity of the steering processes is working in a direction perpendicular to the coast. The surf zone is different from the rest in that processes are strongly depth-controlled; they are driven by an associated loss of momentum and dissipation of energy. Morphodynamic adaptation of the profile is assumed to be relatively fast in the upper part and slow in the lower. Observations suggest timescales of hours around the waterline to millennia near the inner shelf (Stive and De Vriend 1995). The external driving factors in this whole system are sea-level rise and lateral sediment supply.

Long-term analyses have revealed important processes in the large-scale sand balance along the Dutch coast (De Ruig 1997; Van Rijn 1995). They show that serious erosion occurs on the deeper (> 6–8 m below sea level) parts of the shoreface. Sand from the deeper zone is transported to the shallow zone and alongshore northward to the Wadden Sea. The natural supply from the bottom of the North Sea to the deeper part of the shoreface is limited. This means that the offshore slope is steepening.

Functions

The present functions of this subsystem are mainly fishery and recreation. Because the pressure on the coastal zone in Holland is increasing, there are now plans for land reclamation (cf. De Ruig and Hillen 1997; Waterman, Misdorp and Mol 1998; De Ruig 1998). The need for housing and recreation facilities in the dunes is also growing. The densely populated low-lying polder land of the western part of the country, immediately behind the dunes, has increasingly less space. For decades, building in the dunes was restricted because it was thought to endanger the coastal defence function of the dunes, at risk of flooding the polders. Since the 'dynamic preservation' policy has been adopted (see above), this is no longer a valid reason, especially not in the wider dune areas.

It is particularly in the Rijnmond region, the mouth of the Rhine River, that plans for development are being made. The province of South Holland is developing projects for reclamation along the coast between The Hague and Hook of Holland (roughly 4000 ha). These will accommodate urban, recreational and natural areas. The city of Rotterdam is studying the reclamation of land off the coast of Voorne. Here an extension of the Maasvlakte-1 is planned, to give the harbour of Rotterdam more port facilities. Recently, studies were published of plans to build an island in the sea in front of the Dutch mainland coast, which could be used for a new airport (Delft Hydraulics 1997a).

It is interesting to note that in all the plans there is also place for nature development. Studies are being carried out to define the making, management and evaluation of coastal ecosystems which could be located on the new land (LWI 1997). On the other hand, it may be more worth while to consider the development of nature not on new reclamation plans themselves but at other localities which offer better opportunities. In any case, compensation for losses of nature should be made and not in the form of small bits and pieces of land but in areas of substantial size for nature to develop.

Threats

While engineers and policy-planners often think positively about these plans, many other groups, among which are environmental groups and nature conservationists, have more doubts (for example, Janssen 1996).

Threats to valuable ecosystems include the isolation of the existing dune area from the sea resulting, for example, in the reduction of saltspray and the consequent maturing of the plant succession. This may lead to the disappearance of pioneer stages and open dunes, and the increase of shrubs and woodland (compare the development in Voorne since 1950s, well documented by Van Dorp, Boot and Van der Maarel 1985). The foredunes may not receive any more inputs of fresh sand, a precondition for their ecology. Another major influence will be the one on the ecosystem of the North Sea itself. First of all, all marine life will disappear when new land is formed. Second, the digging of sand will affect large areas of North Sea bottom. It is calculated that for 4000 hectares of reclaimed land, approximately 400 million m^3 of sand will be dredged from the sea. Over an area of 200 km^2, 2 m of sand will be taken off (Janssen 1996). New land in the sea will increase the need for infrastructure on land, increasing the pressure on coastal land and its nature. The valuable dune area will be 'squeezed' between the old and the new land.

How can the damage to nature be compensated for by creating new nature on new land, if at all? It is argued that alternatives, like rebuilding and renovating parts of the old suburbs, are a better answer to the need for housing, and that these alternatives should get more attention.

Implications for Management

Recently, some studies were carried out in the Netherlands about land reclamation in the North Sea. They are of interest to the topic of this chapter. They reveal the way of thinking in such large projects, where civil engineers and ecologists work together. How can new land be made? Which form and shape should it ideally have? What will the influence be on the existing coast? Which coastal ecosystems should be created? Is it possible to enhance the ecological quality of the coast as well as its resilience, and at the same time favour the coastal defence quality of the coast? Resilience is the (self-organising) property of the coast to maintain actual and potential functions under changing hydraulic and morphological conditions. This property is based on the morphological and ecological processes operating at the coast (Delft Hydraulics 1997b).

Questions like these illustrate how thinking is developing: (1) plans for reclamation should be seen in the context of all plans for the coast and in an integrated way, i.e. long-term effects on the total coastal system should be considered along with a discussion on the actual needs of specific developments; (2) land could be returned to the sea in order to increase the resilience of the coastal system, thereby decreasing the risk of flooding (De Ruig 1997, 1998).

Case Studies: the Making of New Land

An Island?

Delft Hydraulics has made a study of the feasibility of building an artificial island off the Dutch coast (Delft Hydraulics 1997a). It is meant as a technical contribution concerning the possible location, and lay-out, consequences for morphology and ecology and best options for mitigating negative effects (Figure 9.12). The effect of the island on tides and currents, sand and silt transport, water quality and marine ecosystems were examined.

The conclusion was that a large-scale project such as the construction of an offshore island requires careful consideration not only of its direct effects but also of the implications for the further evolution of the Dutch coast. It was concluded on the basis of the results of the investigation that the construction of an offshore island is technically possible and — if optimally shaped and carefully placed — would have only limited effects on the sea and coastal environment. However, if the island were built, the already limited resilience of the Dutch coast would diminish even further, because more coastal maintenance would be necessary. But if the island were to be combined with other measures, like seaward-protruding sand hooks, the resilience of the coast might improve. Table 9.1 gives an overview of the consequences of construction and presence of an offshore island.

Ecosystems on New Land

The extension of the Rotterdam harbour facilities into the North Sea will include 850 ha earmarked for nature. How can a new dune area be made out of the sea? Basically in three ways, wherein man's interference is decreasing but time is needed for realisation (Löffler and Veer 1999):

- Anthropogenic: sand is deposited in the area where dunes are planned: either 'dunes' are shaped with a shovel or other earth-moving equipment, or we

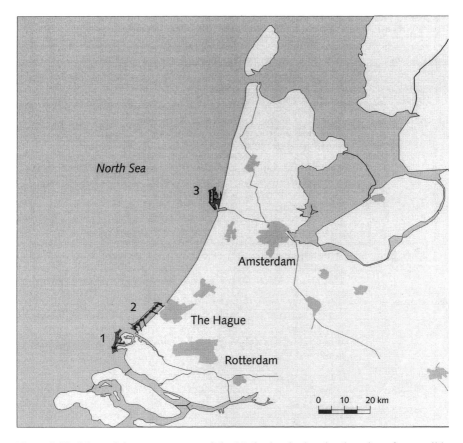

Figure 9.12. Map of the western part of the Netherlands showing locations for possible reclamation of new land. 1 = Maasvlakte II (extension of Maasvlakte I): harbour, industry, nature; 2 = New Holland: housing, recreation, nature; 3 = second international airport (also designed as island at various locations in front of the coast).

wait for dunes to be created by the wind. A dune area can be created within a few years, but the system lacks major natural characteristics.

- Geomorphic: sand is deposited to create a beach and provision is made (e.g. enough fetch) so that foredunes can form themselves by free wind activity. Within 10–50 years a natural dune landscape might be formed which basically possesses the same natural characteristics as a natural landscape.
- Geological: sand is deposited on the foreshore and left there; natural processes of tides, currents and winds are allowed to shape the landscape. A fully natural dune landscape might be formed after several decades, but there is no guarantee.

Table 9.1. Overview of the consequences of construction and presence of an offshore island in front of the Dutch mainland coast (source: Delft Hydraulics 1997a)

Consequences for	Nature and gravity of consequences
Tides and current	Could change in an area of 15 to 40 km around the island No significant influence on large-scale water movements
Morphology	Accretion at the northern side of the island Variation in accretion (*circa* 10 m yr^{-1}) and erosion (*circa* 6 m yr^{-1}) in the lee of the island along the coast
Mud budget	Temporal rise in suspended matter content of sea water during construction of the island (in winter 10–40%, in summer 30–60%; because of large natural variation only observable in summer) Probably no significant effect of the island on mud supply to the Waddensea
Water quality	Around the island up to maximum 5% increase in concentration of PAKs Hardly any effect on the large-scale spread of particles in the sea
Algae growth	Temporary, small (less than 10%) decrease of algae growth during the construction of the island, but few consequences for higher organisms Small increase of algae growth north-west of Texel (within the natural variation)
Marine ecosystems	Temporary loss of up to 2800 tonnes of biomass (ash-free dry weight) due to sand extraction (recovery of population within a few years) Permanent loss of 400 tonnes of biomass following the loss of submarine area Increase of biomass on the hard substrate of the sea defences of 500–800 tonnes and increase of bio-diversity
Birds	Small decrease in feeding area at sea due to the presence of the island New resting and feeding areas on the island New resting area for birds migrating over the North Sea
Coastal dunes	Hardly any effect on wind climate and salt spray Fragmentation of dune area and possible influence on geohydrology due to the construction of a connection with Schiphol Airport
User's functions marine and coastal zone	Hardly any effect on shipping Some hindrance for recreational shipping Damage to fishery due to a loss of fishing grounds and extra damage when fishery is restricted in an area around the island Effect on the openness of the seascape Some noise nuisance for inhabitants and visitors in the coastal zone

Which ecosystems should be developed on newly reclaimed land? The coastal ecosystems that score highly for nature are intertidal areas, dunes and so-called 'slufters'. Slufters are relatively small saltmarshes within areas of dry dune (see above). They drain a primary dune slack through the foredune ridge, or they are formed by a breakthrough of the sea or a parabolic dune through the first foredune ridge. When there is a low-lying area behind the foredune ridge, sea-water invades the dunes. The result is a valuable ecosystem with complex abiotic gradients. Slufters and parabolic dunes for a long time have not been the favourites of Dutch coastal engineers. But the new policy of dynamic preservation gives rise to new possibilities (cf. Hillen and Roelse 1995).

The ecosystems we have mentioned are believed to add to the resilience of the entire coastal system and to its ecological values. The making of such systems in front of the old land is of particular interest at those places where the dune ridge is narrow which means that the defence structure is also thin. In the philosophy of resilience, present agricultural land could even be returned to the sea in order to increase the resilience of the coastal system (see French and Reed, this volume).

DUNE CONSERVATION AND INTEGRATED COASTAL ZONE MANAGEMENT

The coast comprises more than the dune system: so does integrated coastal zone management (ICZM). The coastal zone is an interacting system of physical processes, chemical reactions, biological growth and economic activities. Examples of economic activities are fishery, shipping, recreation, defence, waste disposal, sand and gravel mining, land reclamation, and oil and gas exploration. Economic development in coastal areas depends on sustainable productivity and the viability of coastal resources. Coastal zone management should be based on an understanding of the complexity of the coastal system and its interaction with adjacent urban areas, river catchments in the hinterland, and seas and oceans (Hoozemans et al. 1995). Decisions have to take into account global change and international policy.

It is increasingly recognised that the sustainable development of the coast, and its multiple resources in accordance with their carrying capacity, is possible only by adopting a comprehensive strategy. According to Doody (1997) this will involve integrated action across the many functions of the coast (horizontal approach) and consensus building at the different levels of decision making and policy formulation (vertical approach).

Integrated coastal zone management can be practised at several scales. Van der Meulen and Udo de Haes (1996) sketched the role of nature conservation in ICZM on a global level. At the heart of the problem is the fact that 50–70%

of the Earth's population is concentrated in coastal areas. The competition between the various user categories operating along the coast is fierce, with recreation and tourism as the fastest growers. Combined use causes an ever-increasing tension between safeguarding natural resources, on the one hand, and economic development, on the other. Often short-term economic gains are chosen at the cost of irreversible ecological losses. There are many examples that show that this leads to environmental degradation which eventually backfires on economic development. Van der Meulen and Udo de Haes (1996) argue that sustainable development of the coastal zone is possible only if nature conservation has a strong position in integrated management.

ICZM is clearly a matter of the highest political level. Fortunately this is gradually being realised. In 1994 the Council of Environmental Ministers asked the European Commission to prepare 'a comprehensive strategy on integrated management and planning in the Community coastal zones, providing a framework for its conservation and sustainable use' (Doody 1997).

The tools and techniques developed for ICZM are based on system analysis. It is beyond the scope of this chapter to discuss this approach in any depth. The interested reader is referred to the course on coastal zone management written by Hoozemans *et al.* (1995).

CONCLUSIONS

SOURCES OF INFORMATION

The references to this chapter list the main handbooks for dune management and conservation. Those interested in more information on coastal management issues are referred to the *Journal of Coastal Conservation* (JCC), the *Journal of Coastal Research* (JCR), the *Journal of Coastal Management* (JCM) and *Ocean and Coastal Management* (OCM). The JCC is the official scientific organ of the European Union for Coastal Conservation (who also issue *Coastline*) and is published by Opulus Press AB, Uppsala, Sweden. The JCC focuses on applied research for integrated coastal management with a 'wise use' perspective. The journal has established a policy of cooperation with the JCR which is published in the USA and is more related to pure sciences. The JCM is also published in the USA. OCM is an Elsevier journal which is dedicated to the management also of ocean resources.

FINALLY

Coastal dynamics and management form an ongoing process. Along the Dutch coast we have seen the following phases in the past 10–15 years:

(1) The policy of dynamic preservation created possibilities for soft defence structures instead of hard ones; beach nourishment became an official policy.
(2) Wider coastal zones were thought to have greater resilience, offering more ecological values and greater safety, for example against rising sea level.
(3) This meant that on narrow (a few 100 m wide) coasts, external (i.e. offshore) activities can be beneficial, while on broad coasts (a few km wide) internal activities are opted for; in the latter case, one could think of the creation of slufters or of returning agricultural land to the sea.
(4) Reclamation plans should be seen in this context, taking into account the entire coast and its long-term development along with a discussion on the various present and future functions.

Finding an equilibrium between the interests of socio-economic development and the maintenance of a natural dynamic system is one of the great challenges of our time. The above points illustrate how necessary it is to propose a kind of management that uses natural processes rather than destroying them. The example of the Dutch coast shows what possibilities can be explored when dealing with a soft sedimentary type of coast.

But this is not valid only for the Dutch coast. Many of the world's lowland delta coasts have essentially similar characteristics: two parties competing for the same space; i.e. socio-economy and ecology in a dynamic natural setting. Once the space on land is almost gone, the eye turns towards the sea. These and other experiences can help to develop both on- and offshore space in a responsible way.

ACKNOWLEDGEMENTS

Part of this work was undertaken in the MAST3 INDIA research project funded by the European Commission, Directorate General for Science, Research and Development, under Contract MAS3-CT97-0106. We thank Andrew Warren and Jon French for carefully reading the text.

REFERENCES

Anderson P & Romeril MG (1992) Mowing experiments to restore a species-rich sward on sand dunes in Jersey, Channel Islands, GB. In: RWG Carter, TG Curtis & MJ Sheehy-Skeffington (eds) *Coastal dunes: geomorphology, ecology and management for conservation.* Rotterdam/Brookfield, AA Balkema, 219–34.
Angus S & Elliott MM (1992) Erosion in Scottish machair with particular reference to the Outer Hebrides. In: RWG Carter, TG Curtis & MJ Sheehy-Skeffington (eds)

Coastal dunes: geomorphology, ecology and management for conservation. Rotterdam/ Brookfield, AA Balkema, 93–112.

Anon. (1994) *Natuurlijke dynamiek in de zeereep; een onderzoek naar de mogelijkheden op vier lokaties.* Report Dienst Weg en Waterbouwkunde, Rijkswaterstaat, The Hague, The Netherlands.

Arens SM (1996) Patterns of sand transport on vegetated foredunes. *Geomorphology* **17**, 339–50.

Arens SM & Wiersma J (1994) The Dutch foredunes: inventory and classification. *Journal of Coastal Research* **10**, 189–202.

Bakker TWM (1990) The geohydrology of coastal dunes. In: ThW Bakker, PD Jungerius & JA Klijn (eds) Dunes of the European coast; geomorphology, hydrology, soils. *Catena Suppl.* **18**, 109–19.

Barrère, P (1992) Dynamics and management of the coastal dunes of the Landes, Gascony, France. In: RWG Carter, TG Curtis & MJ Sheehy-Skeffington (eds) *Coastal dunes: geomorphology, ecology and management for conservation.* Rotterdam and Brookfield, AA Balkema, 25–32.

Belsky J (1992) Effects of grazing, competition, disturbances and fire on species composition and diversity in grassland communities. *Journal of Vegetation Science* **3**, 187–200.

Brooks A & Agate E (1986) *Sand dunes; a practical conservation handbook.* Wallingford, British Trust for Conservation Volunteers.

Carter RWG (1988) *Coastal environments; an introduction to the physical, ecological and cultural systems of coastlines.* London, Academic Press.

Carter RWG (1991) Near-future sea level impacts on coastal dune landscapes. In: F Van der Meulen, JV Witter & W Ritchie (eds) Impact of climatic change on coastal dune landscapes of Europe. *Landscape Ecology* **6**, 29–39.

Carter RWG, Hesp PA & Nordstrom KF (1990) Erosional landforms in coastal dunes. In: K Nordstrom, N Psuty & B Carter (eds) *Coastal dunes: form and process.* Chichester, John Wiley, 217–50.

Carter RWG, Curtis TG & Sheehy-Skeffington MJ (eds) (1992) *Coastal dunes; geomorphology, ecology and management for conservation.* Rotterdam/Brookfield, AA Balkema.

Chapman DM (1989) *Coastal dunes of New South Wales: status and management.* University of Sydney, Coastal Studies Unit technical report 89/3.

Davies P, Williams AT & Curr RHF (1995) Decision making in dune management; theory and practice. *Journal of Coastal Conservation* **1**, 87–96.

Delft Hydraulics (1997a) *Eiland in zee, deel van een veerkrachtige kust.* Delft Hydraulics Rapport R3163 (English summary).

Delft Hydraulics (1997b) *Veerkracht van de kust; ontwikkeling en operationalisering van een veerkrachtmeter.* Delft Hydraulics Rapport Z2136.

Depuydt F (1972) *De Belgische strand- en duinformaties in het kader van de geomorfologie der zuidoostelijke Noordzeekust.* Verhandeling van de Koninklijke Academie voor Wetenschappen, Letteren en Schone Kunsten van België, Klasse der Wetenschappen, Jaargang XXXIV, Nr 122. Paleis der Academiën, Brussel (in Dutch, with English summary).

De Raeve F (1989) Sand dune vegetation and management dynamics. In: F Van der Meulen, PD Jungerius & J Visser (eds) *Perspectives in coastal dune management.* The Hague, SPB Academic, 99–109.

De Ruig JHM (1997) Resilience in Dutch coastline management. *Coastline* **2**, 4–8.

De Ruig JHM (1998) Coastline management in the Netherlands: human use versus natural dynamics. *Journal of Coastal Conservation* **4**, 127–34.

De Ruig JHM & Hillen R (1997) Developments in Dutch coastline management: conclusions from the second governmental coastal report. *Journal of Coastal Conservation* **3**, 203–10.

Doing H (1989) Introduction to the landscape ecology of southern Texel. In: F Van der Meulen, PD Jungerius & J Visser (eds) *Perspectives in coastal dune management*. The Hague, SPB Academic, 279–85.

Doody P (1989) Conservation and development of the coastal dunes in Great Britain. In: F Van der Meulen, PD Jungerius & J Visser (eds) *Perspectives in coastal dune management*. The Hague, SPB Academic, 53–67.

Doody P (1997) Coastal zone management in Europe. *Coastline* **4**, 14–15.

Ellenberg H (1978) *Vegetation Mitteleuropas mit den Alpen*. Stuttgart, Verlag Eugen Ulmer.

Favennec J (1995) Coastal management by the French National Forestry Service in Aquitaine, France. In: PS Jones, MG Healy & AT Williams (eds) *Studies in European coastal management*. Cardigan, Samara Publishing Limited, 191–6.

Gibson CWD & Brown VK (1992) Grazing and vegetation change: deflected or modified succession? *Journal of Applied Ecology* **29**, 120–31.

Hesp PA (1988) Morphology, dynamics and internal stratification of some established foredunes in southern Australia. *Sedimentary Geology* **55**, 17–41.

Hesp PA (1989) A review of biological and geomorphological processes involved in the initiation and development of incipient foredunes. In: CH Gimingham, W Ritchie, BB Willetts & AJ Willis (eds) Coastal sand dunes. *Proceedings Royal Society of Edinburgh* **96B**, 181–201.

Hesp PA & Thom BG (1990) Geomorphology and evolution of active transgressive dunefields. In: K Nordstrom, N Psuty & B Carter (eds) *Coastal dunes: form and process*. Chichester, John Wiley, 253–88.

Higgins LS (1933) An investigation into the problem of the sand dune areas on the South Wales coast. *Archaeol. Cambrensis.* **88**, 26–67.

Hillen R & Roelse P (1995) Dynamic preservation of the coastline in the Netherlands. *Journal of Coastal Conservation* **1**, 17–28.

Hoozemans FJM, Klein RJT, Kroon A & Verhagen HJ (1995) *The coast in conflict; an interdisciplinary introduction to coastal zone management*. National Institute for Coastal and Marine Management/RIKZ, The Hague, CZM-Centre Publication 5.

Houston J (1989) The Sefton Coast Management Scheme in north-west England. In: F Van der Meulen, PD Jungerius & J Visser (eds) *Perspectives in coastal dune management*. The Hague, SPB Academic, 249–53.

Janssen M (1996) New plans for reclamation of the North Sea. *Coastline* **3**, 18–19.

Jungerius PD (1989) Geomorphology, soils and dune management. In: F Van der Meulen, PD Jungerius & J Visser (eds) *Perspectives in coastal dune management*. The Hague, SPB Academic, 91–8.

Jungerius PD (1990) The characteristics of dune soils. In: ThW Bakker, PD Jungerius & JA Klijn (eds) Dunes of the European coast: geomorphology, hydrology, soils. *Catena Suppl.* **18**, 155–62.

Jungerius PD & Van der Meulen F (1988) Erosion processes in a dune landscape along the Dutch coast. *Catena* **15**, 217–28.

Jungerius PD & Van der Meulen F (1989) The development of dune blowouts as measured with erosion pins and sequential air photos. *Catena* **16**, 369–76.

Jungerius PD, Koehler H, Kooijman AM, Mücher HJ & Graefe U (1995) Response of vegetation and soil ecosystem to mowing and sod removal in the coastal dunes 'Zwanenwater', the Netherlands. *Journal of Coastal Conservation* **1**, 3–16.

Klijn JA (1990) The Younger Dunes in The Netherlands; chronology and causation. In: ThW Bakker, PD Jungerius & JA Klijn (eds) Dunes of the European coast: geomorphology, hydrology, soils. *Catena Suppl.* **18**, 89–100.

Kooijman AM & de Haan MWA (1995) Grazing as a measure against grass encroachment in Dutch dry dune grassland: effects on vegetation and soil. *Journal of Coastal Conservation* **1**, 127–34.

Kooijman AM & Van der Meulen F (1996) Grazing as a control against 'grass-encroachment' in dry dune grasslands in the Netherlands. *Landscape and Urban Planning* **34**, 323–33.

Löffler MAM & Veer MAC (1999) *Mogelijkheden voor de ontwikkeling van nieuwe duinen bij een Maasvlakte 2.* Report Ministerie van Verkeer en Waterstaat, Directoraat Generaal Rijkswaterstaat, Directie Zuid-Holland, 42pp.

LWI (1997) *Valuation of nature in coastal zones.* Land Water Milieu Informatie Technologie, -project Socio-economic and Ecological Information Systems, Civieltechnisch Centrum Uitvoering Research en Regelgeving, Gouda, Vols I–IV.

Nordstrom KF, McCluskey JM and Rosen PS (1986) Aeolian processes and dune characteristics of a developed shoreline: Westhampton beach, New York. In: WG Nickling (ed.) *Aeolian geomorphology.* Boston, Allen and Unwin, 131–47.

Nordstrom KF, Psuty NP & Carter RWG (1990) Effects of human development. In: K Nordstrom, N Psuty and B Carter (eds) *Coastal dunes: form and process.* Chichester, John Wiley, 337–8.

Paskoff R (1994) *Les Littoraux; impact des aménagements sur leur évolution.* Paris, Masson.

Pickworth Farrow E (1917) On the ecology of the vegetation of Breckland, III; general effects of rabbits on the vegetation. *Journal of Ecology* **5**, 1–18.

Pluis JLA & Van Boxel JH (1993) Wind velocity and algal crusts in dune blowouts. *Catena* **20**, 581–94.

Psuty NP (1989) An application of science to the management of coastal dunes along the Atlantic coast of the U.S.A. In: CH Gimingham, W Ritchie, BB Willetts & AJ Willis (eds) Coastal sand dunes. *Proceedings Royal Society of Edinburgh* **96B**, 289–307.

Rijkswaterstaat (1990) *A new coastal defence policy for the Netherlands.* Ministry of Transport, Public Works & Water Management, The Hague.

Rutin J (1983) *Erosional processes on a coastal sand dune, De Blink, Noordwijkerhout.* Thesis, University of Amsterdam, Publicaties van het Fysisch Geografisch en Bodemkundig Laboratorium van de Universiteit van Amsterdam, nr. 35.

Sherman DJ & Bauer BO (1993) Dynamics of beach-dune systems. *Progress in Physical Geography* **17**, 413–47.

Short AD (1987) Modes, timing and volume of Holocene cross-shore and aeolian sediment transport, southern Australia. In: NC Kraus (ed.) *Coastal Sediments '87.* New York, American Society of Civil Engineers, 1925–37.

Short AD & Hesp PA (1982) Wave, beach and dune interactions in southeastern Australia. *Marine Geology* **48**, 259–84.

Skarregaard P (1989) Stabilisation of coastal dunes in Denmark. In: F Van der Meulen, PD Jungerius & J Visser (eds) *Perspectives in coastal dune management.* The Hague, SPB Academic, 151–61.

Stichting Duinbehoud (1995) *Dynamiek in de kustzone; een advies voor het herstel van dynamische processen in de kustzone van de Schoorlse Duinen* (in Dutch). Leiden Stichting Duinbehoud.

Stive MJF & De Vriend HJ (1995) Modelling shoreface profile evolution. *Marine Geology* **126**, 235–48.

Van Boxel JH, Jungerius PD, Kieffer N & Hampele N (1997) Ecological effects of reactivation of artificially stabilized blowouts in coastal dunes. *Journal of Coastal Conservation* 3, 57–62.

Van der Meulen F (1993) Processes in dune landscapes, the golden fringe of Europe. In: H Synge & E Pongratz (eds) Federation of Nature and National Parks of Europe. *Proceedings of the 1992 FNNPE general assembly and symposium on protecting ecology through natural succession*, 29–36.

Van der Meulen F & Jungerius PD (1989) Landscape development in Dutch coastal dunes; the breakdown and restoration of geomorphological and geohydrological processes. In: CH Gimingham, W Ritchie, BB Willetts & AJ Willis (eds) Coastal sand dunes. *Proceedings Royal Society of Edinburgh* **96B**, 219–29.

Van der Meulen F & Van der Maarel E (1989) Coastal defence alternatives and nature development perspectives. In: F Van der Meulen, PD Jungerius & J Visser (eds) *Perspectives in coastal dune management*. The Hague, SPB Academic, 183–95.

Van der Meulen F & Udo de Haes HA (1996) Nature conservation and integrated coastal zone management in Europe; present and future. *Landscape and Urban Planning* **34**, 401–10.

Van der Meulen F, Jungerius PD & Visser J (eds) (1989) *Perspectives in coastal dune management*. The Hague, SPB Academic.

Van der Meulen F, Kooijman AM, Veer MAC & Van Boxel JH (1996) *Effectgerichte maatregelen tegen verzuring en eutrofiëring in open droge duinen*. Eindrapport Fase 1 1991–1995, FGBL, UvA.

Van der Putten WH (1989) *Establishment, growth and degeneration of Ammophila arenaria in coastal sand dunes*. Doctoral thesis, Agricultural University Wageningen.

Van der Putten WH, Van Dijk C & Peters BAM (1993) Plant specific soil-borne diseases contribute to succession in foredune vegetation. *Nature* **365** (6415), 53–5.

Van Dieren JW (1934) *Organogene Dünenbildung*. The Hague, Martinus Nijhoff.

Van Dijk HWJ (1989) Ecological impact of drinking-water production in Dutch coastal dunes. In: F Van der Meulen, PD Jungerius & J Visser (eds) *Perspectives in coastal dune management*. The Hague, SPB Academic, 163–82.

Van Dorp D, Boot R & Van der Maarel E (1985) Vegetation succession on the dunes near Oostvoorne, The Netherlands since 1934 interpreted from air photographs and vegetation maps. *Vegetatio* **58**, 123–36.

Van Rijn LC (1995) *Sand budget and coastline changes of the central Dutch coast between Den Helder and Hook of Holland*. Delft Hydraulics, report H 2129.

Veer MAC (1998) *Effects of grass-encroachment and management measures on vegetation and soil of coastal dry dune grass lands*. Thesis, University of Amsterdam.

Wanders E (1989) Perspectives in coastal dune management; towards a dynamic approach. In: F Van der Meulen, PD Jungerius & J Visser (eds) *Perspectives in coastal dune management*. The Hague, SPB Academic, 141–48.

Waterman RE, Misdorp R & Mol A (1998) Interactions between water and land in the Netherlands. *Journal of Coastal Conservation* 4, 115–26.

Westhoff V (1989) Dunes and dune management along the North Sea coasts. In: F Van der Meulen, PD Jungerius & J Visser (eds) *Perspectives in coastal dune management*. The Hague, SPB Academic, 41–51.

Williams AT & Randerson PF (1989) Nexus: ecology, recreation and management of a dune system in South Wales. In: F Van der Meulen, PD Jungerius & J Visser (eds) *Perspectives in coastal dune management*. The Hague, SPB Academic, 217–27.

Williams AT & Simmons SL (1996) The degradation of plastic litter in rivers: implications for beaches. *Journal of Coastal Conservation* 2, 63–72.

Wilson P (1990) Coastal dune chronology in the north of Ireland. In: ThW Bakker, PD
    Jungerius & JA Klijn (eds) Dunes of the European coast; geomorphology,
    hydrology, soils. *Catena Suppl.* **18**, 71–9.
Wilson P (1992) Trends and timescales in soil development on coastal dunes in the north
    of Ireland. In: RWG Carter, TG Curtis & MJ Sheehy-Skeffington (eds) *Coastal
    dunes: geomorphology, ecology and management for conservation.* Rotterdam and
    Brookfield, AA Balkema, 153–62.

10 Climate Change and Nature Conservation

CLIVE AGNEW[1] AND SIOBHAN FENNESSY[2]
[1]*University of Manchester, UK*
[2]*Kenyon College, Gambier, Ohio, USA.*

INTRODUCTION

The links between climate and the biosphere have been debated since ancient Greece (Gates 1993). The concept that vegetation reaches equilibrium with climate was used early in the twentieth century to map, and to propose management strategies for, natural resources. Köppen (1931) based his classification of world climates on vegetation distribution (Table 10.1) and Thornthwaite (1948), and later UNESCO (1977), produced climate maps through a water balance calculation of the moisture available to plants. Soil science also developed a system of classification based upon 'zonal' climate. This chapter is not alone, therefore, in assuming a strong association between climate and ecosystems (Prentice *et al.* 1992; Woodward 1987). The older approaches, however, treated climate as spatially rather than temporally variable. In the last half of the twentieth century, with growing concerns over the possible impacts of climate change, the temporal dimension was receiving much more attention. During the last decades of the twentieth century the popular press was active in promoting among the general public the idea of global warming as a significant environmental threat which is likely to lead to changes in lifestyle.

It is now accepted that the climate has changed over the last 100 years, and that this should concern conservationists. This chapter describes these changes briefly, and summarises predictions for this century. It draws heavily upon recent reports by the IPCC (Intergovernmental Panel on Climate Change) and discusses the uncertainties in these predictions before investigating their possible impact on the biosphere and considering strategies for conservation in the face of these threats.

Habitat Conservation: Managing the Physical Environment. Edited by A. Warren and J. R. French.
© 2001 John Wiley & Sons Ltd.

Table 10.1. The Köppen climate classification and associated vegetation (from Briggs and Smithson 1992)

A Tropical climates ($T > 18°C$)
    Af   Wet tropical forest
    Aw   Tropical savanna
    Am   Monsoonal seasonal forest
B Dry climates
    BS   Steppe (P 380 to 760 mm)
    BW   Desert ($P < 250$ mm)
C Warm Temperate climates ($18°C > T_{min} > -3°C$; $T_{max} > 10°C$)
    Cf   Broad-leafed evergreen and tall grassland, no dry season
    Cw   Temperate deciduous, dry winter
    Cs   Sclerophyllous, Mediterranen with dry summer
D Cold climates ($T_{min} < -3°C$; $T_{max} > 10°C$)
           Boreal coniferous forest
E Polar climates ($T_{max} < 10°C$)
    ET   Tundra ($0°C < T_{max} < 10°C$)
    EF   Ice desert ($T < 0°C$)

Note: T = mean monthly temperature.
P = annual precipitation.

CHANGING CLIMATE

Continental drift and tectonic activity have atmospheric effects over geological time. On a somewhat smaller time scale, change is linked to alterations of the energy received at the earth's surface, forced by orbital eccentricities, as explained by the so-called Milankovitch theory (Barry and Chorley 1992; de Blij and Muller 1993; Gordon *et al.* 1998; Houghton 1997; McIlveen 1992). These create cycles over 110 000, 40 000 and 23 000 to 18 800 years. Since the last ice age, mean global temperatures have fluctuated markedly, apparently in response to yet other forces (Figure 10.1). There have been some cool periods, as in the 'Little Ice Age' of the seventeenth century in western Europe, but most interest has been focused on the accelerated warming of the last century or so. The rise since 1880 has been 0.5°C in the northern hemisphere (0.4°C for the world) (Figure 10.2).

This unprecedented temperature increase has given rise to great concern. It has been explained as the impact of the greenhouse gases, in particular the concentration of carbon dixoide (CO_2) emitted in increasing quantities since the onset of the Industrial Revolution. Atmospheric concentrations of CO_2 have risen from 270 ppm (parts per million) around 1750 to 1800, to 355 ppm in the 1990s (though there was no increase during the 1950s and 1960s). In 1995, the IPCC confirmed its belief in the influence on the global climate of these emissions (Houghton *et al.* 1996). This was a much stronger statement than in previous reports.

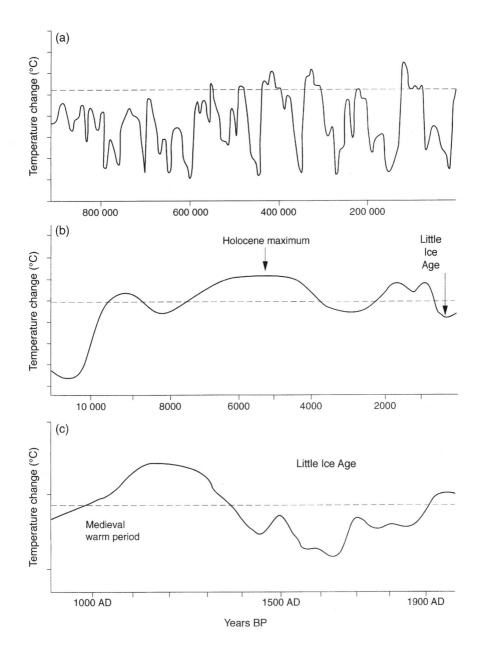

Figure 10.1. Changes in global mean temperature since the Pleistocene on timescales of (a) the last million years; (b) the last 10 000 years and (c) the last 1000 years. From Houghton *et al.* (1991:202), with permission from Cambridge University Press.

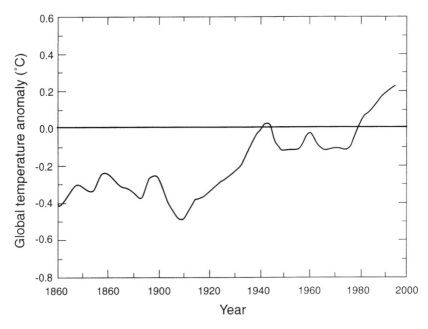

Figure 10.2. A generalised curve representing changes in global average surface temperature rise over last 100 years relative to an average calculated for the period 1961 to 1990. From Houghton (1997:46; with permission from Cambridge University Press).

Many texts now attempt to explain the mechanisms by which greenhouse gases can cause global warming (Gribben and Gribben 1996; Houghton, Callander and Varney 1992; Houghton 1997; Leggett 1990). A major source of CO_2, probably the most important of these gasses, is the burning of fossil fuels. Emissions of carbon increased from 2.5 Gt in 1960 to 6 Gt in 1990 (Houghton, Callander and Varney 1992), mostly from transportation and power generation. Deforestation currently contributes 1.6 (± 1.0) Gt of carbon, if it is assumed that forests are being destroyed at a rate of 1.4 to 2 million ha yr^{-1}. CO_2 is not the only greenhouse gas, however (Figure 10.3), and there have also been rapid rises in concentrations over the last 50 years of other gases, such as the chlorofluorocarbons (CFCs), which also have the ability to absorb terrestrial radiation in large quantities. CFCs are increasing at a rate of around 4% per annum. Given that CFCs are 10 000 more effective than CO_2 at absorbing energy, and that the figures are 2000 times for ozone, 150 times for nitrous oxide and 30 times for methane, there is concern over all atmospheric greenhouse gases, even if their concentrations vary widely. Moreover, the atmospheric lifetimes of most of these gases is 65–200 years, which means their effects will continue for a long time, even if emissions were to be reduced today.

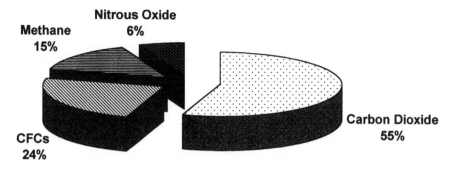

Figure 10.3. The major greenhouse gases, after IPCC 1990 (Houghton, Jenkins and Ephraums 1991:11).

The focus here is upon changes at the decadal timescale, for these could be a major problem for conservation. If it takes decades before a change is apparent we cannot rely simply upon the extrapolation of observed trends to formulate strategies. We need to use models of the atmosphere and the modelling of global climate change has indeed developed into a major international enterprise. The models are so complex and computationally expensive that only a few institutions can resource them. Thus, we have to accept the predictions emanating from places such as the Hadley Centre (UK Meteorological Office) or Geophysical Fluid Dynamics Laboratory (NOAA, USA).

Several types of model have been developed (Henderson-Sellers and McGuffie 1987; Houghton 1997; Wetherald 1991), including energy balance and radiative convective models, but the most commonly cited are General Circulation Models (GCM). These divide the atmosphere vertically and horizontally into a number of cells (typically 100 km horizontal spacing and 1 km vertically) and compute the transfers of energy and moisture which create climate. Gordon *et al.* (1998) identify two types: the simpler equilibrium model that was used in earlier studies and which assumes an instantaneous doubling of CO_2. The other, the transient GCM model, assumes gradual changes in greenhouse gases. It is much more demanding in computational resources and may be coupled to ocean circulations to provide even more reliable results. When this is done the predicted rates of warming are only 60% of those predicted by equilibrium models. However, even these models suffer from limitations such as their inability effectively to deal with cloud cover and vapour pressure changes, and their predictions are not always in agreement. Many authorities, including the IPCC, have pointed to differences in regional forecasts between the models (Parry 1990; Hayes 1991; Wetherald 1991). They can be significant, as Hulme (1996) found for different GCM predictions of southern African rainfall. The UK Hadley Centre model predicts a 2.5–7.5%

fall; the Canadian Climate Centre model, a greater than 7.5% fall; and the Oregon State University model, a rise of 2.5–7.5%, with some parts increasing by 20%. Gordon *et al.* (1998) suggest that GCMs are more useful for reproducing general global conditions, than for detailed regional forecasts. Nevertheless, the general predictions of the models are widely accepted, being based on fundamental laws of physics and fluid motion.

THE ENSO

Discussions of climate change are not complete today without mention of teleconnections (Flohn 1987 and Glantz, Katz and Nicholls 1991), or more precisely, the global interconnectedness of atmospheric and oceanic circulations. It is now widely acknowledged that changes in rainfall are coupled to sea-surface temperature anomalies and GCM modelling is leading to greater understanding of these mechanisms (Robertson and Frankignoul 1990). Glantz (1987) suggested a link between droughts in Africa and Atlantic temperatures, and these now appear to have been successfully modelled (Folland 1987; Owen and Folland 1988). Mason (1995) identified sea-surface temperature changes in the South Atlantic and south-west Indian Oceans that might be significant for rainfall in southern Africa. The most widely discussed coupling is that of the El Niño Southern Oscillation (ENSO) with monsoonal rainfalls (Lockwood 1984). The El Niño or cessation of cold water upwelling in the eastern Pacific is associated with large-scale weather oscillations. Prior to an El Niño there are powerful south-easterly winds which produce a build-up of water in the western Pacific. This is later advected eastwards and interferes with the normal upwelling of cold water off the South American coast (Caviedes 1988). The normal or 'Walker' circulation is reversed and there is a surface westerly wind and persistent high-pressure subsidence in the south-western Pacific. This in turn promotes a weakening of the monsoonal circulation and drought conditions in India and Australia. Reading, Thompson and Millington (1995) noted that during the 1983 El Niño, Australia experienced one of the worst droughts of the twentieth century, followed by unusually heavy rainfall in 1988, when the Walker circulation was re-established. Camberlin (1995) found that rainfalls in north-east Africa correlated well with ENSO, especially droughts in Ethiopia and Uganda. However, Wuethrich (1995) reported that the El Niño phenomenon was not very predictable and that it appeared to follow a more complex pattern than was previously believed.

PREDICTIONS

Global temperatures were expected by the IPCC to increase by 1.4°C by the year 2030 (Houghton, Jenkins and Ephraums 1991) within a range of 0.7–2.0°C. More recent forecasts, based on redefined greenhouse gas emissions and the effects of sulphate aerosols, suggest slightly lower temperature increases by the end of this century (Pearce 1995a). These forecasts are based on assumptions of future greenhouse gas emissions (see below). The UK Countryside Commission (1995) reported that with scenario IS92a (see below for an explanation of this nomenclature) we can expect a rate of global mean warming of 0.2°C per decade. The greatest warming is expected in the higher latitudes. The impact on degree growing days for plants is shown in Table 10.2. Global warming could also cause sea-levels to rise by 20 cm in the next 25 years and 50 cm by the year 2045 (Semeniuk 1994; Warrick, Barrow and Wigley 1993), initially through thermal expansion (Wigley and Raper 1987).

Table 10.2. Changes to degree growing days (temperature > 6°C) for the UK, using global warming scenario IS92a (from Countryside Commission 1995)

	Current	2050
Dumfries	169	192 days
Durham	161	180
Oxford	195	210
Plymouth	210	272

UNCERTAINTY

The evidence in support of global warming is not accepted by all (Balling and Wildavsky 1992; Bate 1998; Bate and Morris 1994; Brown 1996; Courtney 1993; Lindzen 1991; Olstead 1993; Pearce 1995b, 1997; Hoyt 1999), nor did the notion of a warmer world always prevail during the twentieth century. Some 25 years ago, Ponte (1976) was writing about the possible effects of global cooling and the onset of the next ice age (Bryson 1974). More recently, Rahmstorf (1997) has suggested that changes to oceanic circulations may result in significant temperature reductions. The need to be cautious about the predictions arises through:

- Uncertainty over the sensitivity of the climate to greenhouse gas concentrations
- Uncertainty of future greenhouse gas emissions
- Inconsistencies and inadequacies in climate data
- Inadequacies of climate models
- Unforeseen impacts of other climate-forcing mechanisms

Table 10.3 shows the impact of using different climate sensitivities to the doubling of greenhouse gas concentrations. Figure 10.4 plots the temperature

Table 10.3. Estimated increases in global mean air temperatures relative to 1990, using scenario IS92a (from Countryside Commission 1995)

	2020	2050
Low sensitivity (1.5°C)	0.30°C	0.72°C
Best guess (2.5°C)	0.52	1.16
High sensitivity (4.5°C)	0.75	1.70

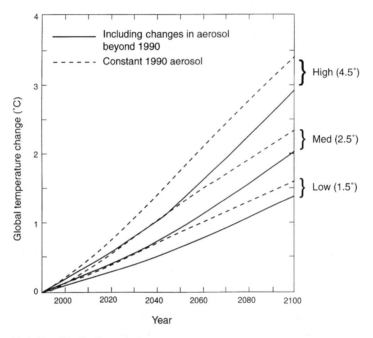

Figure 10.4. Predicted changes in global average surface temperature under IPCC emission scenario IS92a. Three climate sensitivities (response to a doubling of CO_2) are shown and predictions are displayed assuming aerosol concentrations grow (solid line) as under IS92a, in contrast to temperatures if aerosol concentrations remain constant (dashed line). From Houghton (1997:94; with permission from Cambridge University Press).

changes and the predictions for different sensitivities under emission scenario IS92a. Warming is predicted, but its magnitude is far from certain and is likely to be significantly affected by future aerosol concentrations (aerosols are liquid or solid matter suspended in the atmosphere for long periods due to their small size).

There have been many attempts to forecast concentrations of greenhouse gases. The chain of uncertainty is shown in Figure 10.5. Most early climate models were based on the assumption that there would a doubling of pre-industrial CO_2 concentrations (the equilibrium models), but even in 1984 the Royal Commission on Pollution (UK) was reporting that CO_2 concentrations were below expected values. This was explained by greater absorption of CO_2 by the world's oceans than had been expected, estimated by the IPCC (Houghton, Callander and Varney 1992) to be around 2.0 ± 0.8 Gt carbon yr^{-1}.

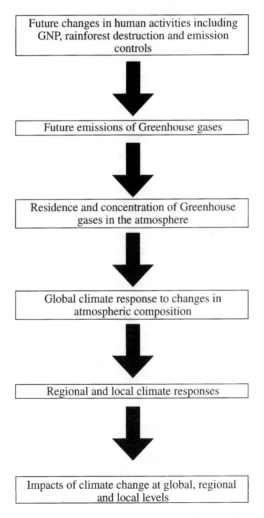

Figure 10.5. The chain of uncertainty concerning global warming impacts.

The degree of this so-called 'atmospheric leaching' is still uncertain as it depends not only on rainfall but also on the temperature of the oceans. There are other sinks for greenhouse gases, such as wetlands, whose destruction, therefore, may be as important for global warming as it is for nature conservation (Franzén 1994). Gribben and Gribben (1996) suggest that roughly half the CO_2 released into the atmosphere each year is absorbed by natural sinks. There is also uncertainty over the future combustion of fossil fuels, which is related to economic activity. In 1990, the IPCC employed a 'business as usual' scenario (SA90) which assumed that no steps were taken to curb greenhouse gas emissions. Releases of carbon were assumed to rise from $6\,Gt\,yr^{-1}$ to over $20\,Gt\,yr^{-1}$ by 2100, and this produced a warming of $0.3°$ per decade. Under this scenario global temperatures would rise by $3°C$ by the year 2100 (a rise of $4°C$ from pre-industrial times). Six alternative scenarios were used after 1992. Scenario IS92a (used below) is comparable to SA90 'business as usual', but includes the curbing of CFC emissions (as agreed in the Montreal Protocols), a revision of population forecasts and some changes to the levels of economic activity. The highest greenhouse gas forecast is in scenario IS92e, which is based on the phase-out of nuclear energy, 3.5% economic growth to 2025 and CO_2 emissions of $16.2\,Gt\,yr^{-1}$. The lowest scenario (IS92c) assumes CO_2 emissions of $8.8\,Gt\,yr^{-1}$ to 2025 and then $4.6\,Gt\,yr^{-1}$, because of fossil fuel shortages.

Even the climate record has been the subject of dispute (Pearce 1997). Air temperatures in the northern hemisphere have not risen continually, and have at times declined, most noticeably in the 1960s. Temperatures have not increased across the world since the 1970s, as is so often assumed, and in the southern hemisphere they have not risen since 1987. Equally puzzling, the reported atmospheric temperature rises have been observed at ground-based sites, yet in the upper atmosphere, temperatures (from satellite data) show no clear warming trend. There is even a suggestion of cooling, but this may have something to do with the Mount Pinatubo eruption in 1991 and the temperature record is too short for full analysis of this event (Gordon et al. 1998). The temperature record has also been criticised for being spatially biased. There are more land-based than ocean-based stations and more in the northern than in the southern hemisphere. Globally, the distribution of climate stations is uneven and in parts of Africa the database is deteriorating. Thus predictions of change, especially over the oceans and developing countries, are super-imposed upon uncertain local conditions.

Furthermore, the GCM results have not wholly reproduced recent climate changes. Earlier models gave a warming trend roughly double that actually observed over the last 100 years and it was recognised over a decade ago that greenhouse gases alone could not explain twentieth-century temperature changes. It is necessary to take into account energy exchanges with and circulations of the world's oceans, sunspot activity and turbidity of the atmosphere due to volcanic eruptions, such as of Mount Pinatubo (Gribben and Gribben 1996). There is, therefore, speculation over IPCC predictions, especially where they are based solely upon the effect of greenhouse gases, and even when they use ocean-coupled GCMs. Furthermore, their over-prediction of rises in temperature is probably related to atmospheric sulphates (*Economist* 1995). The release of SO_2 through the burning of fossil fuels appears to partly counteract the effects of CO_2 released by the same process.

The inadequacies of the climate models have been widely acknowledged. Early models looked only at doubling of CO_2 (equilibrium) but more recent runs include transitional states and are coupled to ocean circulations. Their spatial agreement is limited and regional averages are not predicted with confidence (Parry 1990; Ross 1991). Climate variables besides temperature (precipitation, humidity, windspeed) are modelled with even less certainty. Improvements are leading to greater confidence in regional simulations and most criticism is now levelled at the GCM's inability to simulate feedbacks, in particular the role of clouds.

PREDICTING THE IMPACTS OF CLIMATE CHANGE UPON THE BIOSPHERE

Climate may be a key control on plant and animal life, but it is not the only one, and this makes assessing the impact of climate change very hazardous. The different levels of tolerance to change among species and the complexity of dynamic interactions within ecosystems further complicate assessment. Writing about Africa, Meadows (1996) noted that the contemporary distribution of forest was determined by relatively recent events of which the Quaternary period had been particularly important. Grainger (1996:173) went further, arguing that, 'the biogeography of Africa is now predominately cultural not natural'.

There are two approaches to the assessment of climate change impacts. First, there are analyses based upon fundamental processes. Second, there are empirical studies, often based on historical patterns. This discussion focuses upon vegetation changes for the sake of simplicity. Impacts upon fauna are well covered by Gates (1993). The possible impacts upon wetlands are given special prominence because of their vulnerability to global warming and space is also taken to consider agricultural impacts, because of the potential threats in land-use changes.

FUNDAMENTAL PROCESSES AND VEGETATION CHANGE

CO_2

Climate influences the assimilation rate of biomass in plants through the supply of sunlight, heat and moisture. Growth rate (GR) is a function of the

net assimilation rate of biomass (NAR) and the leaf area index (LAI) where
GR = LAI*NAR (Forbes and Watson 1992).

Photosynthesis produces carbohydrates with solar energy. Respiration, in
which CO_2 and water vapour are released, is equally dependent upon
prevailing atmospheric conditions (Monteith and Unsworth 1990). Three
mechanisms can be identified by which plants assimilate CO_2 during
photosynthesis:

C3   crops such as wheat and rice (CO_2 fixed by a 3-carbon compound
        phosphoglyceric acid)
C4   crops such as maize, sugarcane, millet and the majority of savanna
        grasses (CO_2 fixed by a 4-carbon compound, oxaloacetic acid)
CAM   plants such as cacti and other succulents (CO_2 fixed by crassulacean
        acid)

C3 crops, mostly found in temperate regions, are less efficient converters of
CO_2. Most C4 plants, whose greater efficiency allows them to tolerate higher
temperatures, occur in the tropics (Deshmukh 1986). Enrichment of CO_2 concentrations in the atmosphere could lead to increased plant growth, through a
process known as 'CO$_2$ fertilisation' (Bazzaz and Fajer 1992). C4 plants would
not respond as positively to CO_2 increases as C3 plants (Warrick et al. 1991).
The Countryside Commission (1995) summarised a number of studies to
conclude that cereal yields in eastern England were likely to increase under CO_2
enrichment by 20–40%, but higher respiration rates and water deficiencies in a
warmer world might offset the gains for C3 plants.

Temperature

Temperature affects the spatial distribution of plants (Good 1970; Woodward
1987). The severe winters of high latitudes and altitudes shorten the growing
season and limit germination and growth rates. A warmer world would not,
however, result in a simple poleward migration of species, as individual taxa
would have different response rates and mechanisms (Graham and Grimm
1990; Huntley 1991; Woodward, Thompson and McKee 1991).

Higher temperatures would also not necessarily produce higher growth rates.
As temperature increases so does the rate of chemical reactions in plants, but
growth by cell elongation is most rapid at temperatures around 20°C, and
above 40°C enzyme reactions are reduced (Forbes and Watson 1992). Higher
temperatures produce increased rates of both photosynthesis and respiration,
except in the extremely hot conditions of tropical deserts. The net growth of a
plant depends not merely upon temperature but also on the extent to which
other factors constrain photosynthesis. Leaf temperature needs to be maintained at the optimum through evaporative cooling, and this may not be

possible when water supply is constrained. Goldstein and Sarmiento (1987) report the optimum temperature for two savanna trees to be 25–28°C, while at 35°C, the carbon uptake is only 50% of the maximum. Predicting the impacts of changes in temperature is, therefore, complex, as it needs to take into account deficiencies in moisture. It is for this reason that assessments of cereal production under warming scenarios are often simplified by assuming that the technology is available to overcome deficiencies (namely, irrigation, fertilisers and herbicides; Parry 1990). The issue is further complicated by the fact that average temperatures are less important than extremes, and the number of frost-free days may be a greater constraint than maximum temperatures.

Water

Photosynthesis is not limited by drought directly but by insufficient supply of CO_2. Here, 'supply' refers to the levels of CO_2 found in the plant stomata, not in the atmosphere (because there is normally sufficient CO_2 in the atmosphere). The assimilation of CO_2 is governed by the water status of the plant and Kramer (1983) noted that plant growth and crop yield were reduced by water shortages more than any other environmental variable. The impacts of water shortages on plant growth are well known and reviewed by a number of authors, including Begg and Turner (1976), Kozlowski (1983) and Jones (1992). Many plant processes are affected by moisture deficits, from cell growth to stomatal opening, but the level of tolerance varies between species. Cell enlargement is generally affected at leaf water potentials of -0.2 to -0.4 MPa, and there is usually stomatal closure between -0.8 to -1.0 MPa. Doorenbos and Pruitt (1984) provide a useful tabulation of critical periods, for example for tobacco (knee-high to blossoming), maize (pollination period from tasselling to blister kernel stage) and sugarcane (period of maximum vegetation growth).

As comparatively little of a plant's transpiration requirements is stored in the vegetation, soil moisture is a very important part of the issue. The amount of soil water available to plants (AWC) is normally taken to lie between field capacity (FC, the upper limit of soil water storage when free drainage has ceased) and crop wilting point (WP, around a soil water potential of -1.5 MPa), that is:

$$AWC = FC - WP \text{ (mm of water)} \tag{10.1}$$

The magnitude of the AWC depends upon soil characteristics such as texture, bulk density and organic matter content (Gupta and Larson 1979). Not all the AWC is freely available, and it has been known for some time that the relationship between transpiration rate and soil moisture content is complex (Denmead and Shaw 1962). Jackson (1989) gives examples that demonstrate the significance of AWC to climate, soil type and rooting system. The FAO employ a simple relationship whereby 50% of AWC is assumed to be freely available within the rooting zone, after which soil water is progressively more difficult to extract to the wilting point (Smith 1990, 1992). This is similar to the Penman–Grindley 'root constant' approach (Shaw 1994), but these simple models should not obscure the point that much still needs to be known about the relationships between water shortages and crop growth, in particular the apportioning of energy between above-ground and below-ground biomass. The root/shoot ratio is known to be affected by water deficits (Jones 1992; Monteith 1986), but there are few supporting field data (Lal 1991). In savanna systems, where there are periods of desiccation, plants have a number of 'choices' in their response to water deficits, including larger root/shoot ratios and the control of the osmotic potential in leaves by stomatal regulation (Goldstein and Sarmiento 1987). Tropical plants tend to allocate more production to above-ground biomass than do temperate ones. Tropical forests typically have 90% as above-ground biomass and grasses 50% (Deshmukh 1986). However, Goldstein and Sarmiento (1987) note that early emerging savanna grass species have a higher root/shoot ratio, which is to say a greater development of subsurface biomass than grasses emerging in wetter conditions. There are thus a number of ways in which vegetation may respond to constraints in water supply in a warmer world.

The empirical Water Use Efficiency (WUE) concept is widely used to quantify the relationship between crop production and water consumption:

$$\text{WUE} = \text{yield per unit area/water used to produce yield} \qquad (10.2)$$

or

$$\text{WUE} = Y/T \text{ (see Table 10.4 for examples)} \qquad (10.3)$$

Table 10.4. Examples of WUE for selected crops (adapted from Jones 1992). (Note greater WUE of C4 plants)

	$(Y/T)*10^{-3}$
C4 PLANTS	
Millet	2.72–3.88
Sorghum	2.63–3.65
Maize	2.67–3.34
C4 range	2.41–3.88
C3 PLANTS	
Wheat	1.93–2.20
Rice	1.47
Alfalfa	1.09–1.60
Pulses	1.33–1.76
C3 range	0.88–2.65

where

Y = yield

T = crop transpiration (evapotranspiration for ecological studies)

A more complete analysis of WUE reveals that the rate of transpiration is dependent upon atmospheric resistance to water vapour diffusion such that (Lal 1991):

$$N = KT/(es - e) \qquad (10.4)$$

where

N = total shoot–dry matter production

T = crop transpiration

$(es - e)$ = average vapour pressure deficit during daylight hours

K = constant

For a given saturation deficit $(es - e)$, N/T is constant (Gregory 1989), so that WUE is amenable to improvement if transpiration increases and this is a function of water supply. Many agricultural practices recognise this relationship and, if adopted, could offset the adverse impacts of global warming. Examples include water harvesting (Gowing and Young 1996), efficient plant spacing to maximise water uptake, using varieties with deeper rooting systems, reducing water losses by fallowing, weed control and mulches. It can also be seen that an enrichment of atmospheric CO_2 might enhance WUE through greater transpiration efficiency, even allowing plants to occupy more arid tropical regions.

This brief summary of the fundamental relationships between CO_2, temperature and water supply to plant growth illustrates the complexity of the interactions. There have been attempts to develop biogeographical models based on these kinds of relation into more comprehensive models for complete ecosystems. For example, Prentice et al. (1992) used knowledge of growth constraints, such as degree growing days in their BIOME model, and this was used in turn by Hulme (1996) to investigate the impacts of climate change in southern Africa. Vegetation changes were estimated using moisture supply, the temperature of the coldest month and whether or not the WUE (see above) was improved by higher CO_2 concentrations. The model showed significant shifts in most biomes, but the nature of the changes was dependent upon the different climate models and upon the assumptions adopted. For example, grasslands could decrease by 50% but might also increase with greater WUE, deserts could expand using the UK Hadley Centre model or they might decrease using the Oregon State University model. It was noted by Hulme (1996:28) that 'unambiguous conclusions about the effect of climate change on southern Africa vegetation cannot be made'.

There are still many areas for research. WWF (1993) noted the need for more understanding of community–ecology interactions, ecophysiology, population dynamics, dispersal and migration rates, colonisation, succession, to mention but a few from their list. Scientific analyses are constrained by the quality of the climate change predictions, and the diversity among the predictions of the GCM models (Ross 1991). It is not surprising to find studies that seek to simplify the impacts of climate change by looking for empirical associations or using historical analogues.

EMPIRICAL STUDIES OF VEGETATION CHANGE

Empirical studies use observed associations between ecosystems and climate. They may also take into account indirect effects of climate change through land-use shifts or sea-level rise (Roberts 1994). Holdridge (1947; Figure 10.6)

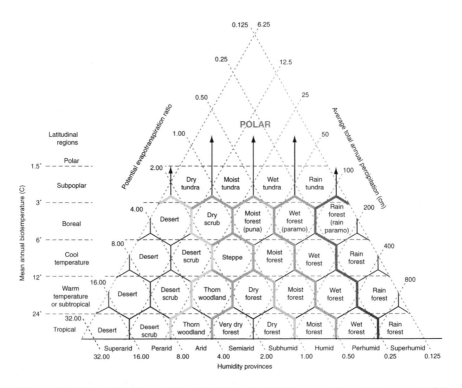

Figure 10.6. Simplification of the Holdridge's (1947) associations between rainfall, evaporation and global vegetation distribution, as used in Figure 10.7. Average biotemperature is the annual average temperature computed with values below zero set to equal zero. The PEt ratio is the ratio of Potential Evapotranspiration divided by average annual rainfall. Reproduced with permission from *Science*.

mapped the main global vegetation patterns as they were related to rainfall and evaporation and his approach was used by Wyman (1991) to examine possible latitudinal shifts based on a doubling of CO_2 concentrations. Wyman's analysis predicts a poleward shift of vegetation of 500–1000 km over the next 100 years. He believes that amphibians are the best faunal indicators of climate change, because of their sensitivity to the environment, but concludes (p. 152) that 'there are too many unknowns regarding the degree and extent of changes that will occur to allow for accurate predictions of likely effects of global climate change'.

Figure 10.7 shows predicted changes under a CO_2 doubling for North America (after Emanuel, Shugart and Stevenson 1985). A latitudinal shift of vegetation of 200–300 km sees the temperate grasslands expanding at the expense of boreal forest and tundra. In tropical regions, the boundary between savanna and tropical forest is perhaps the most sensitive to change, although desert margins are also highly variable. Basing predictions only on temperature changes, a doubling of CO_2 might increase the area occupied by tropical climates by over 25%, with a commensurate decline in subtropical and boreal forests (Bolin et al. 1991; Table 10.5). It is anticipated that global warming will favour an expansion of grasslands with a decrease in deserts (Archibold 1995). These predictions, like those discussed above, are affected by the disagreements between different GCM forecasts. They are also influenced by assumed rates of change, in other words the efficiency of migratory mechanisms (Peters 1991). The migration rate of tree species in the past (Houghton et al. 1996) is believed to be around 4–200 km per century. Global warming could result in latitudinal shifts ten times faster, perhaps 100 km in ten years (DoE 1991).

Archibold (1995) reported that predictions of changes in forest cover were far from clear, the predictions for dry forest ranging from zero to 71% of present area, and for moist forest between -10% to $+11\%$. There are further uncertainties about changes in the topography and soils of the areas onto which vegetation might migrate. As Stott (1994, p. 299) notes, 'there are still large uncertainties in all our estimates of global change, especially at the regional or biome levels, because our basic understanding of the relevant physics remains incomplete . . . and our actual predictions of green-house gas emissions remain approximate'.

Sinha (1991) is also cautious, because:

- CO_2 concentrations have increased over the last 100 years, yet there has been no noticeable effect on vegetation.
- Increases in cloudiness in the tropics with higher temperatures and humidities may have deleterious effects, as on rice yields.
- Monsoonal circulation changes are uncertain.
- Too little is known about tuber crops.
- Pest and disease attacks can offset any gains.

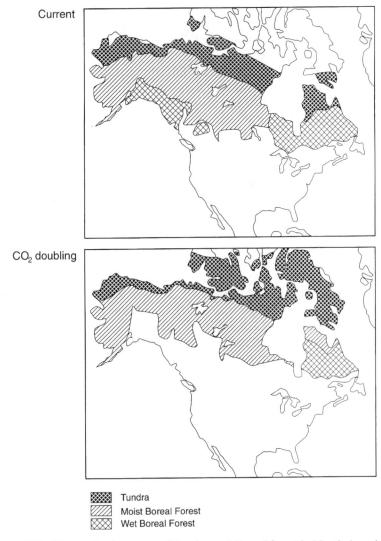

Current

CO$_2$ doubling

▨ Tundra
▨ Moist Boreal Forest
▨ Wet Boreal Forest

Figure 10.7. Changes in the extent of Tundra and Boreal forest in North America under a doubling CO$_2$ scenario (adapted from Emanuel, Shugart and Stevenson 1985).

There are further dangers of accepting the global patterns of vegetation established by climatological variables. Agnew (1995) discusses this problem in the demarcation of desert boundaries, and Mather (1974) has criticised Köppen's (1931) climatically derived maps of global vegetation for failing to take adequate account of water supply. Moreover, because temperature and rainfall are only generally related to major vegetation changes, Mather suggests

Table 10.5. Vegetation changes under a CO_2 doubling scenario (from Bolin *et al.* 1991)

	% of land coverage	
	Present	Future
Tundra	3.3	0
Grasslands	17.7	28.9
Desert	20.6	23.8
Forests	58.4	47.4
Tropical forest	25	40 (as fraction of forest total)
Boreal forest	23	< 1
Warm temp. forest	21	25
Subtropical forest	16	14
Cool temp. forest	15	20

that any correct results from Köppen's vegetation maps were fortuitous. Gates (1993) also discusses the problems of empirical bioclimatic models and notes that they reproduce observed global vegetation patterns with accuracies of only about 40%.

Dissatisfaction with the oversimplification of empirical approaches has led to the search for historical analogues. This approach is also hazardous, as environmental conditions are rarely comparable for different parts of the world at different historical periods, but Cannel and Pitcairn (1993) examine the impacts of mild winters and hot summers during 1989 and 1990 in the UK as an analogue of future conditions. Here, the very mild winters helped the early emergence of plants and animals. Blackthorn and horse chestnut were several weeks early. They showed some concern about seeds that required chilling prior to germination, such as those of hogweed and cowparsely, but they could find no clear effects, although mountain ash might have been affected by frost damage in late spring. Frogs were found to be spawning earlier, and the spring numbers of butterflies and moths increased in northern England. Aphids also increased significantly and there was an outbreak in 1989 of green spruce aphid on sitka and Norway spruce. Caterpillars were also more abundant. Hence insects and amphibians were able to take advantage of the milder conditions, but there was also some evidence that predators, such as parasitic wasps, also increased in numbers, while the dry, hotter summers presented problems for species dependent on ponds. The summer conditions had adverse impacts on the musk orchid, ferns and other shallow-rooting plants, and plants with deep taproots appeared to have flourished (such as ribwort and knapweed).

Wetlands and Aquatic Ecosystems

If global warming were to alter the supply and distribution of water, wet ecosystems would also be altered. Changes may range from a conversion to terrestrial systems to an increase in the wetland area. Carpenter *et al.* (1992) summarise a range of water balance studies, which predict changes in runoff for different regions of North America. In the currently arid West, a 10% decline in precipitation would lead to a decrease in annual average runoff volumes of between 40% and 76%. Predictions for the wetter Delaware River basin range from a decrease of 51% (resulting from a 20% decrease in precipitation) to an increase of 31% (with a 20% increase in precipitation).

However, changes in the timing, frequency and magnitude of extreme rainfall events are probably more important to wetlands than changed mean conditions (Carpenter *et al.* 1992; Poff 1991). These could change river channel geomorphology, wetland hydroperiod or lake turnover times. The rise and fall of water affects annual net production of the biotic community and also temperature regimes, which in turn control the rate of nutrient uptake, decomposition rates and the denitrification processes (Bayley 1995). Small and shallow habitats such as ponds, headwater streams and marshes respond most quickly to reduced water inputs (Carpenter *et al.* 1992), and one consequence is higher water temperatures. High water temperatures may eliminate cold-water fisheries. The capacity of both inland and coastal wetlands to provide nursery areas and food for populations of macroinvertebrates and fish may also be diminished. Water-quality changes may result as well, for not only may salts be concentrated but also the oxygen-holding capacity of the water may decline.

Groundwater is an important source of water in many wetlands, and changes in the balance between precipitation and evapotranspiration may well affect the groundwater supply. Peatlands are particularly vulnerable, especially those associated with permafrost in high latitudes. Warming may shift the zone of peat formation poleward. The balance between these two processes, precipitation and evapotranpiration, is not known (Roots 1989), but the potential of peatlands to sequester CO_2 or release it as the peat decomposes puts them in a unique position of providing a feedback to the global warming cycle. Sequestration is offset to a degree by the production of methane (CH_4). If peatland shrinks, methane production will decrease. A rapid change in global temperatures, if it occurs, may cause the degradation of southern peatlands more quickly than the build-up of northern ones (Gorham 1991).

Coastal wetlands are particularly vulnerable to climate change and to sea-level rise. In the USA alone, there are about $13\,000\,km^2$ of dry land within 50 cm of high tide (of which $10\,000\,km^2$ is currently undeveloped). Projections made by the US EPA are for a 15 and 34 cm rise during this century (Titus and Narayanan 1995), with up to a 65 cm rise possible (see also French and Spencer, 2001). A 50 cm rise would eliminate between 17% and 43% of the wetlands, and half of the loss would occur in Louisiana (Titus and Narayanan 1995) which along with the southern Atlantic coast is subsiding. Titus *et al.* (1991) estimated that an area approximately the size of Massachusetts would need to be abandoned if coastal wetlands were to be allowed to migrate inland as the water-levels rise. Migration would be possible in some areas, but the total wetland area would shrink, because slopes inland of most wetlands are steeper than in the present wetlands themselves.

Agriculture

As so much of the Earth's surface is cultivated, changes in agriculture are very relevant to nature conservation, but the likely changes here are far from clear. Various scenarios have been investigated (Parry 1990; Rosenweig and Parry 1993, 1994), and the general conclusion is that global cereal production is likely to fall by 5%. The most marked decline may be in developing countries (9–11%), and there may even be an increase of 10% in developed countries. Tropical agriculture is particularly vulnerable as crops are often grown near their tolerances of heat and water supply. Yields here would decline, unless water use efficiency were to be increased. Irrigation or fertilisation could achieve this. Rainfall may increase in places like northern Australia and this might encourage more rangeland production, but in many arid areas, where the major limiting factor is phosphorus, an improvement in climate will not necessarily lead to greater productivity (Stott 1994). In the UK, a temperature rise of 1.6°C is expected to reduce cereal yields by 5–10% because of a reduction of the maturation period. Root crops might not be affected in the same way, and would benefit from higher temperatures, to give 5–10% increases in yield, providing water does not become limiting. The potential reduction in cereal yields would probably be offset by increases by CO_2 fertilisation and technological advances, and overall wheat yields are expected to rise from 6.94 t ha^{-1} in 1986/90 to 10 t ha^{-1} by 2050.

A study by the US Department of Agriculture (Schimmelpfennig *et al.* 1995) concluded that even a 3–5°C warming would not significantly affect agricultural production, if farmers were willing to adopt new farming strategies. Changes in crops, fertiliser application and growing seasons could maintain current yields. Calculations made without considering CO_2 fertilisation show that yields could fluctuate between −24% to 24% by the time atmospheric CO_2 doubles. The CO_2 fertilisation effect, which remains controversial, has been estimated to produce gains of between US$119 billion to US$197 billion over the same period. In many regions, however, there may be a decline in the availability of water for irrigation. Regional dislocations might occur if crops and livestock production areas were shifted northwards.

IMPLICATIONS FOR CONSERVATION

Responses to the threats and opportunities posed by global warming depend first upon the confidence placed upon predictions of climate change. The chain of uncertainty (Figure 10.5) leaves plenty of scope for discussion. However, the threat is real enough to make a review of conservation strategies wise, not least to ensure that ecosystems are resilient enough to adapt.

Strategies may be directed to the causes of climate change, mainly the emissions of greenhouse gases, but also to the management of nature itself. It is the latter with which this chapter is primarily concerned. Given the global scale of climate change it is appropriate for management strategies to be developed at the international scale. Table 10.6 lists recent targets set at the Kyoto Conference in Japan. These are intended to limit emissions from the industrialised nations. Despite the uncertainties at the national level, this is the only realistic scale for action.

THE NEED FOR NATIONAL STRATEGIES

The IPCC has produced a summary of possible changes to ecosystems because of global warming (Houghton *et al.* 1996). These include:

- A poleward shift of vegetation zones by 150–550 km and an altitude shift of 150–500 m.
- Drought-sensitive plants will be vulnerable despite the higher CO_2 levels.
- Secondary effects through soils, fires, pests and disease may be as great as direct climatic impacts.
- The geographical position of wetlands will shift.
- Warmer rivers and lakes will affect cool-water species.
- Permafrost in high latitudes will shrink.
- Coastal ecosystems are particularly at risk.

Table 10.6. Commitments to limit or reduce emissions of equivalent CO_2 from 1990 to 2010

Countries/regions[1]	Allowed (%) 1990–2010[1]	Observed (%) 1990–95	World % of CO_2 Emissions in 1990[2]
European Union	−8	−1	
UK	−8	−4	2.78
Portugal	−8	+49	
Germany	−8	−9	4.6
OECD (excluding EU)	−6	+8	
USA	−7	+7	23.45
Japan	−6	+8	4.93
Economic Transition	−1	−29	
Developing		+25	

[1] After Bolin (1998).
[2] After Anon. (1994).

Walker and Steffan (1997) also noted that:

- Species will have different response rates, and this will produce new communities.
- Vegetation changes predicted by the newer transient GCMs are different from those predicted by the older equilibrium models.
- Local conditions may strongly influence changes.
- Invasion of non-native species is likely to be a serious problem.

We take two examples of national strategies that are responding to these threats, in the UK and in the USA.

NATURE CONSERVATION STRATEGIES IN THE UK

Natural ecosystems in Europe are highly vulnerable to climate change because they are established on the poorer soils, are highly fragmented and species in mountainous areas have nowhere to migrate (Watson et al. 1998). In 1991 the UK Department of the Environment compiled a report on the possible impacts of climate change in the UK which is summarised here (DoE 1991). The basis for their analysis was a 'business as usual scenario' with CO_2 concentrations rising to 450 ppm by 2030. In the UK this would produce warmer winters than at present (1.4–2.5°C), with a more marked warming in the north, as in the Highlands of Scotland. In summer, the warming would be around 1.4°C (2.1°C by 2050). Precipitation was expected to rise in winter by around 5% (8% by 2050), but there was uncertainty over summer rainfalls which might increase or decrease by up to 16%. The incidence of droughts was expected to rise. It was the change in the occurrence of extreme events such as drought that was most worrying. Summers as hot and dry as that of 1976 might increase in frequency by a hundredfold to 1 year in 10. It was expected that this rate of climate change would have adverse impacts on 50% of statutory protected areas (NNR and SSSIs). Grime (1990) noted that vegetation changes were most likely to take place as a series of spurts in response to extreme weather conditions, as after hot dry summers or stormy winters.

Concern is greatest for montane communities, saltmarshes and other coastal ecosystems, isolated ecosystems and wetlands. The DoE predictions, however, lack detail for specific regions, partly because rainfall intensity or storm frequency, which would affect some regions more than others, could not be predicted with any certainty. But these general predictions are under review. Since the 1991 report, a number of other studies have attempted greater spatial detail. In 1995, the Countryside Commission reported on impacts in the year 2050, using the IPCC 1S92a emissions scenario. This study predicts major change in south-eastern England, and much less in Scotland and the north-west of England. Temperature rises are expected to be greatest in the south-east, where maximum winter temperatures might rise by 1.5°C over those of 1990 (0.8°C in Scotland). The mean UK temperature, 8.3°C (1961–90), is expected to rise to 9.6°C by 2050, equivalent to a latitudinal movement of 200–300 km. Conditions similar to those experienced in Cornwall today would be found across most of England. These predictions and predictions of drought are not as high as they were in earlier reports. The frequency of droughts of the magnitude of the 1976 event is expected to rise from once every 357 years to once every 14 years by 2050. There is less certainty about precipitation. In general, winters are predicted to get wetter and summers drier (see Table 10.7).

A number of possible impacts have been noted for the UK (Countryside Commission 1995; DoE 1991; Grime 1990):

- Warmer springs would affect individual tree species, because of the chilling requirement for bud burst. Early emerging plants such as the snowdrop might suffer from more competition.
- Higher temperatures would threaten alpine plants and fauna such as the mountain hare.
- Saltmarshes and coastal communities might not be able to retreat in the face of sea-level rises or might be adversely affected by higher loads of suspended sediments (see also, French and Reed, 2001). The inhabitants of mudflats such as the redshank might suffer.
- Birds and mammals, being more mobile than plants, should be able to respond, provided there is a suitable environment for them and migratory corridors.
- Species in isolated islands (for example, those surrounded by agriculture) would need the creation of dispersal corridors for survival.
- Heaths might be subject to more frequent burning.
- Fenlands might dry out, especially with higher summer demands for water and overabstraction.
- Moisture stress might make plants more susceptible to pests and diseases.
- More droughts would impact most upon shallow rooting plants such as woodland flowers, as in the impacts observed on dog's mercury in the dry summer of 1990.

Table 10.7. Predicted UK changes in precipitation from 1990 to 2050 (source: Countryside Commission 1995)

	% change	Level of uncertainty
Winter	6.8	3.8 to 9.8 % change
Summer	2.3	−3.7 to +8.1 % change

- There would be greater potential for the invasion of alien weeds, pests and diseases.
- Greater competition between species can be expected, favouring fast-growing 'woody' species and less effective competitors would suffer.
- Trees and other species with a slow migration rate would not be able to keep pace with the movement of the isotherms. Trees that normally come early in the succession process should be less vulnerable.

Several authors have raised the last of these points and voiced their concerns over slow migration rates in the face of rapid warming, yet Walker and Steffan (1997) argue that many species can migrate fast enough in response to the predicted changes, though only if allowed to, as along dispersal corridors. Some species, such as the fast-growing perennials, could benefit from a milder climate. Amphibians and reptiles might be favoured by warmer winters, providing summer droughts were not too severe. Birds such as the stone curlew and woodlark, which are now at the northern limits of their range, might find their ranges extended.

In all the reports on the possible impacts of global warming on the UK countryside, there is a note of caution that human activities may be much more significant than climate change as a determinant of species survival. Herrington (1996) calculated that domestic water demands would rise by 4% by the year 2021 under a higher temperature regime, but that changes such as household composition and an ageing population could produce an increase of 25%. The potential damage caused by overabstraction of surface and groundwater and associated problems of effluent disposal are likely to be more immediate and serious than those posed by climate change.

All the predictions imply the need to maintain or enhance the resilience of existing ecosystems. The best strategy for nature conservation in the face of uncertainty is to maintain diversity and vitality. The need for an integrated approach is also widely recognised, managing changes across both regional and international boundaries. Here the problems are as much political as scientific.

NATURE CONSERVATION STRATEGIES IN THE USA

In 1992, the US National Academy of Sciences issued a study entitled 'Policy Implications of Greenhouse Warming'. This concluded that, despite the uncertainty over the consequences of global warming, its potential threat was great enough to warrant response. Taking action would not only serve as protection against the uncertainties it would also help to mitigate the possibilities of surprises.

Much of the ability to conserve ecosystems in the face of global warming depends on their resilience, and their capacity to change position as climatic conditions shift. In large part this is a function of the dispersal abilities of individual species. If the rate of warming is faster than a species can migrate then extinction is possible (Webb 1992). The average migration rate for North American trees has been estimated to be between 20 and 40 km per century, much slower than the anticipated rate of shift of ecosystem boundaries in response to climate change. Davis and Zabinski (1992) speculate that, while many commercially important species might be re-established by foresters or agriculturalists, the survival of many plant and animal species is uncertain, unless efforts are taken to transplant them on a large scale. The principles of ecosystem restoration will have to be applied to move these ecosystems to new climate zones, and new reserves are called for to prevent the wholesale extinction of many species, particularly late successional ones.

A recent report by the IPCC (Watson *et al.* 1998) on the regional impacts of climate change summarises the key impacts on both ecological and socio-economic systems in North America. The predicted degree of vulnerability to the effects of climate change varies significantly from region to region, implying a need for variable response strategies. Specific impacts alluded to in the report include:

- The occurrence of landslides and debris flows in unstable areas of the Rocky Mountains could become more common as winters become wetter and glaciers retreat.
- The area of arid lands in the south-west may increase.
- The southern boundary of permafrost areas may shift northwards by 500 km by 2050. This would affect ecosystems, wildlife and infrastructure.
- Fish populations in large, deep lakes may respond to warmer water temperatures with increased growth and survival rates. Increased food availability from, for example, increased plankton growth, could increase fish production. Higher temperatures in smaller water bodies, however, may limit the distribution of cold water species, including salmonids.
- Marine mammals and turtles may have lower survival rates and health levels because of changes in food supply, the extent of sea ice, and a reduction in the area of breeding habitats.
- Sea-level rise, as discussed above, may lead to the flooding of significant portions of North American coastal wetlands.

Any decision to abandon a coastal area is dependent on a complex mix of environmental, economic, social and governmental factors. In an exploration of these issues, Titus *et al.* (1991) advocated the abandonment of coastal areas and the permission of coastal ecosystems to migrate inland. Although giving up valuable coastal property may not be necessary for several decades, planning for it now is wise. It will be easier to steer development away from vulnerable areas now rather than abandoning the investments later, and abandoning currently developed areas will be easier if there is sufficient planning. Another option is the establishment of so-called 'rolling easements' which allow development to occur on the condition that the property will not be protected from rising waters in the future (through diking, bulkhead construction, etc.). In this way development has to give way to migrating ecosystems (Titus *et al.* 1991). The state of Maine has adopted rolling easements to protect wetlands in areas developed after 1987.

CONCLUSIONS

The IPCC (Houghton *et al.* 1996) has argued that ecosystems need to be protected against the threats of global warming as they:

- Provide food, fibre and fodder
- Process and store carbon
- Assimilate wastes
- Are a gene reservoir
- Maintain natural resources such as water and soil
- Have intrinsic existence values.

The present discussion has pointed to many of the conservation implications of a warmer world. Forests and wetlands are particularly vulnerable. Some ecosystems may have greater productivity and diversity, but possibly only with intervention. The WWF (1993) recommends that the key requirements of any strategy to combat global warming should be based upon:

- Better information through improved monitoring and scientific research
- Flexibility (enhancing the resilience of species and ecosystems)
- Creating new habitats, as well as maintaining existing ones
- A more holistic (integrated) approach.

These points are reiterated by Walker and Steffen (1997) and they, the IGBP (1990), and Watson *et al.* (1998) all emphasise the need for greater understanding of the biological consequences of climate change. In particular, the following research needs are identified:

- Better baseline data in order to assess change
- Better regional climate-change scenarios
- Greater understanding of the interactive effects of ecosystem physiology and structure
- Better quantitative understanding of the relations between resilience and complexity

- Understanding how ecosystem adaptation could be enhanced.

In response to the threats posed by global warming the WWF (1993) proposes a number of strategies:

- Nature reserves should be large enough to enable spatial shifts to take place.
- Reserves should contain a range of altitudinal and latitudinal zones.
- Coastal sites should be able to cope with sea-level rise.
- Connecting dispersal corridors should support reserves.

REFERENCES

Agnew CT (1995) Desertification, drought and development in the Sahel. In: A Binns (ed.) *People and environment in Africa.* Chichester, John Wiley, 137–49.

Anon. (1994) *Climate change in the UK programme.* London, HMSO.

Archibold OW (1995) *Ecology of world vegetation.* London, Chapman & Hall.

Balling RC & Wildavsky A (1992) *The heated debate. Greenhouse predictions vs. climate reality.* San Francisco, Pacific Research Institute for Public Policy.

Barry RG & Chorley RJ (1992) *Atmosphere, weather and climate* (6th edn). London, Routledge.

Bate R (ed.) (1998) *Global warming: the continuing debate.* Cambridge, European Science and Environment Forum.

Bate R & Morris J (1994) *Global apocalypse or hot air?* London, IEA.

Bayley PB (1995) Understanding large river–floodplain ecosystems. *Bioscience* **45**, 153–63.

Bazzaz FA & Fajer E (1992) Plant life in a CO_2 rich world. *Scientific American* **266**(1), 18–23.

Begg J & Turner N (1976) Crop water deficits. *Advances in Agronomy* **28**, 161–217.

de Blij HJ & Muller PO (1993) *Physical geography and the global environment.* Chichester, John Wiley.

Bolin B (1998) The Kyoto negotiations on climate change: a science perspective. *Science* **279**, 330–1.

Bolin B, Doos BR, Jager J & Warrick RA (1991) *The greenhouse effect, climate change and ecosystems.* Chichester, John Wiley.

Briggs D & Smithson P (1992) *Fundamentals of physical geography.* London, Routledge.

Brown, P. (1996) *Global warming: can civilization survive?* London, Blandford.

Bryson RA (1974) A perspective on climatic change. *Science* **184**, 753–60.

Camberlin P (1995) June–September rainfall in North East Africa and atmospheric signals over the tropics. *International Journal of Climatology* **15**, 773–83.

Cannel M & Pitairn C (1993) *Impacts of mild winters and hot summers 1988–1990.* London, Department of the Environment, HMSO.

Carpenter SR, Fisher SG, Grimm N & Kitchell JF (1992) Global change and freshwater ecosystems. *Annual Review of Ecology and Systematics* **23**, 119–39.

Caviedes CN (1988) The effects of ENSO events in some key regions of the South American continent. In: S Gregory (ed.) *Recent climatic change.* London, Belhaven Press, 252–66.

Countryside Commission (1995) *Climate change, air pollution and the English countryside, summary of potential impacts.* Walgrave, Countryside Commission.

Courtney RS (1993) The end of the world is not nigh. *Coal Transactions* **8**(3), 40–2.

Davis MB & Zabinski C (1992) Changes in geographical range resulting from greenhouse warming: effects on biodiversity in forests. In: RL Peters & TE Lovejoy (eds) *Global warming and biological diversity*. New Haven, Yale University Press, 297–308.

Denmead OT & Shaw RH (1962) Availability of soil water to plants as affected by soil moisture content and meteorological conditions. *Agronomy Journal* **54**, 385–90.

Deshmukh I (1986) *Ecology and tropical biology*. London, Blackwell.

Department of the Enviroment (DoE) (1991) *The potential effects of climate change in the UK*. London, HMSO.

Doorenbos J & Pruitt WO (1984) Crop water requirements. *FAO Irrigation and Drainage Paper 24*. Rome, FAO.

Economist (1995) Reading the patterns: the evidence that greenhouse gases are changing the climate. **143**(2), 109–11.

Emanuel WR, Shugart HH & Stevenson MP (1985) Climate change and the broad scale distribution of terrestrial ecosystems complexes. *Climate Change* **7**, 29–43.

Flohn H (1987) Rainfall teleconnections in northern and eastern Africa. *Theoretical and Applied Climatology* **38**, 191–7.

Folland CK (1987) Sea temperatures predict African drought. *New Scientist* **116**(1580), 25.

Forbes JC & Watson RD (1992) *Plants in agriculture*. Cambridge, Cambridge University Press.

Franzén LG (1994) Are wetlands the key to the ice-age cycle enigma? *Ambio* **23**, 300–8.

French JR & Reed DJ (2001) Physical contexts for saltmarsh conservation. In: A Warren & JR French (eds) *Habitat conservation: managing the physical environment*. Chichester, John Wiley, 179–228.

French JR & Spencer T (2001) Sea-level rise. In: A Warren & JR French (eds) *Habitat conservation: managing the physical environment*. Chichester, John Wiley, 305–47.

Gates DM (1993) *Climate change and its biological consequences*. Sunderland, Mass., Sinauer Associates Inc.

Glantz MH (1987) Drought in Africa. *Scientific American* **256**(6), 34–40.

Glantz MH, Katz RW & Nicholls N (eds) (1991) *Teleconnections: linking worldwide climate anomalies*. Cambridge, Cambridge University Press.

Goldstein G & Sarmiento G (1987) Water relations of trees and grasses and their consequences for the structure of savanna vegetation. In: BH Walker (ed.) *Determinants of tropical savannas*. Oxford, IRL Press, 13–38.

Good R (1970) *The geography of flowering plants*. New York, John Wiley.

Gordon A, Grace W, Schwerdtfeger P & Bryon-Scott R (1998) *Dynamic meteorology*. New York, John Wiley.

Gorham E (1991) Northern peatlands: role in the carbon cycle and probable response to climatic warming. *Ecological Applications* **1**, 182–95.

Gowing JW & Young MDB (1996) *Evaluation and promotion of rainwater harvesting in semi-arid areas*. Final Technical Report, University of Newcastle Upon Tyne.

Graham R & Grimm E (1990) Effects of global climate change on the patterns of terrestrial biological communities. *Trends in Ecology and Evolution* **5**(9), 89–292.

Grainger A (1996) Forest environments. In: W Adams, AS Goudie & AR Orme (eds) *The physical geography of Africa*. Oxford, Oxford University Press, 173–95.

Gregory P (1989) Water use efficiency of crops in semi-arid tropics. *ICRISAT workshop on soil, crop and water management systems for rainfed agriculture in the Sudano-Sahelian zone at Niamey, Niger*. Andhra Pradesh, ICRISAT, 85–98.

Gregory S (1988) El-Niño years and the spatial pattern of drought over India 1901–70. In: S Gregory (ed.) *Recent climatic change*. London, Belhaven Press, 226–36.

Gribben J & Gribben M (1996) The greenhouse effect. *New Scientist* **151**(2037), 1–4.

Grime JP (1990) The impact of climate change on British vegetation. In: RW Battarbee & S Patrick (eds) *The greenhouse effect: consequences for Britain*. ECRC, University College London, 31–7.

Gupta SC & Larson WE (1979) Estimating soil water retention characteristics from particle size distribution, organic matter percent and bulk density. *Water Resources Research* **15**, 1633–5.

Hayes JT (1991) Global climate change and water resources. In: RL Wyman (ed.) *Global climate change and life on Earth*. London, Routledge, 18–42.

Henderson-Sellers A & McGuffie K (1987) *A climate modelling primer*. London, John Wiley.

Herrington P (1996) *Climate change and the demand for water*. London, HMSO.

Holdridge LR (1947) Determinations of world plant formations from simple climatic data. *Science* **105**, 367–8.

Houghton JT (1997) *Global warming: the complete briefing*. Cambridge, Cambridge University Press.

Houghton JT, Jenkins GJ & Ephraums JJ (eds) (1991) *Climate change, the IPCC 1990 scientific assessment*. Cambridge, Cambridge University Press.

Houghton JT, Callander BA, & Varney SK (1992) *Climate change: IPCC 1992, supplementary report*. Cambridge, Cambridge University Press.

Houghton JT, Meira Filho LG, Callander BA, Harris N, Kattenberg A & Maskell K (1996) *Climate change: IPCC 1995, the science of climate change*. Cambridge, Cambridge University Press.

Hoyt D (1999) Greenhouse warming: fact, hypothesis, or myth? http://users.erols.com/dhoyt1/index.html (accessed 5/8/1999).

Hulme M (1996) *Climate change and Southern Africa*. Norwich, CRU/WWF.

Huntley B (1991) How plants respond to climate change: Migration rates, invididualism and the consequences for plant communities. *Annals of Botany* **67**, 15–22.

International Geosphere Biosphere Programme (1990) Global change: the initial core projects. *IGBP Report No. 12*, Stockholm.

Jackson IJ (1989) *Climate, weather and agriculture in the tropics*. Harlow, Longman Scientific and Technical.

Jones HG (1992) *Plants and microclimate*. Cambridge, Cambridge University Press.

Köppen W (1931) *Die Klimate der Erde*. Berlin.

Kozlowski TT (ed.) (1983) *Water deficits and plant growth*. New York, Academic Press.

Kramer PJ (1983) *Water relations of plants*. Orlando, Academic Press.

Lal R (1991) Current research on crop water balance and implications for the future. In: MVK Sivakumar, JS Wallace, C Renard & C Giroux (eds) *Soil water balance in the Sudano-Sahelian zone*. IAHS Publication 199, Wallingford, 31–44.

Leggett J (1990) *Global warming*. Oxford, Oxford University Press.

Le Houérou HN & Popov GF (1981) *An ecological classification of inter-tropical Africa*. Rome, FAO.

Lindzen RS (1991) Review of climate change. *Quarterly Journal of the Royal Meteorological Society* **117**, 651–2.

Lockwood JG (1984) The southern oscillation and El Niño. *Progress in Physical Geography* **8**, 102–10.

McIlveen R (1992) *Fundamental of weather and climate*. London, Chapman & Hall.

Mason SJ (1995) Sea surface temperatures and South African rainfall associations. *International Journal of Climatology* **15**, 119–35.

Mather JR (1974) *Climatology, fundamentals and applications.* New York, McGraw-Hill.

Meadows ME (1996) Biogeography. In: W Adams, AS Goudie & AR Orme (eds) *The physical geography of Africa.* Oxford, Oxford University Press, 161–72.

Monteith JL (1986) How do plants manipulate the supply and demand of water? *Philosophical Transactions of the Royal Meteorological Society of London* **A316**, 245–59.

Monteith JL & Unsworth MH (1990) *Principles of environmental physics.* London, Edward Arnold.

National Academy of Sciences (1992) *Policy implications of greenhouse warming: mitigation, adaptation, and the science base.* Washington, DC, National Academy Press.

Olstead, J. (1993) Global warming in the dock. *Geographical Magazine* **65**(9), 12–17.

Owen JA & Folland CK (1988) Modelling the influence of sea surface temperatures on tropical rainfall. In: S Gregory (ed.) *Recent climatic change.* London, Belhaven Press, 141–53.

Parry M (1990) *Climate change and world agriculture.* London, Earthscan.

Pearce F (1995a) Fiddling while earth warms. *New Scientist* **145**(1970), 14–15.

Pearce F (1995b) Global warming: the jury delivers guilty verdict. *New Scientist* **148**(207), 14.

Pearce F (1997) Global warming chills out over the Pacific. *New Scientist* **153**(2070), 16.

Peters RL (1991) Consequences of global warming for biological diversity. In: RL Wyman (ed.) *Global climate change and life on Earth.* London, Routledge, 99–118.

Poff NL (1991) Regional hydrologic response to climate change: an ecological perspective. In: P Firth & G Fisher (eds) *Climate change and freshwater ecosystems.* New York, Springer-Verlag, 88–115.

Ponte L (1976) *The cooling.* Englewood Cliffs, NJ, Prentice-Hall.

Prentice IC, Cramer W, Harrison SP, Leemans R, Monserud RA & Solomon AM (1992) A global biome model based on plant physiology and dominance soil properties an climate. *Journal of Biogeography* **19**, 117–34.

Rahmstorf S (1997) Ice cold in Paris. *New Scientist* **153**(2068), 26–30.

Reading A, Thompson RD & Millington AC (1995) *Humid tropical environments.* Oxford, Blackwell.

Roberts N (1994) The global environmental future. In: N Roberts (ed.) *The changing global environment.* Oxford, Blackwell, 3–21.

Robertson AW & Frankignoul C (1990) The tropical circulation: a simple model versus a general model. *Quarterly Journal of the Royal Meteorological Society* **116**, 69–87.

Roots EF (1989) Climate change: high latitude regions. *Climate Change* **15**, 223–53.

Ross S (1991) Atmospheres and climate change. In: PM Smith & K Warr (eds) *Global environmental issues.* London, Hodder & Stoughton, 73–120.

Rozenzweig C & Parry, ML (1993) Potential impacts of climate change on world food supply. In: HM Kaiser & TE Drennen (eds) *Agricultural dimensions of global climate change.* Delray Beach, St Lucie Press.

Rozenzweig C & Parry ML (1994) Potential impacts of climate change on world food supply. *Nature* **367**(6459), 133–8.

Schimmelphennig DJ, Lweandrowski J, Reilly M, Tsigas, J & Perry I (1995) *Agricultural adaptation to climate change: issues of longrun sustainability.* AER-740, Natural Resources and Environment Division, Economic Research Service, US Department of Agriculture.

Semeniuk V (1994) Predicting the effect of sea level rise on mangroves in North-Western Australia. *Journal of Coastal Research* **10**, 1050–76.

Shaw E (1994) *Hydrology in practice.* London, Chapman & Hall.

Sinha SK (1991) Impacts of climate change on agriculture. In: J Jager & HL Ferguson (eds) *Climate change: science, impacts and policy*. Cambridge, Cambridge University Press, 99–107.

Smith M (1990) CROPWAT A computer program for irrigation planning and management. *FAO Irrigation and Drainage Paper 46*. Rome, FAO.

Smith M (1992) *Expert consulation on revision of FAO methodologies for crop water requirements*. Rome, FAO.

Stott P (1994) Savanna landscapes and global environmental change. In: N Roberts (ed.) *The changing global environment*. Oxford, Blackwell, 287–303.

Thornthwaite CW (1948) An approach towards a rational classification of climate. *Geographical Review* **38**, 55–94.

Titus JG & Narayanan V (1995) *The probability of sea level rise*. US Environmental Protection Agency Publication #230-R-95-008, Washington DC.

Titus JG, Park RA, Leatherman SP, Weggel JR, Greense MS, Mausel PW, Brown S, Gaunt C, Trehan M & Yohe G (1991) Greehouse effect and sea level rise: the cost of holding back the sea. *Coastal Management* **2**, 171–210.

UNESCO (1977) *Map of the world distribution of arid regions*. MAB Technical Note 7, Paris.

Walker B & Steffen W (eds) (1997) *A synthesis of GCTE and related research. The terrestrial biosphere and global change: implications for natural and managed ecosystems*. IGBP Report No. 1, Stockholm.

Warrick RA, Shugars HH, Antionovsky MJ, Tarrant JR & Tucker CJ (1991) The effects of increased CO_2 and climate change on terrestrial ecosystems. In: B Bolin, BR Doos, J Jager, & RA Warrick (eds) *The greenhouse effect, climate change and ecosystems*. Chichester, John Wiley, 363–92.

Warrick RA, Barrow E & Wigley T (eds) (1993) *Climate and sea level change: observations, projections and implications*. Cambridge, Cambridge University Press.

Watson RT, Zinyowera MC, Moss RH & Dokkan DJ (eds) (1998) *The regional impacts of climate change: IPCC 1998*. Cambridge, Cambridge University Press.

Webb T (1992) Past changes in vegetation and climate: lessons for the future. In: RL Peters & TE Lovejoy (eds) *Global warming and biological diversity*. New Haven, Yale University Press, 59–75.

Wetherald RT (1991) Changes in temperature and hydrology caused by an increase of atmospheric carbon dioxide as predicted by GCMs. In: RL Wyman (ed.) *Global climate change and life on Earth*. London, Routledge, 1–17.

Wigley TML & Raper SCB (1987) The global expansion of sea water associated with global warming. *Nature* **330**(6144), 127–31.

Woodward F (1987) *Climate and plant distribution*. Cambridge, Cambridge University Press.

Woodward F, Thompson G, & McKee I (1991) The effects of elevated concentrations of carbon dioxide on individual plants, populations, communities and ecosystems. *Annals of Botany* **67**, 23–38.

WWF (1993) *Some like it hot*. Gland, World Wildlife Fund.

Wuethrich B (1995) El-Niño goes critical. *New Scientist* **145**(1963), 32–5.

Wyman RL (ed.) (1991) *Global climate change and life on earth*. Andover, Chapman & Hall.

11 Sea-level rise

J. R. FRENCH[1] AND T. SPENCER[2]
[1]*University College London, UK*
[2]*University of Cambridge, UK*

INTRODUCTION

Sea-level is of importance for the definition of shoreline position and, with superimposed tidal, wave and meteorological influences, as a crucial determinant of landform and habitat extent in coastal and estuarine settings. Modern sea-level research commenced in the nineteenth century following the acceptance within the geological science community that repeated large-amplitude fluctuations in sea-level have occurred in conjunction with glacial episodes, and of the importance of sea-level as a fundamental control on continental sedimentation and stratigraphy. The emergence of Quaternary science as a recognisable subdiscipline stimulated increasingly detailed investigations of regional and local sea-level history, partly in support of glacial studies, but also as a means of elucidating the evolution of specific landform–habitat assemblages (for a recent review, see Carter 1992). Holocene studies have been of particular value in providing a context for the interpretation and utilisation of modern coastal and estuarine environmental settings (Braatz and Aubrey 1987; Devoy 1987; Shennan 1987).

The period since the mid-nineteenth century has also witnessed the establishment, expansion and technological sophistication of sea-level monitoring networks. In particular, the archive of world tide gauge records maintained since 1933 by the UK Permanent Service for Mean Sea-Level (Woodworth 1991) constitutes an invaluable dataset to support the analysis of historical sea-level trends (Gornitz 1995). Since the 1970s oceanographers and geophysicists have provided new insights into the very nature of sea-level that add an important spatial dimension to these temporal, and essentially *coastal*, studies. Satellite altimetry of the oceans, now possible to an accuracy of a few cm (Robinson 1985), reveals a complex topography which reflects the close relationship between average sea-level and the surface of gravitational and rotational potential which constitutes the global geoid (Mörner 1980). One of the more significant findings has been the discovery of regionally coherent monthly and inter-annual variations in mid-ocean sea-level, notably in the

Habitat Conservation: Managing the Physical Environment. Edited by A. Warren and J. R. French.
© 2001 John Wiley & Sons Ltd.

tropical Pacific, where they have been used to monitor and forecast the development of El Niño events (Wyrtki 1985). Sea-level is thus a potentially sensitive indicator of climate and ocean dynamics as well as a major environmental forcing variable.

Since the mid-1980s, well-publicised scientific research into effects of 'greenhouse' gas emissions upon climate and ocean behaviour has established global sea-level rise as one of the key issues driving contemporary environmental debates (Barth and Titus 1984; National Research Council 1987; Houghton *et al.* 1996). The prospect of an accelerated rise in sea-level over this century and beyond has been viewed by conservationists as tangible evidence of the scale of human impacts and as a threat to coastal habitats already diminished in extent and quality through continuing development. These environmental issues are invariably embedded in broader economic and societal concerns. From a planning perspective, even a small increase in mean sea-level represents a serious threat to an estimated 60% of the world's population, and their economic and cultural activities, which are presently located within the 'coastal zone' (as broadly defined by UNEP 1990). Coastal and estuarine lowlands exhibit a delicate balance between their geological legacy, present physical and ecological processes, and constraints imposed by centuries of engineering intervention for the purposes of stabilisation and flood protection. The emergence of a global component of sea-level rise, accelerating in rate over this century and beyond, would almost inevitably exacerbate *existing* engineering and habitat loss problems and would *extend* their geographical scope to areas presently unaffected.

Following the terminology adopted by the Intergovernmental Panel for Climate Change (IPCC) in 1990, there are three main elements of the sea-level rise issue:

• Processes
• Impacts
• Response measures

Leaving aside the doubts expressed in some quarters concerning the validity of the greenhouse gas models and the interpretation of climate model results in the light of historical temperature records (Lindzen 1990; Bate 1998; Agnew and Fennessy 2001), much uncertainty still surrounds the *processes* by which sea-level responds to atmospheric temperature changes. For a specific climate change scenario, different model assumptions concerning, *inter alia*, the thermal expansion of the upper ocean and the rate and timing of ice melt, give rise to a considerable spread of projected sea-level curves. A common element in recent studies, however, is the sharp acceleration in the rate of rise projected for the middle of this century. This is significant, not merely in terms of the ability of coastal sedimentary landforms and associated ecosystems to accommodate high rates of environmental forcing, but also in that it reduces the time interval available for responsive mitigative action (such as a switch between alternative greenhouse gas emissions policies; Gornitz, 1995).

Assessments of the *impacts* of global sea-level rise have been strongly influenced by projections from modelling studies published in the mid-1980s, some of which envisaged a *maximum* rise of several metres by 2100 (see, for example, Hoffman 1984). In the developed world, attention has focused on the increased flood risk to land protected by seawalls, and morphodynamic adjustments which could lead to more rapid and more widespread erosion and damage to coastal property. Major changes in the location, extent and character of intertidal habitats have been identified as an important, though secondary, issue. In the tropics, media attention has been directed towards the ability of lowlying reef islands to maintain their physical and ecological integrity in the face of not just sea-level rise but also increasing stress due to more frequent storm events and higher sea-surface temperatures (Buddemeier and Smith 1988; Spencer 1995). These concerns have evolved in parallel with an enhanced appreciation of the functions, values and fragility of key habitat types (Mitsch and Gosselink 1993; see also Thompson and Finlayson 2001) and the extent to which natural habitats continue to be degraded and lost through human activity. It is difficult to isolate environmental and anthropogenic effects. However, Nicholls, Hoozemans and Marchand (1999) estimate that up to 70% of the world's coastal wetlands could be lost by the 2080s, and that about 22% could be lost through the effects of sea-level rise alone.

Responses to sea-level rise are likely to be driven first and foremost by the need to mitigate the effects of flood risk and erosional adjustments which impinge upon population centres and sites of high economic value, and to do so at tolerable cost to both central government and local communities. For key locations — major population centres and industrial facilities — these concerns will likely take precedence over local environmental concerns. However, there is a growing sense of shared purpose among coastal engineers, geomorphologists and conservationists. This partly reflects the greater value placed by society on 'environmental capital'. Significantly, there is also an increasing awareness of the role of beaches, dunes and saltmarshes as an integral part of more cost-effective and sustainable defences. Within areas subject to relative sea-level rise of geological origin, engineers, geomorphologists and conservation organisations are already working together in the interests of sustainability. More informed approaches to the management of sea-level rise can thus provide *opportunities* for conservation (Arens, Jungerius and van der Meulen 2001; French and Reed 2001). These are well exemplified by 'soft engineering' approaches to erosion control (for example, Child 1996; French *et al.* 2000) and by adaptive flood defence strategies based upon coastal realignment (Titus 1991; Agriculture Select Committee 1998). Thus, although economic considerations are likely to prevail, habitat conservation and recreation can be expected to play a significant role in future flood defence and erosion management schemes (Dixon, Leggett and Weight 1998).

From a conservation perspective, there are several obstacles to effective planning for sea-level rise. First, there is the ambiguous nature of the evidence for change and the uncertainty inherent in models of environmental change and their outputs. Second, complex scientific arguments concerning the linkage between greenhouse gas emissions and climate and sea-level responses are often poorly represented by the media and popular press. Third, many important habitats remain poorly understood in terms of the physical and biotic processes which combine to produce distinctive habitat characteristics. All these are linked, in that journalism thrives on the contradictions presented by the extremes of scientific opinion. Equally, environmental interest groups have been quick to utilise the upper bounds of scientific projections in order to galvanise public support for their message. In reality, sea-level changes are likely to be more subtle, but the geomorphological and ecological responses more complex, than implied in the popular literature.

The aim of this review is to equip conservationists and environmental managers with the up-to-date appreciation of the realities and subtleties of sea-level rise that is necessary to inform such planning and decision making. The emphasis is primarily on the *processes* of sea-level rise, although some consideration is given to the generalities of its *physical impacts*. Specifically, the following questions are addressed:

The nature of sea-level rise and its variability
- What *is* 'mean sea-level' and how is it measured?
- What variability and trend in sea-level is evident from the instrumental record?

Accelerated sea-level rise and climate change
- Is there any evidence for globally coherent sea-level rise due to climate change?
- Is there a consensus over predicted global sea-level rise for this century?

Physical impacts of sea-level rise
- How does sea-level rise operate as an agent of coastal change?
- What kinds of physical habitat changes are likely under current 'best-estimate' scenarios?

Those involved in coastal conservation and management need to be cognisant of these aspects of the *physical* environment, not least, in order that the right kinds of questions are asked of those involved in the prediction of environmental change.

SEA-LEVEL AND ITS VARIABILITY

PROCESSES DRIVING SEA-LEVEL CHANGE

Over 'geological time', global sea-level has experienced fluctuations of several hundred metres in amplitude, with major highstands up to 350 m above present level occurring in Ordovician and Cretaceous time (Haq, Hardenbol and Vail 1987). In this context, the repeated fluctuations in sea-level associated with the Quaternary glaciations appear modest: at the last glacial maximum (approximately 18 000 years BP), average sea-level stood around 120 m lower than at present (Matthews 1990). Numerous processes combine to produce such variations at both local and regional scales: some of the more important of these are summarised in Table 11.1. Conventionally, a distinction is made between

Table 11.1. Major processes resulting in secular trend and interannual variability in Mean Sea Level (modified from Gornitz 1995; with additional information from Shennan 1993; Mörner 1983; Sahagian, Schwartz and Jacobs 1994)

Process	Rate (mm yr^{-1})	Timescale (yr)
SECULAR TRENDS		
Eustasy		
Tectono-eustasy	± 0.001–0.1	10^3–10^8
Glacio-eustasy	± 1–10	10^3–10^5
Regional (100–1000km) land movements		
Glacio-isostasy	± 1–10	10^4
Lithospheric cooling and sediment loading	0.03	10^7–10^8
Local (< 100 km) land movements		
Neotectonic uplift/subsidence	± 1–10	10^2–10^4
Shelf sedimentation; delta plains	1–5	10–10^4
Anthropogenic processes		
Water impoundment (reservoirs)	-0.5–0.75	< 100
Groundwater extraction (via river runoff)	0.4–0.7	< 100
Deforestation and wetland loss	0.2	< 100
Subsidence due to water, hydrocarbon, mineral extraction (very local)	1-5 +	< 100

	Amplitude (cm)	Period (yr)
INTERANNUAL VARIABILTY		
Geostrophic currents	1–100	1–10
Low-frequency atmospheric forcing	1–4	1–10
El Niño	10–50	1–3

eustatic and isostatic processes. *Eustatic* effects, resulting from changes in the volume of ocean water, are most obviously associated with glacial/interglacial cycles. However, tectono–eustatic changes result from physical alterations in ocean volume due, for example, to variation in the rate of seafloor spreading (Hays and Pitman 1973; Mörner 1983). *Isostatic* effects frequently dominate relative sea-level rise as recorded at the coast. Very slow, long-range movements are associated with sedimentation on continental shelves and with cooling (and sinking) of the ocean lithosphere. More rapid isostatic changes result from deformation of the crust during glacial loading and unloading.

Changes in sea-level over long periods of geological time are only peripherally relevant to the practical task of managing sea-level rise over 'human' timescales, measured in decades or centuries. However, it is important to note that research into Quaternary (and particularly Holocene) sea-levels undertaken over the last 20 years has fundamentally altered our perception of the sea-level. At one level, a debate has raged over the nature and definition of eustasy (see, for example, Mörner 1986; Fairbridge 1989). Whereas scientists previously sought to define a truly global eustatic curve, it is now realised that the sea surface closely approximates the geoid (Mörner 1980) and, as such, is subject to complex spatial and temporal perturbations that preclude identification of a universal eustatic curve. Eustatic changes are now considered to be global in scope (i.e. due to changes in the amount of water in the ocean), but regionally varied in magnitude (Pirazzoli 1996). At another level, attention has shifted to the reconstruction of ever more refined regional and local relative sea-level histories (well exemplifed by the various contributions in Tooley and Shennan 1987). The interaction of eustatic and isostatic factors can be particularly complex in areas directly influenced by glacial and deglacial tectonics (see, for example, Quinlan and Beaumont 1981). These interactions can be generalised at a regional scale (Figure 11.1) to give a series of characteristic relative sea-level signatures (Clark, Farrell and Peltier 1978) against which to evaluate more detailed reconstructions of coastal response.

Holocene sea-level change is of special relevance in that it exerts a first-order control on the large-scale evolution of coastal landforms and their associated habitats. Climatic forcing is the key control at this scale, with changes in global ice-balance producing rapid fluctuactions in sea-level which have a direct influence on the physical (Anderson and Thomas 1991) and ecological (for example, Woodroffe and Grindrod 1991) dynamics of the coastal zone. The rate and timing of these effects is strongly conditioned by geographical variations in the history of deglaciation, of the kind envisaged by Clark, Farrell and Peltier (1978). Thus, some coastlines have experienced long periods of relatively stable sea-level, while others are still adjusting to the continued effects of post-glacial rise or fall.

Both eustatic and isostatic processes contribute significant variation in sea-level over human timescales (here measured in decades to centuries). The link

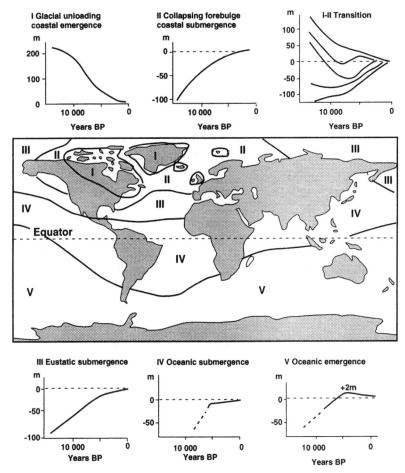

Figure 11.1. Regionalised Holocene sea-level curves resulting from contrasting deglaciation histories (modified from Clark, Farrell and Peltier 1978, with permission of Academic Press).

with climate is strong and includes interannual variability which often dominates instrumental records. However, anthropogenic activity can also lead to small, though quantifiable (Table 11.1) changes in global sea-level which complicate the direct investigation of the climate–ocean linkage.

MEAN SEA-LEVEL

At human timescales, direct insights into variations in sea-level can be obtained from coastal gauges deployed at fixed locations (UNESCO 1985). Such records provide a basis for establishing current trends and, potentially, diagnostic evidence of climate change. The interpretation of instrumental records necessarily leads to an emphasis on the properties of sea-level time-series, and it is instructive to consider the ways in which a value for mean sea-level are obtained from such data. Pugh (1987) provides a convenient representation of observed sea-level, $X(t)$, as a time dependent function of the form:

$$X(t) = Zo(t) + T(t) + S(t) \qquad (11.1)$$

where $Zo(t)$ is the mean-sea level, $T(t)$ is the tidal variation in water level, $S(t)$ is the variation in level due to meteorological influences (often referred to as the surge component), and t is time. T and S contribute, respectively, periodic and aperiodic variability at relatively short timescales (but see also below). Zo is also a time-dependent function, since the mean sea-level itself exhibits variability both within and beyond the duration of an instrumental record. Within-record variation in Zo is usually small in comparison with that due to T and S, although clearly this is not true over the geological time scales referred to above. This chapter is concerned with progressive variation (i.e. secular trend) in Zo.

The term 'mean sea-level' is commonly deployed in two slightly different senses. First, mean sea-level at a particular location is determined from averaging or filtering hourly measurements over a period of at least a year and, where possible, over a period of at least 19 years in order to average the 18.6-year nodal variation in tidal level (Doodson and Warburg 1941; Pugh 1987). Mean sea-level, in this sense, serves as a convenient reference for navigation and survey purposes, and provides a hypothetical level for the sea surface in the absence of tidal action. Second, mean sea-level is derived from monthly or annual averaging in order to specifically highlight seasonal and interannual variability and longer-term trends in level. Monthly values often exhibit quite complex variability due to low frequency tidal components (including 18.6-year nodal tide and seasonal variations), meteorological effects (related to air pressure and its gradients) and residuals due to measurement errors and unmodelled oceanographic factors (such as seasonal changes in temperature and salinity).

An important consequence of time-averaging is that the accuracy of sea-level measurement is much improved. As Pugh (1987) observes, a single hourly measurement of water level may be accurate to, say, 0.01 m (relative to the local tidal benchmark). Averaging even a hundred hourly values, will reduce this error to 0.001 m. Thus it is meaningful to search for very subtle *relative* variability and trend within the instrumental record. However, *absolute* accuracy is still constrained by the accuracy and stability of the local benchmark. Disturbances to benchmarks, and to the gauges themselves, present a major impediment to the recovery of continuous records over periods of more than a few decades.

VARIABILITY AND TREND IN THE INSTRUMENTAL RECORD

Tide Gauge Data

The first modern tide records were established in Amsterdam in 1682 (Van Veen 1954). These measurements, and others made at Stockholm (since 1704) and Brest (since 1807), relied on visual observations of a graduated tide pole: the first self-recording tide gauge was installed in 1831, at Sheerness, on the Thames estuary (Pugh 1987). Subsequently, thousands of tide gauges have been established worldwide, primarily for navigational and port engineering purposes, rather than for science. Since 1933, the UK Permanent Service for Mean Sea Level (PSMSL; Woodworth 1990) has maintained a database of records from more than 1800 stations around the world. The PSMSL screened dataset, based on stations where datum stability and record integrity has been checked, currently includes data from around 1000 stations.

Statistically, the PSMSL dataset is affected by a number of quality problems. One of the most critical evaluations is that by Gröger and Plag (1993) who identify a number of deficiencies which militate against the inference of global sea-level trends from tide gauge data. First, many records are short and longer records, in particular, typically contain one or more gaps. Various minimum record lengths have been proposed for the identification of sea-level trend, with 20 years being a practical minimum (Gornitz 1995) in order to reduce the effects of interannual variability. In contrast, Pirazzoli (1989) considers that 50 years is more realistic: such a threshold reduces the number of eligible stations to around 230. Shorter records are valuable, however, especially as a basis for investigating interannual as well as spatial variability in sea-level (Tsimplis and Woodworth 1994). Second, there is a strong geographical bias in favour of the northern hemisphere, with the longest records being concentrated in Europe and North America. Third, substantial interannual variability complicates the extraction of secular trends and, especially, any climate-induced signal.

Tide gauges record variations in *relative* sea-level, in which the 'global' eustatic component is often dominated by vertical land movements of more local origin. Only within the last two decades have advances in satellite-based positioning and laser range-finding made it possible to control for these effects. There are at least three main avenues for research here. First, high-resolution gravimeters can now be used to detect minute changes in gravity (Carter *et al.* 1994). Promising results have been obtained from Hudson Bay, Canada, where gravity changes since 1987 and a 50-year tide gauge record both indicate uplift at $11 \, \text{mm} \, \text{yr}^{-1}$. Second, Global Positioning System (GPS) coordinates can be used to fix the position of the tide gauge benchmarks, thereby controlling for relative land movements. Preliminary results from such a programme covering the Mediterranean have recently been presented by Zerbini *et al.* (1996). Third, satellite altimetry (Robinson 1985) can be used to obtain repeat coverage of the ocean surface. Of particular importance is the TOPEX/POSEIDON programme, which can achieve several mm of precision and broad spatial coverage (Nerem 1995). In all cases, it will be some years before sufficient data exist to allow separation of underlying trends from the interannual variability. In the meantime, the PSMSL dataset constitutes the major resource for analyses of sea-level variability and trend.

Interannual Variability

Examples of longer MSL series from around the world are presented in Figure 11.2. All the series of mean annual levels show a combination of interannual variability superimposed on varying forms of progressive trend. In shorter records, the magnitude of interannual variability contributes background 'noise' that complicates the extraction of any long-term trend. However, such variability does provide some important insights into ocean response to short-term atmospheric forcing and into the nature of the climate–ocean linkage.

Close inspection of the same records reveals an interannual variation that is coherent between neighbouring stations (for example, between Newlyn and Brest), but differs in structure between more distant pairs of stations. Pugh

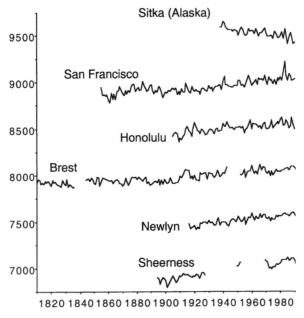

Figure 11.2. Records of mean annual sea-level from selected stations, indicating various degrees of interannual variability and direction and magnitude of trend (data from UK Permanent Service for Mean Sea Level).

(1987) provides an authoritative discussion of the various processes contributing to monthly and annual variability in mean level. These include direct meteorological effects, related to atmospheric pressure and associated wind fields, and density effects, caused by changes in temperature and/or salinity. Meteorological influences are typically very important at higher latitudes, especially where storms are more common. By way of example, Figure 11.3(a) shows the relationship between MSL and mean annual air pressure at Newlyn (south-west UK). There is a strong negative correlation which corresponds very closely to the theoretically anticipated 'inverse barometer' effect (Pugh 1987). A slight discrepancy can be attributed to wind effects which are correlated with pressure. The effect of applying such a barometric correction to the Newlyn record is illustrated in Figure 11.3(b).

The lack of direct coherence observed between more distant stations implies that gradients in sea-level are maintained over periods of years, and that these are associated with major ocean dynamics rather than local weather. Insights into the both the temporal and spatial character of these processes can be obtained from comparison of simultaneous monthly levels across a network of tide gauges (Wyrtki 1985), aided by the synoptic coverage provided by satellite altimetry (Nerem 1995).

Some of the most important findings to date have concerned the behaviour of the tropical Pacific. On average, the surface waters of the tropical Pacific are characterised by a relatively deep (100–200 m) western 'warm pool' (29–30°C) and a shallow (< 75 m) 'cool tongue' (22–24°C) in the east. The latter results from upwelling of cold, nutrient-rich, deep water driven by the Trade Winds converging on the warm pool. This pattern of sea surface temperatures, east to west deepening of the thermocline, and dominant atmospheric circulation represents a state of quasi-equilibrium in the ocean–atmosphere system. This state can be described by surface air pressure differences between Darwin, North Australia (indicative of ascending warm and moist air over the western warm pool), and Tahiti (influenced by dry descending air): this difference, in standardised form, is known as the Southern Oscillation Index (SOI). Major perturbations to the ocean–atmosphere it represents are known as El Niño Southern Oscillation (ENSO) events (Philander 1990). Warm phase pertubations, when ocean warming occurs across the entire equatorial Pacific Ocean, are known as El Niño events. These last 18–24 months and occur every 2–10 years. An El Niño event is often followed by unusually cool conditions, in form of a La Niña event, characterised by vigorous upwelling along the coast of South America and along the Equator (Palmer and Webster 1997).

El Niño and La Niña events are accompanied by sea-level changes. Wind stress due to the action of the Trade Winds over the great distances of the tropical Pacific results in an east-to-west slope on the sea surface, with the warm pool being on average 45 cm higher than the eastern ocean margin.

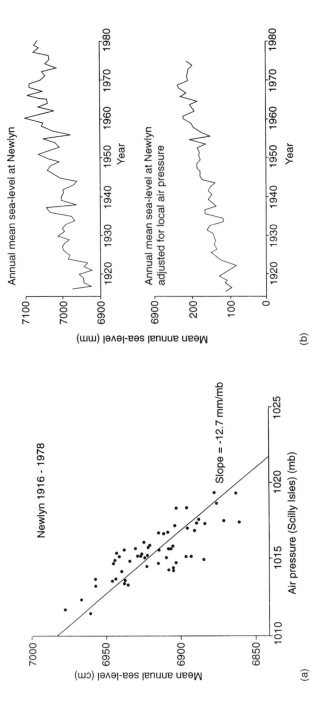

Figure 11.3. (a) Correlation between annual mean sea-level and annual mean air pressure at Newlyn, south-west UK, 1916–78; (b) Effect of local pressure correction on 'noise' level within Newlyn record (both reproduced from Pugh 1987).

When the Trades weaken with the onset of El Niño, this differential is lost. Typical western Pacific sea levels of -8 to -36 cm and -20 to -60 cm below MSL were recorded during the 1976–7 moderate and the 1972–3 strong El Niño events respectively (Wyrtki 1979).

Perturbations of sea-level can travel as far north as Alaska. Enfield and Allen (1980, 1983; Komar and Enfield 1987; Figure 11.4) have identified coherent sea-level anomalies along the eastern Pacific coast, from Alaska to Chile. Large positive anomalies (higher than average sea-level) were associated with the El Niño events of 1957–8, 1965–6 and 1972–3. North of San Francisco, the anomalies become strongly correlated with local wind stress. Sea-level variability due to ocean dynamic behaviour is, therefore, much modified at the coast. Such variations in sea-level have been implicated in the triggering erosional episodes along the coasts of California (Flick and Cayan 1985) and Oregon, where the strong El Niño of 1982–3 caused a seasonal increase in sea-level of 0.6 m, some 0.35 m greater than the average seasonality (Komar 1986).

Historical Trends

There have been numerous attempts to deduce 'global' trends in mean sea-level from tide gauge data (Table 11.2). The more recent studies have benefited from the ever-increasing length of the instrumental record and have also deployed more sophisticated techniques for the extraction of any eustatic signal from records 'contaminated' by varying amounts of vertical land movement.

The most widely referenced recent analyses are those by Barnett (1988) and Gornitz and Lebedeff (1987). Both derive curves of global mean sea-level change over the last 100 years (Figure 11.5), but utilise different approaches to the extraction of a eustatic signal. Barnett relies on averaging out a broad geographical distribution of records (while avoiding locations obviously dominated by glacial rebound and tectonic activity) so that the net contribution from land movements tends to zero. Gornitz and Lebedeff filter out vertical land movements by reference to late Holocene sea-level histories which include both eustatic and isostatic/tectonic effects. The basic idea here is that any recent eustatic signal appearing over the last 100 years or so should stand out from much lower-frequency land movements over the last few thousand years. Further filtering can be achieved by averaging by region. Aside from specialised statistical investigations (Nakiboglu and Lambeck 1991; Gröger and Plag 1993), there is also a trend towards the adoption of tighter quality-control procedures and an emphasis on a small number of relatively long records (Douglas 1991).

Almost all the studies in Table 11.2 suggest a rise in global MSL of between 1 and 1.5 mm yr^{-1}, over the last century. However, it has been argued that such agreement is partly fortituous and reflects the use of an essentially common (if

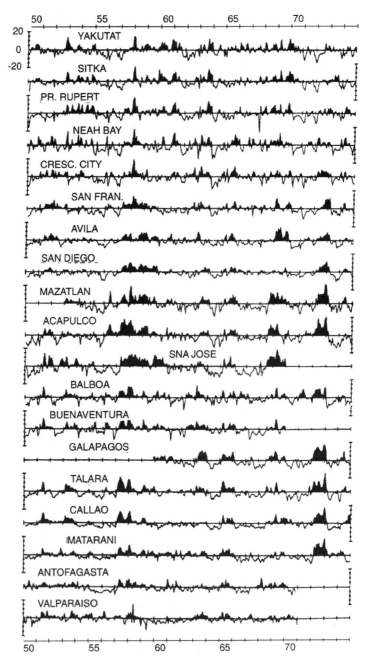

Figure 11.4. Time series of monthly mean sea-level anomalies along the east coast of America, 1950–74. Positive anomalies are shaded black (after Enfield and Allen 1983).

Table 11.2. Successive estimates of global eustatic sea-level rise based upon analysis of tide gauge data (studies selected from more comprehensive lists in from Pirazzoli 1989; and Gornitz 1995)

Author(s)	Period (region)	Stations	Rate (mm yr^{-1})
Gutenberg (1941)	1807–1937	69	1.1 ± 0.8
Valentin (1952)	1807–1947	253	1.1
Emery (1980)	1850–1978	247	3.0
Gornitz, Lebedeff and Hansen (1982)	1880–1980 (14 regions)	193	1.2 ± 0.1*
Barnett (1984)	1881–1980	152	1.2 ± 0.3*
Gornitz and Lebedeff (1987)	1880–1982	130	1.2 ± 0.3*
	1880–1982 (11 regions)	130	1.0 ± 0.1*
Barnett (1988)	1880–1996	155	1.15
Pirazzoli (1989)	1881–1980 (Europe)	58	0.9–1.2
Douglas (1991)	1880–1980	21	1.8 ± 0.1
Nakiboglu and Lambeck (1991)	1807–1990 (10° × 10° blocks)	655	1.15 ± 0.4
Emery and Aubrey (1991)	1807–1996	517	Not determined
Shennan and Woodworth (1992)	1901–88 (UK + North Sea)	33	1.0 ± 0.15
Gröger and Plag (1993)	1807–1992	854	Not determined

Rates quoted together with standard deviation, except where indicated * (95% confidence interval) and (standard error).

incrementally extending) dataset. Pirazzoli (1989) has challenged the assumption that, by averaging over a large enough sample of sites, the contribution from local land movements can be reduced to zero. Of the 229 records which have a duration of at least 30 years, very few have trends which lie within the consensus region of 1 to 1.5 mm yr^{-1}. Instead, many sites are dominated by uplift or subsidence. There are reasons to suppose that subsidence is likely to be more prevalent than uplift in coastal regions (Pirazzoli 1989), resulting in overestimation of actual eustatic rise by analyses of this kind. This can contaminate the temporal trend deduced from records of differing lengths. Thus, the suggestion by Barnett (1988; Figure 11.5(a)) that a higher rate of rise has occurred since the 1930s, is attibuted by Pirazzoli (1989) to the inclusion, after 1930, of new stations on the subsiding east coast of the USA. This problem is substantially overcome by a combination of regionalisation and explicit filtering of isostatic and tectonic effects. The regionally derived estimates in Table 11.2 are slightly lower and have narrower confidence intervals. Coastal locations are frequently affected by very local subsidence or uplift; this is difficult either to represent through regional sea-level histories based on geographically sparse sea-level index points, or to simulate, using geophysical glacial rebound models that are themselves calibrated with reference to those same data! Equally, vertical land movements are not always

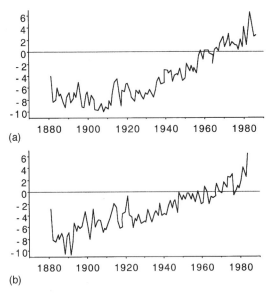

Figure 11.5. Global average sea-level rise over the last century, according to analyses by (a) Barnett (1988) and (b) Gornitz and Lebedeff (1987). Changes are given relative to the 1951–70 average. Figures redrawn from Gornitz (1995).

low-frequency in nature — rapid co-seismic uplift may create relative sea-level rise that is comparable in magnitude with that due to other processes operating over long periods (Bilham and Barrientos 1991).

   None of the published studies proposes that sea-level is either stable or falling. Also, the estimated sea-level rise is significantly higher than the trend implied by geological studies covering the last few thousand years (Gornitz and Seeber 1990): this implies that the data *do* contain a recent eustatic signal. A rise of between 1.0 and 1.2 mm yr^{-1} over the last century seems most pausible, although the latest IPCC summary (Houghton *et al.* 1996) quotes a range of 1.0 to 2.5 mm yr^{-1}.

CLIMATE CHANGE AND SEA-LEVEL RISE

CLIMATIC CONTROLS ON SEA-LEVEL RISE

Global warming is expected to effect a rise in global mean sea-level through two sets of processes:

(1) Thermal expansion of the upper ocean
(2) Melting of land ice (valley glaciers, ice caps and the major ice sheets).

There is a broad consensus that sea-level rise through thermal expansion, due to the negative dependence of density on temperature, will be an important consequence of global climate warming. Since the adjustment involves internal heat and mass redistribution within the ocean, there are significant spatial and temporal lags in ocean response to temperature forcing. Consequently, sea-level rise through thermal expansion will be more rapid in some areas than others (Warrick et al. 1996). Different assumptions concerning the nature of heat redistribution lead to a wide range of model results. These are difficult to relate to hydrographic data, which are sparse and include few long-term surveys of temperature versus depth. There are, however, some regional studies (for example, Roemmich 1992) which suggest that recently observed thermal expansion is consistent with sea-level rise recorded by nearby tide gauges.

The fate of valley glaciers and small ice caps has received considerable attention since, although these are volumetrically insignificant compared to the major ice sheets, their response times are much faster. Numerous studies have pointed to a long-term ice loss from glaciers since the nineteenth century (Meier 1993). The processes are complex, and the interaction between mass balance and meteorological forcing is likely to involve a series of poorly understood feedbacks. Again, there is a heavy reliance upon modelling, augmented by limited observations, and extended in time with reference to longer climatic datasets. Resulting sea-level changes are typically estimated through empirical models of glacier sensitivity to changes in air temperature (Oerlemans and Fortuin 1992).

Most of the non-oceanic water on Earth resides in the Antarctic and Greenland ice sheets, and the bulk of this is stored above present mean sea-level (Warrick et al. 1996). Even a slight change in the mass balance of the ice sheets will have a significant effect on sea-level. The two great ice sheets differ in that the latter has no major floating ice shelves. There are also differences in temperature regime. The Greenland ice sheet experiences significant melting (of the same order of magnitude as iceberg calving; Reeh 1991). Thus, global warming is likely to produce a relatively rapid change in ice mass balance with an immediate consequence for sea-level. In contrast, the Antarctic ice sheet is characterised by large ice shelves. Although these do not affect sea-level directly, they form part of a closely coupled ice flow system and much media attention has been generated by the prospect of large-scale breakup of these shelves. However, it is extremely difficult to disaggregate the effects of natural variability from longer-term trends in either ice flow dynamics or overall mass balance on the basis of extremely short observational records. As Warwick et al. (1996) acknowledge, this means that a heavy reliance is placed upon numerical simulations, such as those by Huybrechts (1994) of very long-term ($-125\,000$ to $+25\,000$ years relative to present) ice sheet dynamics. For the Antarctic ice sheet, these indicate a strong response to the last glacial–interglacial transition, and a mass balance that is currently negative. In contrast, the response of the Greenland ice sheet is relatively weak, and its mass balance is close to zero.

CLIMATIC AND ANTHROPOGENIC CONTRIBUTIONS TO HISTORICAL SEA-LEVEL RISE

A crucial question is whether these climatic controls on sea-level can be used to explain the rise in sea-level over the last century that is apparent from the instrumental record. Attempts to answer this question have involved either the direct comparison of observed sea-level change with modelled change due to climate forcing or the search for statistically significant accelerations in the sea-level record that might indicate a signal of climatic rather than a geological origin.

Table 11.3(a) (derived from the latest IPCC report, Warwick *et al.* 1996) synthesises published estimates of the contributing climatic effects into low, middle and high scenarios. These provide some indication of the overall modelling uncertainty, and can be compared with the range of observed sea-level rise values. The Table also includes estimates for sea-level changes due to alterations in surface water storage. These are largely a consequence of anthropogenic activities. Some of these (such as reservoir storage) tend to lower sea-level, while others (groundwater depletion, wetland loss, and deforestation) have a positive effect on sea-level (Sahagian, Schwartz and Jacobs 1994). Overall, there is clearly a problem in reconciling the observed rise with the combined effect of the contributing factors. In part, this difficulty arises from large uncertainties in component contributions. There are significant differences in the magnitude and direction of change attributed to the behaviour of the Antarctic ice sheet, and the smaller, though still important changes in surface and groundwater storage. Thus, the range of variation in contributions (-19 to 37 cm equivalent sea-level rise over the last century) easily spans the narrower range of sea-level rise estimates from the instrumental record (10–30 cm). However, it is misleading to structure the comparison in this way. It is possible, for instance, that actual climate forcing might be best represented by a mixed scenario of the component terms. Equally, the mid-range climate-forcing scenario might be more appropriately compared with the more plausible low-range estimate for the observed sea-level rise: the two estimates of 8 and 10 cm of sea-level rise are then quite close.

All the estimates of global sea-level rise over the last century imply a significant acceleration compared to the changes occurring in the recent geological past. This apparently commenced sometime around the middle of the nineteenth century, although the exact timing remains uncertain. Such an interpretation is based on careful analysis of a small number of long records, often assembled from composite sources (Smith 1980; Maul and Martin 1993). However, there is little evidence, within such records, of a significant

Table 11.3. (a) Estimated contribution of climatic and anthropogenic factors to sea-level rise over the last century (after Warrick et al. 1996). All values in cm

Range	Thermal expansion	Glaciers and small ice caps	Greenland ice sheet	Antarctic ice sheet	Surface water storage	Total contribution	Observed rise
Low	2	2	−4	−14	−5	−19	10
Middle	4	3.5	0	0	−3.5	8	18
High	7	5	4	14	7	37	25

(b) Selected projections of future sea-level rise (cm, to 2100) including, where possible, the contribution from specific climatic controls. Modified from similar tables in Gornitz (1995) and Warrick et al. (1996)

Author(s)	Thermal expansion	Glaciers and small ice caps	Greenland ice sheet	Antarctic ice sheet	Range	Best estimate
Hoffman, Wells and Titus (1986)	28–83	12–37	6–27	12–220	58–367	—
Thomas (1986)	35–48	12–42	13–34	20–80	80–204	110
Warrick and Oerlemans (1990) (IPCC 1990)	43	18	10	−5	31–110	66
Wigley and Raper (1992)	—	—	—	—	15–90	48
Titus and Narrayanan (1996)	21	9	5	−1	5–77	34
Warrick et al. (1996) (IPCC 1996)	28	16	6	−1	20–86	49
Warrick et al. (1996)	15	12	7	−7	—	27

acceleration in the rate of sea-level rise over the twentieth century (Woodworth 1990; Douglas 1992). The search for such an acceleration, which might be particularly diagnostic of the early phase of climatic and ocean warming, is greatly complicated by the inherent limitations of the global tide gauge network and by the strong variability in sea-level at interannual and inter-decadal timescales. These include Pacific ENSO events (Komar and Enfield 1987), and low-frequency variation in the North Atlantic (Maul and Hanson 1991).

PROJECTED SEA-LEVEL RISE

Projections have been made of the overall rise in sea-level likely to result from global climate change and are summarised in Table 11.3(b). None incorporates an explicit surface-water storage contribution, since water management strategies and patterns of water usage under a changing climate are hard to model. Rather, they are based on fairly simple climate models, linked to separate submodels for thermal expansion and ice melt. All these simulations envisage an accelerating rate of sea-level rise over the next century, although the nature and rate of this rise has been significantly revised since the earliest studies of the mid-1980s. The most recent projections incorporate advances in the understanding of atmosphere–ocean interactions and ice sheet dynamics, and include additional feedbacks, notably the cooling effect of increased aerosols of anthropogenic, biological and volcanic origin. The effect of such developments has been to reduce the spread of individual projections, and to lower the 'best estimate' of sea-level rise. As noted above, the anticipated response of the Antarctic ice sheet is now very different from that envisaged in the 1980s. Rather than mass wasting leading to a rapid sea-level rise of several metres, current thinking envisages increased snow accumulation under warming temperatures, leading to a weakly negative contribution to overall sea-level rise.

The most influential projections have been those produced in 1990 and 1995 under the auspices of the IPCC. The IPCC 1990 report (Houghton, Jenkins and Ephraums 1990; Warrick and Oerlemans 1990) utilised three values for the sensitivity of climate to radiative forcing (1.5, 2.5 and 4.5°C for a doubling of CO_2), together with a mix of energy use and emissions scenarios (of which the 'business as usual', or BAU, scenario is the most often quoted). A significant update was produced in 1992 (Wigley and Raper 1992), which incorporated the effects of the Montreal protocol on halocarbon (CFC) emissions and some additional feedback mechanisms including stratospheric ozone depletion and aerosols. The latest IPCC report (Houghton et al. 1996; Warrick et al. 1996) leaves the earlier projections qualitatively unchanged, but incorporates results from an improved coupled ocean–atmosphere model of thermal expansion, more sophisticated parameterisation of glacier and ice sheet response, and a wider range of simulated aerosol feedbacks.

The 'best estimate' curves in Figure 11.6(a) are part of a wide spectrum of projections which result from various combinations of assumed climate sensitivity, future emission levels and the selection and parameterisation of the component models. The IPCC 1996 projections encompass six emissions

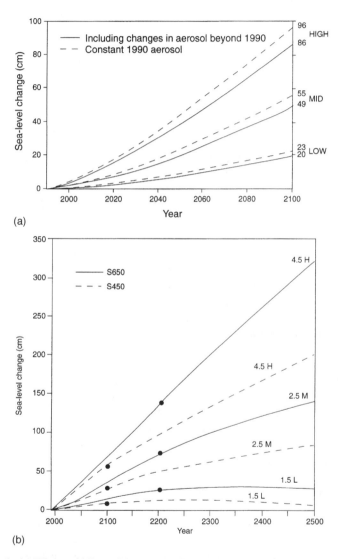

(a)

(b)

Figure 11.6. (a) High, middle and low projections for the period 1990 to 2100 for IPCC Scenario IS92a (redrawn from Warrick *et al.* 1996); (b) long-term projections produced by Raper, Wigley and Warrick (1996), assuming stabilisation of CO_2 levels at 450 ppm (dashed lines) and 650 ppm (solid lines).

scenarios (IS92a to f), three climate sensitivities (low, mid and high), and extreme projections derived from combinations of scenarios and sensitivities chosen to maximise or minimise sea-level response. Simulations are also performed with and without the effect of future changes in aerosol levels. Finally, an alternative set of simulations is presented, which utilise a completely different set of models driven by the same forcing conditions as the IPCC models. The best estimate here (Table 11.3(b)) is significantly lower than for the equivalent IPCC projection, largely due to a much smaller rate of thermal expansion and a more strongly negative contribution from Antarctica.

The current best-estimate projection from the IPCC envisages a rise of 49 cm by 2100, with extreme projections of 94 and 13 cm (Warrick *et al.* 1996). The climate model was run from 1765, with ice melt effects being introduced from 1880. It is interesting to compare the modelled sea-level change over this period with the separately estimated and observed overall sea-level over this period. The modelled changes (Table 11.3) lie with the range of those estimated separately from the processes considered above: this is not surprising, given that similar scientific understanding is deployed in both cases. There is clearly a discrepancy with the instrumental record. Part of this arises from the omission of anthropogenic effects on surface water storage. Also, it is probably more reasonable to take the low end of the tide gauge record to be the best estimate. The results from these different approaches are then in much closer accord.

Detection of a true climate-induced trend in mean sea-level requires the isolation of low-frequency signals from tide gauge records that are both noisy and short in duration. That there are trends with tide gauge records is not in doubt. Far less certain is whether there is any *acceleration* in the data that is consistent with the modelled projections. Both Woodworth (1990) and Douglas (1992) show that there is only limited empirical evidence for such accelerations. The record from Newlyn (Cornwall, UK) exhibits a linear trend of 1.72 mm yr^{-1}, similar to that of the globally averaged data. Using a simple numerical experiment, Woodworth (1990) shows that current interannual variability is much less than the IPCC 1990 'best estimate' of sea-level rise for the next century. However, the accelerating component of sea-level rise would not be detectable (at the 95% confidence level) until at least 2010. This 'detection horizon' is clearly somewhat later for the revised best estimate projection of Warrick *et al.* (1996).

A significant aspect of the IPCC 1996 report is the inclusion of longer-term simulations which consider sea-level changes over the period up to 2500 (Figure 11.6(b)). These reveal strong lag effects, such that sea-level rise continues to occur long after a stabilisation of CO_2 levels (Raper, Wigley and Warrick 1996). The key lesson here is that, irrespective of governmental action on greenhouse gas emissions, a substantial potential for sea-level rise may already be contained within the oceans.

PHYSICAL IMPACTS OF SEA-LEVEL RISE

SEA-LEVEL RISE AS AN AGENT OF COASTAL CHANGE

Oscillations in sea-level occurring over 'geological' timescales have produced major shifts in shoreline position and composition. Given the importance of sea-level as a first-order control on coastal evolution (Carter and Woodroffe 1994), it is not surprising that numerous attempts have been made to conceptualise and model the influence of sea-level rise on coastal morphodynamics. No universal model has been developed, nor is one likely to emerge in the foreseeable future. A major complication is the multiplicity of factors determining coastal behaviour (Figure 11.7(a)) and the range of the scales at which these operate (Figure 11.7(b)). This makes it difficult to extract a clear sea-level signal from coastal morphological changes. Instead, sea-level simply provides a moving plane of reference upon which the dynamic effects of coastal processes, subject to strong mediation by sediment supply, are superimposed.

Many mid- to high-latitude coasts have been profoundly influenced by the reworking of glacially influenced sediment accumulations by rising Holocene sea-levels. Within this context, modern coastal behaviour is often dominated by human influences which operate primarily through interference with local and regional sediment budgets (Komar 1999). Most are negative in impact and often far outweigh the effects of sea-level rise. Thus Allen's (1981) analysis of the erosion problems at Sandy Hook recreational beach (New Jersey, USA), shows that only 1% of the erosion since 1953 is attributable to sea-level rise: the rest is due to a combination of sediment starvation downdrift of a major groyne installation and an increase in the frequency and magnitude of major storms. More recently, Gaillot and Piégay (1999) have shown that recent beach erosion within Calvi Bay (Corsica) has followed a reduction in sediment delivery from coastal catchments due to mining of fluvial gravels for construction purposes. Although this activity is now prohibited by law, abandoned extraction pits are likely to trap a significant proportion of river sediment load for several decades to come.

IMPACTS OF SHORT-TERM SEA-LEVEL VARIABILITY

A further complication arises from the addition of dynamic behaviour due to short-term ocean and climate variability. At decadal scales, for example, perturbations associated with El Niño events have significant impacts along the Pacific coast of North America. These include coastal flooding from elevated water levels in El Niño years; landsliding associated with high rainfall storms; and episodes of severe beach erosion by storm waves (Richmond and Gibbs 1997). All these phenomena have been well documented for recent El Niño events, notably for the strong 1982–3 event, which produced an elevation in

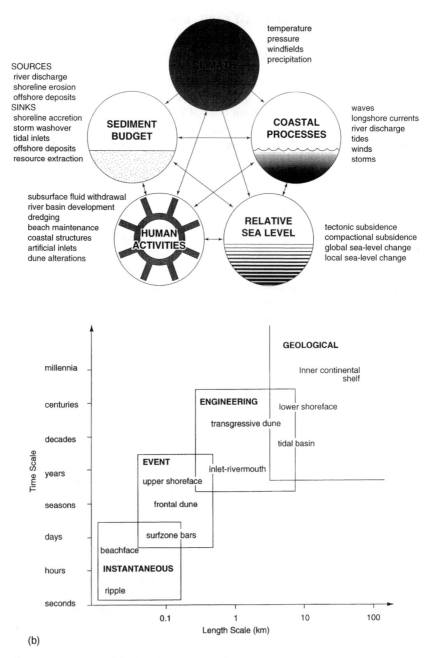

Figure 11.7. (a) Factors affecting coastal environments (modified from Williams, Dodd and Gohn 1991); (b) spatial and temporal scales in coastal evolution (after Cowell and Thom, 1994, with permission of Cambridge University Press).

sea-level of up to 0.35 m along the Oregon coast (Komar and Enfield 1987). During the same event, heavy rainfall along the ocean coast south of San Francisco Bay led to gully erosion on the subaerial cliffs, and basal cliff retreat of 10 to 20 m (Lajoie and Mathieson 1997). Giant breakers up to 6–7.5 m in height were experienced in Oregon during the 1982–3 event: similar storm waves during the 1997–8 El Niño resulted in up to 4 m of vertical beach lowering in Monterey Bay, California (Richmond and Gibbs 1998). Such changes also alter the sediment budget: in southern California, it is estimated that beach erosion during the 1997–8 El Niño amounted to $3.8 \times 10^3 \, m^3$ of material (Anon. 1998).

These aspects of short-term coastal dynamic behaviour provide insights into the effects of a sudden increase in mean sea-level. However, they also serve to illustrate both the interdependence of the factors depicted in Figure 11.7 and the importance of lag effects which can have longer-lasting consequences for management and conservation. The latter are especially evident in the behaviour of sandy pocket beaches and coastal spits. Along the Oregon coast, seasonal variations in wave approach redistribute sediments with these embayments, although the net transport of sand is negligible. However, in the 1982–3 El Niño event, storms to the south generated considerable northward transport, leading to major accumulations of sand at the northern ends of embayments and beach erosion (often to bedrock) at their southern margins. In the case of some spit complexes, sediments usually subject to seasonal migration between beaches and offshore bars were moved northwards and swept into backbarrier bays. This net sediment loss from the system has led to erosion of the oceanside spit beaches, which has continued in some cases, for three to six years after the El Niño event. Artificial beach nourishment is one strategy proposed to alleviate this longer-term problem, but sand dredging from within the backbarrier bays may damage the ecologically important wetlands contained therein (Komar and Good 1991).

El Niño-related variability also includes negative sea-level anomalies. These have impacts which may affect longer-term coastal behaviour and the viability of intertidal habitats under future sea-level rise. For example, typical western Pacific sea-level anomalies of up to −0.36 m and −0.60 m were recorded during the moderate 1976–7 and the strong 1972–3 El Niño events respectively (Wyrtki 1979). ENSO-related fluctuations of 0.60 m occurred at Jarvis and 0.80 m at Nauru in 1981–3 (Lukas, Hayes and Wyrtki 1984). Several studies have shown that these short-term falls in sea-level are associated with reef dieback, especially of large coalescing colonies of flat-topped corals, or 'microatolls', (for example, Woodroffe and McLean 1990; McLean 1996). At Tongareva atoll, northern Cook Islands, it has been estimated that recovery from die-back and regrowth of coral to previous levels takes up to 20 years (Spencer et al. 1997). This delayed recovery arises, in part, from the effects of multiple subsequent sea-level events. Since a healthy reef flat is extremely effective at dissipating wave energy around islands composed on largely unconsolidated sands and gravels, lagged ecological recovery from these brief sea-level falls actually contributes to the problem of adjustment to more sustained sea-level rise.

PHYSICAL EFFECTS OF SUSTAINED SEA-LEVEL RISE

The principal physical effects of a more sustained rise in mean sea-level are reasonably well established. The most general effects include:

- Inundation of lowlying terrain
- Increased erosion and shoreline recession
- Intrusion of saltwater within estuaries and into coastal acquifers.

The relative significance of these changes, their mode of operation and their consequences for habitats of conservation importance, will vary between contrasting coastline types.

Inundation and Related Processes

Assessments of the threat posed by accelerated sea-level rise have tended to emphasise the possibility of widespread inundation of coastland lowlands (Barth and Titus 1984; Boorman, Goss-Custard and McGrorty 1989; Alcock 1991; Tooley and Jelgersma 1992) and the threat to lowlying oceanic island populations (Nunn 1990; Pernetta and Hughes 1990). This reflects the increasing concentration of human activity on the coastal plains of developed regions (see, for example, Cendrero and Charlier 1989) and the economic value of lowlying areas which are already at risk from coastal flooding (Peerbolte, De Ronde and Baarse 1991; Frassetto 1991; Yohe et al. 1996). These assessments have been greatly aided by the development of Geographical Informations Systems (GIS) capable of integrating information on lowland topography, population density, and land use (Shennan and Sproxton 1990; ElRaey et al. 1997). GIS-based analyses can also be used to delimit vulnerable habitats of conservation significance, such as freshwater grazing marshes and brackish reed swamps, as well as potential sites for habitat restoration (Hollis, Thomas and Heard 1989). By coupling terrain information with ecological models, it is also possible to simulate the habitat transitions that might be expected under given sea-level rise scenarios (for example, Lee, Park and Mausel 1992; Percival, Sutherland and Evans 1998).

In the case of coastal lowlands where inundation is a major element of the response to sea-level rise, further complications arise from the interaction between large-scale morphology and coastal hydrodynamics. Based on the application of a three-dimensional tide model, Gehrels et al. (1995) estimated the contribution of changes in mean tidal range to post-glacial sea-level rise in the Gulf of Maine and Bay of Fundy. Of 5–8 m of sea-level rise during the last 5000 years, between 0.4 and 0.5 m can be attributed to an increase in the range of the lunar semi-diurnal (M_2) tide. When the IPCC92 best-estimate scenario of 0.48 m of sea-level rise to 2100 AD is applied to the present coastline, their model predicts a small additional contribution of around 0.01 m from the M_2 range (i.e. around 0.02 m increase in *range*). This kind of feedback between morphology and process is hard to model using GIS-based systems which do not incorporate any explicit representation of physical processes.

Of the habitats most obviously susceptible to inundation, particular attention has been paid to the fate of coastal wetlands. The processes affecting coastal freshwater systems are considered by Thompson and Finlayson (2001). Freshwater habitats are especially vulnerable where these have acquired conservation significance within landscapes created through historical land claim and where their character is dependent upon the maintenance of sea defences and drainage systems. Here, the risk is from catastrophic inundation following failure of the defences protecting these areas from tidal flooding (RSPB *et al.* 1997). Although such failures can be repaired, the saltwater incursion may exert a longer-term effect upon species composition, especially where complex assemblages of plant species are developed along fresh to brackish transitions. In the case of saltmarshes, much has been made of the maintenance of elevations and inundation regimes by sedimentation. As French and Reed (2001) emphasise, the nature of saltmarsh response to sea-level forcing varies between distinctive wetland types as well as between contrasting geological settings. Impacts such as these, where isolated events force lagged ecological transitions, or where sustained environmental change forces a dynamic system response, are difficult to simulate using simple inundation models.

Some of the most promising developments concerning the prediction of ecosystem-level consequences of sea-level rise have occurred in the field of spatial landscape modelling. Whereas earlier ecological modelling focused on temporal changes, a new generation of spatially explicit coastal ecological landscape models is being developed to simulate large-scale and long-term wetland habitat changes under joint management and environmental change scenarios. These models are dynamic in that feedbacks are included between processes and landscape types, thereby allowing physical and successional habitat transitions to occur in response to external environmental forcing (Sklar, Costanza and Day 1985). Costanza, Sklar and White (1990) have used this approach to model habitat dynamics in the actively forming Atchafalaya delta plain in coastal Louisiana. In this setting, the balance between wetland sedimentation and deltaic subsidence is critical to the maintenance of ecosystem structure. In particular, changes in freshwater inflow and sediment supply give rise to major successional shifts in habitat at decadal timescales and extending over hundreds of km². Their model is computationally-intensive: 3000 1 km² cells were simulated using 1248 weekly time-steps using input data for 1956–78. But the predictive power is high, with 68% of the 1978 habitat states being correctly modelled. Subsequent studies have harnessed more powerful computational resources to achieve more ambitious spatial resolution of habitat dynamics and a fuller parameterisation of physical and biotic processes. Sklar *et al.* (1994) present results from a 10 000-cell simulation of estuarine inundation and salinity intrusion within maritime forest lowlands in a South Carolina estuary. Sea-level data for 1970 to 1990 provided the primary environmental forcing. Key elements of their modelling approach include (1) the use of a GIS for spatial data management; (2) modelling software to develop and test the geomorphological, hydrological and ecological submodels at the level of a single cell; and (3) specialised software to handle the computation across the whole array of landscape cells, possibly involving parallelisation of the computation across an array of machines. This kind of approach is becoming more feasible as a successor to the standard GIS-based habitat impact assessments of the past. Although the representation of important hydrodynamic and hydrochemical processes is necessarily simplified, an important feature of spatially explicit models is their ability to detect non-linear responses to slight changes in initial conditions or rates of enviromental change. For example, Sklar *et al.* (1994) found little evidence of saline intrusion using the historical sea-level rise of 2–3 mm yr^{-1}. However, when the model was run with a sea-level rise of 3 mm yr^{-1}, a 20% increase in inundation area was predicted for high-tide conditions.

In more spatially restricted settings, the notion of 'coastal squeeze' (Titus 1991) has become enshrined in the coastal management literature. Along European and North American coasts protected by seawalls or backed by steep terrain (Kana *et al.* 1988; Boorman, Goss-Custard and McGrorty 1989), remaining saltmarshes are fragmented and are often experiencing rapid areal loss through marginal erosion. As accelerated erosion reduces marsh area (Phillips 1986), and increased inundation leads to a shift towards lower marsh species, the lack of any natural saltmarsh–brackish–freshwater transition zone (Burd 1995) means that diverse high marsh habitat will inevitably be lost. Under these circumstances, conservationists have to direct their attention away from assessments of habitat vulnerability towards the application of the best possible scientific understanding to the problem of wetland restoration (French and Reed 2001).

Erosion and Shoreline Recession

In practice, it is difficult to isolate the effects of inundation and erosion, since a change in the rate and pattern of erosion will almost invariably accompany the inundation of coastal terrain (Barth and Titus 1984; Kaplin 1989). Rocky

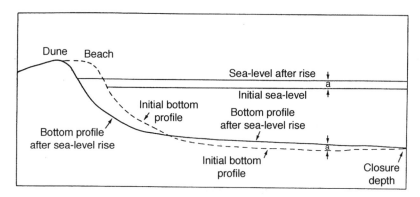

Figure 11.8. Schematic illustration of the Bruun model of shoreline erosion in response to sea-level rise.

shorelines, formed of resistant materials, may undergo very little erosional adjustment in the short term: any shift in shoreline position is likely to occur primarily through inundation, with coastal slope being the determining factor. At the other extreme, soft rock cliffs and low dune barriers, already characterised by high erosion rates, may recede at an accelerated rate, with increased erosion being the dominant agent of change (Leatherman 1987; Bray and Hooke 1997).

The adjustment of sandy shorelines to sea-level rise is often modelled with reference to the concept of the equilibrium shoreface profile. In a two-dimensional sense, this refers to a hypothetical long-term average profile achieved under a given wave climate and in a particular set of materials (Schwartz 1982). Such profiles rarely, if ever, exist in nature, yet they provide the basis for the widely applied Bruun Rule of shoreline erosion (Figure 11.8; Bruun 1962, 1983). This envisages a balance between the volume of material eroded from the beach and deposition on the immediately adjacent shoreface, at least in the absence of substantial gradients in longshore sediment flux. A key element of the Bruun model is the inclusion of a 'closure depth', at which there are no measurable elevational changes before and after storms which are morphologically active over the landward part of the profile: this is rather akin to the concept of the 'surf base' (Dietz 1963). The choice of the closure depth is somewhat arbitrary. Bruun (1962) adopted a depth of 18 m, whereas more recent studies have tended to use a depth of around 9 m (Pilkey et al. 1993). Bruun (1988) suggested that field identification could be made on the basis of sedimentological boundaries, whilst Nicholls, Birkemeier and Lee (1998) calculate closure depth as a function of nearshore wave statistics.

The Bruun model is intuitively attractive and remains a widely used tool for predicting shoreline migration under a given rise in sea-level. In the original model, the form of the assumed equilbrium profile means that shoreline recession will be in the region of 100 × the rise in sea-level. The most successful applications have been in restricted fetch settings where the depth of closure is limited. Examples include Rosen's (1978) investigation of erosion along the Chesapeake Bay shore; Hands' (1983) study of Lake Michigan, where sandy beaches exhibit erosional and depositional responses to climatically induced fluctuations in lake level; and Kaplin and Selivanov's (1995) analysis of shoreline responses to recent historical water-level fluctuations in the Caspian Sea.

Conceptually, however, the Bruun Rule has severe limitations. Critiques have been provided by Dubois (1992) and Pilkey et al. (1993) who raise fundamental objections to the use of equilibrium profiles as a basis for modelling shoreline change. Actual coastal profiles are conditioned by far more numerous hydrodynamic and geological factors than are admitted by the Bruun model. Many coastal morphologies bear a strong imprint of past processes in the form of resistant outcrops, and/or inherited sediment bodies which are of limited extent and which have strongly three-dimensional geometry: they do not reflect a simple balance between wave energy and sand transport. The Bruun model becomes especially unrealistic at a shoreface scale and where sediment transport pathways are strongly three-dimensional (Dubois 1992). A series of modifications have been proposed to take account of losses of fine material under storm conditions (Hands 1983), sediment washover from barrier beaches into adjacent lagoons (Dean and Maurmeyer 1983; Everts 1985) and sediment storages within coastal foredunes (Weggel 1979). These typically have the effect of increasing the predicted shoreline retreat for a given increment of sea-level rise. As Bird (1993) observes, this has immediate implications for the fate of sandy beaches facing developed shorelines, some of which could be completely lost through erosion if they are not artificially nourished. In modified form, the Bruun model has also provided a starting point for understanding the erosional response of coastal saltmarshes (Phillips 1986) and soft-rock cliffs (Bray and Hooke 1997). It has also formed the coastal process model at the heart of GIS-based analyses of coastal change (see, for example, ElRaey et al. 1999).

There is empirical evidence for a widespread recent tendency towards erosion along the world's sedimentary shorelines (Bird 1985). Given the scale of interventions in coastal sediment budgets, this is clearly not a simple consequence of historical sea-level rise (Williams, Dodd and Gohn 1991; Komar 1999). However, where human interference has been limited, sea-level trends can account for a significant proportion of observed erosion (see, for example, studies by Everts 1985; and Inman and Dolan 1989, of the Outer Banks of North Carolina). Elsewhere, for example at Assateague Island, Maryland (Leatherman 1987), dramatic erosion of the barrier beach system has occurred downdrift of the artificially stabilised Ocean City inlet. Here, engineering intervention accounts for virtually all the recent erosion, which peaked at $9\,\mathrm{m\,yr^{-1}}$ prior to alterations in the layout of the inlet training jetties.

More recently, Leatherman, Zhand and Douglas (2000) have attempted to derive a Bruun-type relationship between records of sea-level rise and shoreline position for the US east coast in areas not obviously influenced by coastal engineering works (Figure 11.9). Here, the observed horizontal recession with actively eroding sections ranges between 110 and 180 × vertical sea-level rise.

Although coastal erosion is typically portrayed as a threat to habitat conservation, there are circumstances in which the maintenance of active erosion may be beneficial. An emerging aspect of coastal erosion management is the increasing interest in geological and archaeological as opposed to purely biological conservation (Hooke 1998). This is particularly evident in the case of soft rock cliffs, the vast majority of which (in developed countries at least) have been defended to some degree. Often this has arisen through the piecemeal efforts of private landowners and developers, or in reaction to catastrophic events (in the UK, for example, following the 1953 North Sea storm surge; Clayton 1989). Although cliff-top residents exert strong pressure on local government authorities to control erosion, current engineering practice recognises the detrimental effect that cliff toe protection (in the form of rock armour, timber pallisades, groynes and detached breakwaters) can have on adjacent beaches (Brampton 1998). Instead, schemes are being designed which manage, rather than halt, the erosion (Powell 1992). Such schemes have been strongly supported by geological conservationists, who are concerned at the loss of active stratigraphic exposures (McKirdy 1987) including important reference sections. However, erosion control requires careful engineering

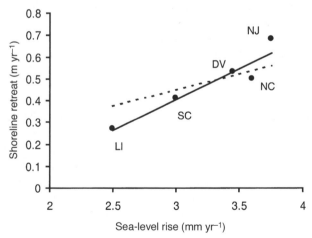

Figure 11.9. Relationship between the rate of sea-level rise and shoreline retreat for actively eroding sections of the US east coast not influenced by coastal engineering works. LI = Long Island; NJ = New Jersey; DV = Delmarva; NC = North Carolina; and SC = South Carolina. Solid line is best linear fit; dashed line is forced through the origin (i.e. zero shoreline change for zero sea-level rise). Redrawn using data in Leatherman, Zhang and Douglas (2000).

intervention in order to sustainably balance the needs of conservation against the consequence of continued cliff-top recession (Barton 1998). Further, as Regnauld, Lemasson and Dubreuil (1998) argue, erosion management strategies must be formulated with respect to local conservation needs. For example, archaeological conservation is sometimes best served by the protection of key sites, and sometimes by continued erosion and exposure of new artefacts. Further, the nature of the conservation interest, whether geomorphological or historical, may change fundamentally as discrete features are sequentially consumed. Accelerating sea-level rise and an increased tendency towards erosion will make it more difficult to achieve some of these compromises, and may serve to increase tensions between geological and biological interests—for example, as faster rates of exposure alter the dynamics of unique cliff-face plant communities.

Salinity Effects

Physical environmental changes will also arise through saline intrusion within estuaries and into coastal aquifers (Sherif and Singh 1999). The extensively engineered estuarine systems of the Netherlands well illustrate the sensitivity of both physical and ecological habitat characteristics to changes in sea-level and tidal regime, and to the engineering strategies implemented to manage them. Deepening of the Rotterdam approach channels has resulted in the saltwater limit migrating 35 km upstream over the last 60 years (Jelgersma 1992). More recently, the construction of surge barriers in the Eastern Scheldt has substantially altered the marine influence and the physical and biological functioning of tidal flats and fringing saltmarshes (DeLeeuw *et al.* 1994; DeJong, DeJong and Mulder 1994). Salinity has increased in the estuary, owing to a restriction on freshwater inflow. Tidal flushing has decreased, with a 10–15% reduction in average range. The ecosystem has proved resilient, however, with many of the changes being close to natural levels of interannual variability (Nienhuis and Smaal 1994). Following the completion of the Eastern Scheldt barrier in 1987, widespread erosion of tidal shoals and estuarine marsh cliffs has occurred (Louters, van der Berg and Mulder 1998). The ecological impacts of these physical changes are complex, and not always negative. Meire (1996), for example, has documented the successful adaption of oystercatcher (*Haematopus ostralegus*) populations, and their main prey species (*Cerastoderma edule* and *Mytilus edulis*), to a reduction in the area of intertidal habitat.

Mitigation of salinity changes is one of the main problems facing managers of shallow lagoonal and reedbed habitats associated with backbarrier shingle and dune systems. The conservation value (i.e. habitat character; biodiversity) of these environments is critically dependent upon the maintenance of gradients in abiotic conditions, such as soil moisture, groundwater level and salinity (Grootjans *et al.* 1997; see also Thompson and Finlayson, 2001). Such environments tend to be naturally ephemeral and comprise complex mosaics of brackish and freshwater habitats (Smith and Laffoley 1992) that are vulnerable to the various effects of tidal flooding and saline intrusion. Once intrusion has occurred, it may take years to flush out the salt water and for invertebrate communities to recover (see, for example, Andrews and Burgess 1990).

In the UK, particular concern is being expressed over the fate of wetlands developed behind late-Holocene gravel barriers. Reversion to tidal conditions may result in new saltwater habitat creation to offset losses through erosion. However, a net loss of freshwater habitat is envisaged, partly owing to the lack of natural 'refuge' sites, but also as a consequence of wider management plans formulated around coastal defence concerns. For example, the current generation of Shoreline Management Plans (SMPs) advocate a non-interventionist strategies along large stretches of East Anglian coastline where the cost of engineered protection is difficult to justify on economic grounds (Figure 11.10). As a result, some 550 ha of grazing marsh and 223 ha of reedbed are at risk from saltwater seepage and occasional flooding (RSPB *et al.* 1997). These areas include some of the most important sites recognised under the EU Birds and Habitats Directives.

Elsewhere in the UK, Orford and Jennings (1998) present an intriguing case study of the recent and likely future morphodynamic integrity of the gravel barrier at Porlock Bay, Somerset. Twentieth-century interventions, including groyne construction to protect tidal sluices and grading and nourishment of the barrier crest, have created an artificial freshwater backbarrier wetland that is vulnerable to episodic saline incursion associated with mesoscale variation in storminess. There is political pressure (from both residents and conservation agencies) in favour of the engineering intervention required to maintain barrier integrity. As Orford and Jennings argue, this is unsustainable in the face of sustained future sea-level rise, and has the effect of increasing the magnitude of unplanned incursions of the sea. A more enlightened strategy would be to allow natural landward barrier migration, with intermittent washover and/or breaching occurring such that backbarrier habitat changes are *progressive* rather than catastrophic. This is consistent with the notion of preparing coastal margins for sea-level rise prior to future realignment of sea defences (Burd 1995; Dixon, Leggett and Weight 1998).

The identification and set-aside of new sites for freshwater wetland creation must a priority if such approaches are to succeed. However, interim approaches may be necessary in order to 'buy time' while long-term solutions are sought. For example, at Benacre Broad, in Suffolk (Figure 11.10), dieback of freshwater reed-bed became evident in 1984, following overtopping and breaching of the narrow shingle barrier-beach. Engineering intervention to enhance the stability of the barrier would be expensive and might have detrimental impacts on the geomorphological functioning of the coastal sedimentary cell, and on

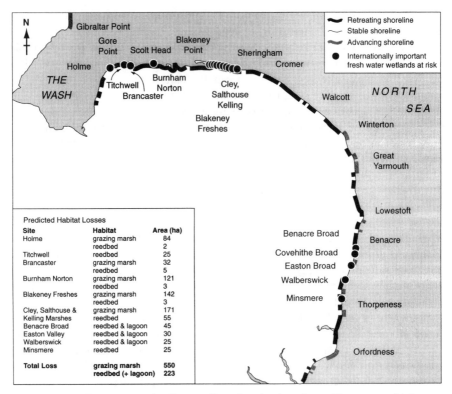

Figure 11.10. The UK East Anglia coastline, showing locations of important freshwater wetland habitats at risk from future acceleration in the rate of erosion, as envisaged in the current generation of Shoreline Management Plans. Redrawn from RSPB *et al.* (1997).

neighbouring locations of both economic and nature capital value. Kaznowska and Waller (1993) outline the implementation of a low-cost interim strategy designed to avoid the need for abandonment and to maximise the retention of wildlife interest. A series of low earth bunds have been constructed by English Nature in order to spatially restrict, rather than prevent, tidal flooding. The bunds are constructed at increasing crest elevations, allowing higher levels of freshwater to be maintained in winter to counteract the influence of over-topping surge tides. Given anticipated rates of sea-level rise, the long-term survival of the reed-bed habitat still depends, of course, on the purchase or management of new sites.

CONCLUSIONS

A more informed debate concerning the actual, as opposed to publicly perceived, threat posed by sea-level rise must be rooted in an appreciation of the relevant physical processes and the natural limits to their variability. Only then can an appropriate balance be struck between the contrasting perspectives of flood defence and coastal protection, scientifically informed conservation and public opinion. As this review has shown, the world's tide gauge records contain unambiguous evidence of both variability and trend in mean sea-level over timescales of decades to centuries. Although vertical land movements often account for a large proportion of the trend over the last century, there is a consensus among sea-level researchers that mean sea-level has risen globally at a average of between 1 and 2 mm yr^{-1} over this period. It is likely that a proportion of this rise is the result of climatic change, but a link with anthropogenic greenhouse gas forcing cannot be conclusively demonstrated.

Fundamental research into the nature of the climate–ocean linkage has resulted in more refined models of sea-level response to natural and anthropogenic climate forcing. These refinements have led to a significant downward revision of global sea-level predictions for the this century. In place of the dramatic scenarios that were common in the mid-1980s is a current 'best estimate' of between 0.4 and 0.5 m of global sea-level rise by 2100. Despite the strong scientific consensus behind this forecast, the interannual variability in sea-level is such that it will still be 10–20 years before any climate-driven acceleration in the rate of sea-level rise will be unambiguously detectable within the longest of the instrumental records.

The impacts of sea-level rise on coastal landforms and habitats are qualitatively understood, through the reconstruction of Holocene changes, the study of analogues such as rapidly subsiding deltas or fluctuating lake levels, and the application of conceptual and numerical models. Empirical isolation of a clear sea-level signal from past erosional or ecological trends is complicated by the sheer extent of anthropogenic influences, especially in terms of coastal sediment budgets. These effects underlie many of the world's coastal erosion problems. However, forecast sea-level rise is still sufficient to exacerbate existing erosion and habitat loss problems and can be expected to extend their geographical range into areas and environments which are presently unaffected. These impacts can only be evaluated within a local geomorphological and ecological context which incorporates knowledge of past human impacts. Many coastal systems are too far from any idealised equilibrium state to permit the extraction of meaningful answers from models such as the Bruun Rule. From a conservation perspective, it is especially important to appreciate the vulnerability of key habitats to subtle increases in sea-level and the associated hydrological and hydrochemical changes. As some of the other contributions in this volume demonstrate, opportunities for the conservation of wetland and dune habitats exist through managed habitat migration and restoration within emerging policies and strategies for accommodating sea-level rise. Viewed positively, the spectre of sea-level rise thus supplies a powerful motivation for better understanding the physical dynamics of the habitats upon which it threatens to impinge.

REFERENCES

Agnew C & Fennessy S (2001) Climate change and nature conservation. In: A Warren & JR French (eds) *Habitat conservation: managing the physical environment*. Chichester, John Wiley, 273–304.

Agriculture Select Committee, House of Commons (1998) *Sixth Report: Flood and coastal defence, Volume 1*. London, HMSO.

Alcock G (1991) Effects of sea-level rise on the UK coast. In: RW Battarbee & ST Patrick (eds) *The greenhouse effect: consequence for Britain*. London, ENSIS Publishing, 26–30.

Allen JR (1981) Beach erosion as a function of variations in the sediment budget, Sandy Hook, New Jersey, USA. *Earth Surface Processes & Landforms* 6, 139–50.

Anderson JB & Thomas MA (1991) Marine ice-sheet decoupling as a mechanism for rapid eustatic sea level change: the record of such events and their influence on sedimentation. *Sedimentary Geology* 70, 87–104.

Andrews JT & Burgess ND (1990) Some examples of brackish and saltwater habitat creation for birds. In: CM Finlayson & T Larsson (eds) *Wetland Management and Restoration Workshop Proceedings*, 45–55.

Anon. (1998) California's beaches: out to sea. *The Economist* 20 June, 46.

Barnett TP (1984) The estimation of 'global' sea level change: a problem of uniqueness. *Journal of Geophysical Research* 89C, 7980–8.

Barnett TP (1988) Global sea level. In: National Climate Program Office, *Climate variations over the past century and the greenhouse effect*. Washington DC, NCPO/NOAA.

Barth MC & Titus JG (eds) (1984) *Greenhouse effect and sea-level rise*. New York, Van Nostrand Reinhold.

Barton, ME (1998) Geotechnical problems associated with the maintenance of geological exposures in clay cliffs subject to reduced erosion rates. In: J Hooke (ed.) *Coastal defence and earth science conservation*. London, Geological Society, 32–45.

Bate R (ed.) (1998) *Global warming: the continuing debate*. Cambridge, European Science and Environment Forum.

Bilham R & Barrientos SE (1991) Sea-level rise and earthquakes. *Nature* 350, 386.

Bird ECF (1985) *Coastline changes: a global review*. Chichester, John Wiley.

Bird ECF (1993) *Submerging coasts — the effects of a rising sea level on coastal environments*. Chichester, John Wiley.

Boorman LA, Goss-Custard JD & McGrorty S (1989) *Climate change, rising sea level and the British coast*. Research Publication No. 1, Institute of Terrestrial Ecology, London, HMSO.

Braatz B & Aubrey DG (1987) Recent relative sea-level change in eastern North America. In: D Nummedal, OH Pilkey & JD Howard (eds) *Sea-level fluctuation and coastal evolution*. Tulsa, SEPM, 29–46.

Brampton AH (1998) Cliff conservation and protection: methods and practices to resolve conflicts. In: J Hooke (ed) *Coastal defence and earth science conservation*. London, Geological Society, 21–31.

Bray M & Hooke JM (1997) Prediction of soft-cliff retreat with accelerating sea-level rise. *Journal of Coastal Research* **13**, 453–67.

Bruun P (1962) Sea-level rise as a cause of storm erosion. *ASCE Journal of the Waterways and Harbours Division* **88**(WW1), 117–30.

Bruun P (1983) Review of conditions of use for the Bruun Rule of erosion. *Coastal Engineering* **7**, 77–89.

Bruun P (1988) The Bruun Rule of erosion by sea-level rise: a discussion on large-scale two- and three-dimensional usages. *Journal of Coastal Research* **4**, 627–48.

Buddemeier RW & Smith SV (1988) Coral reef growth in an era of rapidly rising sea-level: predictions and suggestions for long-term research. *Coral Reefs* **7**, 51–6.

Burd F (1995) *Managed retreat: a practical guide.* Peterborough, English Nature.

Carter RWG (1992) Sea-level changes: past, present and future. *Quaternary Proceedings* **2**, 111–32.

Carter RWG & Woodroffe C (1994) Coastal evolution: an introduction. In: RWG Carter & CD Woodroffe (eds) *Coastal evolution: Late Quaternary shoreline morphodynamics.* Cambridge, Cambridge University Press, 1–31.

Carter WE & 13 others (1994) New gravity meter improves measurements. *EOS Transactions, American Geophysical Union* **75**, 90–2.

Cendrero A & Charlier RM (1989) Resources, land use and management of the coastal fringe. *Geolis* **3**(1–2), 40–60.

Child M (1996) Soft engineering: a firm future? *Proceedings of Institution of Civil Engineers* **114**, 145–6.

Clark JA, Farrell WE & Peltier WR (1978) Global changes in postglacial sea level: a numerical calculation. *Quaternary Research* **9**, 265–78.

Clayton K (1989) Sediment input from Norfolk cliffs — a century of protection and its effect. *Journal of Coastal Research* **5**, 433–42.

Costanza R, Sklar FH & White ML (1990) Modelling coastal landscape dynamics. *Bioscience* **40**, 91–107.

Cowell PJ & Thom BG (1994) Morphodynamics of coastal evolution. In: RWG Carter & CD Woodroffe (eds) *Coastal evolution: late Quaternary shoreline morphodynamics.* Cambridge, Cambridge University Press, 33–86.

Dean RG & Maurmeyer EM (1983) Models for beach profile responses. In: PD Komar (ed.) *Handbook of coastal processes and erosion.* Boca Raton, CRC Press, 151–66.

DeJong DJ, DeJong Z & Mulder JPM (1994) Changes in area, geomorphology and sediment nature of salt marshes in the Oosterschelde Estuary (SW Netherlands) due to tidal changes. *Hydrobiologia* **283**, 303–16.

Deleeuw J, Apon LP, Herman PMJ, de Munck W & Beeftink WG (1994) The response of salt-marsh vegetation to tidal reduction caused by the Oosterschelde storm-surge barrier. *Hydrobiologica* **283**, 335–53.

Devoy RJN (1987) Sea-level applications and management. *Progress in Oceanography* **18**, 273–86.

Dietz RS (1963) Wave-base, marine profile of equilibrium and wave built terraces: a critical appraisal. *Bulletin of the Geological Society of America* **74**, 971–90.

Dixon AM, Leggett DJ & Weight RC (1998) Habitat creation opportunities for landward coastal re-alignment: Essex case studies. *Journal of the Chartered Institution of Water and Environmental Management* **12**, 107–12.

Doodson AT & Warburg HD (1941; reprinted 1980) *Admiralty Manual of Tides.* London, HMSO.

Douglas BC (1991) Global sea-level rise. *Journal of Geophysical Research* **96**, 6981–92.

Douglas BC (1992) Global sea-level acceleration. *Journal of Geophysical Research* **97**, 12699–706.

Dubois RN (1992) A re-evaluation of Bruun's Rule and supporting evidence. *Journal of Coastal Research* **8**, 618–28.

ElRaey M, Fouda Y & Nasr S (1997) GIS assessment of the vulnerability of the Rossetta area, Egypt, to impacts of sea level rise. *Environmental Monitoring and Assessment* **47**, 59–77.

ElRaey M, Frihy O, Nasr SM & Dewidar KH (1999) Vulnerability assessment of sea level rise over Port Said Governorate, Egypt. *Environmental Monitoring and Assessment* **56**, 113–28.

Emery KO (1980) Relative sea levels from tide-gauge records. *Proceedings of the National Academy of Science USA* **77**, 6968–72.

Emery KO & Aubrey DG (1991) *Sea levels, land levels, and tide gauges*. New York, Springer-Verlag.

Enfield DB & Allen JS (1980) On the structure and dynamics of monthly mean sea level anomalies along the Pacific coast of North and South America. *Journal of Physical Oceanography* **10**, 557–78.

Enfield DB & Allen JS (1983) The generation and propagation of sea level variability along the Pacific coast of Mexico. *Journal of Physical Oceanography* **13**, 1012–33.

Everts CH (1985) Sea level rise effects on shoreline position. *Journal of Waterway Port Coastal and Ocean Engineering* **11**, 985–99.

Fairbridge RW (1989) Crescendo events in sea-level changes. *Journal of Coastal Research* **5**, ii–vi.

Flick RE & Cayan DR (1985) Extreme sea levels on the coast of California. *Proceedings of the 19th Coastal Engineering Conference. American Society of Civil Engineers*, 886–98.

Frassetto R (ed.) (1991) *Impacts of sea level rise on cities and regions*. Proceedings, First International Conference on Cities and Water. Venice, Marsilio Editori.

French JR, Watson CJ, Möller I, Spencer T, Dixon M & Allen R (2000) Beneficial use of cohesive dredgings for foreshore recharge. Keele, *35th MAFF Conference of river and coastal engineers — Innovation Forum*, 11.10.1–4.

Gaillot S & Piégay H (1999) Impact of gravel-mining on stream channel and coastal sediment supply: example of the Calvi Bay in Corsica (France). *Journal of Coastal Research* **15**, 774–88.

Gehrels WR, Belknap DF, Pearce BR, & Gong B (1995) Modeling the contribution of M_2 tidal amplification to the Holocene rise of mean high water in the Gulf of Maine and the Bay of Fundy. *Marine Geology* **124**, 71–85.

Gornitz V (1995) Sea-level rise: a review of recent past and near-future trends. *Earth Surface Processes and Landforms* **20**, 7–20.

Gornitz V & Lebedeff S (1987) Global sea-level changes during the past century. In: D Nummedal, OH Pilkey & JD Howard (eds) *Sea level fluctuation and coastal evolution*, SEPM Special Publication **41**, 3–16.

Gornitz V, Lebedeff S & Hansen J (1982) Global sea level trend in the past century. *Science* **215**, 1611–14.

Gornitz V & Seeber L (1990) Vertical crustal movements along the East Coast, North America, from historic and late Holocene sea level data. *Tectonophysics* **178**, 127–50.

Gröger M & Plag HP (1993) Estimations of a global sea level trend: limitations from the structure of the PSMSL global sea level data set. *Global and Planetary Change* **8**, 161–9.

Grootjans AP, Jones P, van der Meulen F & Paskoff R (eds) (1997) Ecology and restoration perspectives of soft coastal ecosystems. *Journal of Coastal Conservation* **3**, 3–102.

Gutenberg B (1941) Changes in sea level, postglacial uplift, and mobility of the Earth's interior. *Bulletin of the Geological Society of America* **52**, 721–72.

Hands EB (1983) The Great Lakes as a test model for profile responses to sea-level change. In: PD Komar (ed.) *Handbook of coastal processes and ersoion*. Boca Raton, CRC Press, 167–89.

Haq BU, Hardenbol J, & Vail PR (1987) Chronology of fluctuating sea-levels since the Triassic. *Science* **235**, 1156–66.

Hays JD & Pitman WC (1979) Lithospheric plate motion, sea-level changes and climate and ecological responses. *Nature* **246**, 16–22.

Hoffman JS (1984) Estimates of sea-level rise In: MC Barth & JG Titus (eds) *Greenhouse effect and sea-level rise*. New York, Van Nostrand Reinhold, 245–66.

Hoffman JS, Wells JB & Titus JG (1986) Future global warming and sea level rise. In: G Sigbjarnarson (ed.) *Proceedings Iceland Coastal and River Symposium '85*. Reykjavik, National Energy Authority, 246–66.

Hollis GEH, Thomas D & Heard S (1989) *The effects of sea-level rise on sites of conservation value in Britain and northwest Europe*. London, University College London. Report to WWF, Project 120/88.

Hooke J (ed.) (1998) *Coastal defence and earth science conservation*. London, Geological Society.

Houghton JT, Jenkins GJ & Ephraums JJ (eds) (1990) *Climate change: The IPCC Scientific Assessment*. Cambridge, Cambridge University Press.

Houghton JG, Meira Filho LG, Calander BA, Harris N, Kattenberg A & Maskell K (eds) (1996) *Climate Change 1995: The science of climate change*. Cambridge, Cambridge University Press.

Huybrechts P (1994) The present evolution of the Greenland ice sheet: an assessment by modelling. *Global and Planetary Change* **9**, 39–51.

Inman DL & Dolan R (1989) The Outer Banks of North Carolina: budget of sediment and inlet dynamics along a migrating barrier system. *Journal of Coastal Research* **5**, 193–237.

Jelgersma S (1992) Vulnerability of the coastal lowalnds of the Netherlands to a future sea-level rise: In: MJ Tooley & S Jelgersma (eds) *Impacts of sea-level rise on European coastal lowlands*. Oxford, Blackwell, 94–123.

Kana TW, Eiser WC, Baca BJ & Williams ML (1988) Charleston case study. In: JG Titus (ed.) *Greenhouse effect, sea level rise and coastal wetlands*. Washington DC, US EPA, 37–59.

Kaplin PA (1989) Shoreline evolution during the twentieth century. In: A Ayala-Castanares, W Wooster & A Yanez-Arancibia (eds) *Oceanography 1988 Proceedings*. Mexico City, Mexico Autonomous University, 59–64.

Kaplin PA & Selivanov AO (1995) Recent coastal evolution of the Caspian Sea as a natural model for coastal responses to the possible accleration of global sea-level rise. *Marine Geology* **124**, 161–75.

Kaznowska S & Waller C (1993) Buying time for Benacre. *Enact* **1**, 10–11.

Komar PD (1986) The 1982-83 El Niño and erosion on the coast of Oregon. *Shore & Beach* **54**, 3–12.

Komar PD (1999) Coastal change — scales of processes and dimensions of problems. In: NC Kraus & WG McDougal (eds) *Coastal Sediments 99: Proceedings of the 4th International Symposium on Coastal Engineering and Science of Coastal Sediment Transport Processes*, **1**, 1–17.

Komar PD & Enfield DB (1987) Short-term sea level changes and coastal erosion. In: D

Nummedal, OH Pilkey & JD Howard (eds) *Sea level fluctuation and coastal evolution.* SEPM Special Publication **41**, 17–27.

Komar PD & Good JW (1991) Long term erosion impacts of the 1982–83 El Niño on the Oregon coast. In: OT Brown (ed.) *Coastal Zone '89: Proceedings of the Sixth Symposium of Coastal and Ocean Management.* New York, ASCE, 3785–94.

Lajoie KR & Mathieson SA (1997) 1982-83 coastal erosion: San Mateo County, California. http://elnino.usgs.gov/SMCO-coast-erosion/introtext.html

Leatherman (1987) Beach and shoreface response to sea-level rise: Ocean City, Maryland. *Progress in Oceanography* **18**, 139–49.

Leatherman SP, Zhang K & Douglas BC (2000) Sea level rise shown to drive coastal erosion. *EOS* **81**(6), 55–7.

Lee JK, Park RA & Mausel PW (1992) Application of geoprocessing and simulation modelling to estimate impacts of sea-level rise on the northeast coast of Florida. *Photogrammetric Engineering and Remote Sensing* **58**, 1579–86.

Lindzen RS (1990) Some coolness concerning global warming. *Bulletin of the American Meteorological Society* **71**, 288–99.

Louters T, van den Berg JH & Mulder JPM (1998) Geomorphological changes of the Oosterschelde tidal system during and after the implementation of the delta project. *Journal of Coastal Research* **14**, 1134–51.

Lukas R, Hayes SP & Wyrtki K (1984) Equatorial sea level response during the 1982–83 El Niño. *Journal of Geophysical Research* **89**, 10435–50.

McKirdy AD (1987) Protective works and geological conservation. In: MG Culshaw (ed.) Planning and engineering geology. *Geological Society of London Special Publication* **4**, 81–5.

McLean RF (1996) Biological indicators of sea level rise.: the use of coral microatolls. In: TH Aung (ed.) *Proceedings of the Ocean Atmosphere Pacific International Conference.* Adelaide, Flinders University of South Australia, 1–6.

Matthews RK (1990) Quaternary sea-level change. In: National Research Council, *Sea level change.* Washington DC, 88–103.

Maul GA & Hanson K (1991) Interannual coherence between North Atlantic atmospheric surface pressure and composite southern USA sea level. *Geophysical Research Letters* **18**, 653–6.

Maul GA & Martin DM (1993) Sea-level rise at Key West, Florida, 1846–1992: America's longest instrumental record? *Geophysical Research Letters* **20**, 1955–8.

Meier MF (1993) Ice, climate and sea level: do we know what is happening? In: WR Peltier (ed.) NATO ASI Series 1: *Global Environment Change*, **12**, Heidelberg, Springer-Verlag, 141–60.

Meire A (1996) Feeding behaviour of oystercatchers *Haematopus ostralegus* during a period of tidal manipulations. *Ardea* **84**, 509–24.

Mitsch WJ & Gosselink JG (1993) *Wetlands.* New York, Van Nostrand Reinhold.

Mörner NA (1980) *Earth rheology, isostasy and eustasy.* Chichester, John Wiley.

Mörner NA (1983) Sea levels. In: R Gardner & H Scoging (eds) *Mega-geomorphology.* Oxford, Clarendon Press, 73–91.

Mörner NA (1986) The concept of eustasy: a redefinition. *Journal of Coastal Research, Special Issue* **1**, 49–51.

Nakiboglu SM & Lambeck K (1991) Secular sea level change. In: R Sabadini, K Lambeck & E Boschi (eds) *Isostasy, sea-level and mantle rheology.* Dordrecht, Kluwer Academic, 237–58.

National Research Council (1987) *Responding to changes in sea-level: engineering implications.* Washington DC, National Academy Press.

Nerem RS (1995) Global mean sea level variations from TOPEX/POSEIDON altimeter data. *Science* **268**, 708–10.

Nicholls RJ, Birkemeier WA & Lee G-H (1998) Evaluation of depth of closure using data from Duck, NC, USA. *Marine Geology* **148**, 179–201.

Nicholls RJ, Hoozemans FMJ & Marchand M (1999) Increasing flood risk and wetland losses due to global sea-level rise: regional and global analyses. *Global Environmental Change* **9**, 69-87.

Nienhuis PH & Smaal AC (1994) The Oosterschelde estuary: a case study of a changing ecosystem — an introduction. *Hydrobiologica* **283**, 1–14.

Nunn PD (1990) Recent environmental changes on Pacific islands. *Geographical Journal* **156**, 127–40.

Oerlemans J & Fortuin JPF (1992) Sensitivity of glaciers and small ice caps to greenhouse warming. *Science* **258**, 115–17.

Orford J & Jennings S (1998) The importance of different time-scale controls on coastal management strategy: the problem of Porlock gravel barrier, Somerset, UK. In: J Hooke (ed.) *Coastal defence and earth science conservation.* London, Geological Society, 87–102.

Palmer TN & Webster PJ (1997) The past and the future of El Niño. *Nature* **390**, 562–4.

Peerbolte EB, De Ronde JG & Baarse G (1991) *Impact of sea level rise on society: a case study for the Netherlands.* UNEP and Government of the Netherlands.

Percival SM, Sutherland WJ & Evans PR (1998) Intertidal habitat loss and wildfowl numbers: applications of a spatial depletion model. *Journal of Applied Ecology* **35**, 57–63.

Pernetta JC & Hughes PJ (1990) *Implications of expected climate changes in the the South Pacific region: an overview.* UNEP Regional Seas Reports and Studies 128.

Philander SGH (1990) *El Niño, La Niña and the Southern Oscillation.* San Diego, Academic Press.

Phillips JD (1986) Coastal submergence and marsh fringe erosion. *Journal of Coastal Research* **2**, 427–36.

Pilkey OH, Young RS, Riggs SR, Smith AWS, Yu W & Pilkey WD (1993) The concept of shoreface profile of equilibrium: a critical review. *Journal of Coastal Research* **9**, 255–78.

Pirazzoli PA (1989) Present and near-future global sea-level changes. *Palaeogeography, Palaeoclimatology, Palaeoecology (Global and Planetary Change Section)* **75**, 241–58.

Pirazzoli PA (1996) *Sea-level changes. The last 20 000 years.* Chichester, John Wiley.

Powell KA (1992) Engineering with conservation issues in mind. In: MG Barrett (ed.) *Coastal zone planning and management.* London, Thomas Telford, 237–49.

Pugh DT (1987) *Tides, surges, and mean sea-level. A handbook for engineers and scientists.* Chichester, John Wiley.

Quinlan G & Beaumont C (1981) A comparison of observed and theoretical post glacial relative sea-level rise in Atlantic Canada. *Canadian Journal of Earth Sciences* **8**, 1146–63.

Raper SCB, Wigley TML & Warrick RA (1996) Global sea level rise: past and future. In: JD Milliman (ed.) *Rising sea level and subsiding coastal areas.* Dordrecht, Kluwer.

Reeh N (1991) Parameterization of melt rate and surface temperature on the Greenland Ice Sheet. *Polarforschung* **59**, 113–28.

Regnauld H, Lemasson L & Dubreuil V (1998) The mobility of coastal landforms under climatic changes: issues for geomorphological and archaeological conservation. In: J Hooke (ed.) *Coastal defence and earth science conservation.* London, Geological Society, 21–31.

Richmond B & Gibbs A (1997) Coastal impacts of an El Niño season. http://elnino.usgs.gov/richmond.html

Richmond B & Gibbs A (1998) El Niño coastal monitoring program. http://elnino.usgs.gov/coastal/coastal.html

Robinson IS (1985) *Satellite oceanography*. Chichester, Ellis Horwood.

Roemmich D (1992) Ocean warming and sea level rise along the southwest U.S. coast. *Science* **257**, 273–5.

Rosen PS (1978) A regional test of the Bruun Rule of shoreline erosion. *Marine Geology* **26**, 7–16.

RSPB, National Trust, Norfolk Wildlife Trust, Suffolk Wildlife Trust & English Nature (1997) *Coasts in crisis: world famous wetlands at risk in Norfolk and Suffolk*. Norwich, RSPB.

Sahagian DL, Schwartz F & Jacobs DK (1994) Direct anthropogenic contributions to sea-level rise in the twentieth century. *Nature* **367**, 54–7.

Schwartz ML (1982) *The encyclopedia of beaches and coastal environments*. Stroudsburg, Hutchinson Ross.

Shennan I (1987) Holocene sea-level changes in the North Sea. In: MJ Tooley & I Shennan (eds) *Sea-level changes*. Oxford, Blackwell, 109–51.

Shennan I (1993) Sea-level changes and the threat of coastal inundation. *Geographical Journal* **159**, 148–56.

Shennan I & Sproxton I (1990) Possible impacts of sea-level rise: a case study from the tees estuary, Cleveland County. In: JC Doornkamp (ed.) *The greenhouse effect and rising sea levels in the UK*. Nottingham, M1 Press, 109–33.

Shennan I & Woodworth P (1992) A comparison of late Holocene and twentieth-century sea-level trends from the UK and North Sea region. *Geophysical Journal International* **109**, 96–105.

Sherif MM & Singh VP (1999) Effect of climate change on sea water intrusion in coastal aquifers. *Hydrological Processes* **13**, 1277–87.

Sklar FH, Costanza R & Day JW (1985) Dynamic spatial simulation modelling of coastal wetland habitat successions. *Ecological Modelling* **29**, 261–81.

Sklar FH, Gopu KK, Maxwell T & Costanza R (1994) Spatially explicit and implicit dynamic simulations of wetland processes. In: WJ Mitsch (ed.) *Global wetlands: old world and new*. Amsterdam, Elsevier, 537–54.

Smith BP & Laffoley DA (1992) *A directory of saline lagoons and lagoon-like habitats in England*. English Nature Science Series 5. Peterborough, English Nature.

Smith RA (1980) Golden Gate tidal measurements: 1854–1978. *Journal of the Waterway, Port, Coastal and Ocean Division, American Society of Civil Engineers* **106**, 407–10.

Spencer T (1995) Potentialities, uncertainties and complexities in the response of coral reefs to future sea-level rise. *Earth Surface Processes and Landforms* **20**, 49–64.

Spencer T, Tudhope AW, French JR, Scoffin TP & Utanga T (1997) Reconstructing sea level change from coral microatolls, Tongareva (Penrhyn) Atoll, Northern Cook Islands. *Proceedings 8th International Coral Reef Symposium, Panama* 1, 489–94.

Stevenson JC, Ward LG & Kearney MS (1986) Vertical accretion in marshes with varying rates of sea-level rise. In: DA Wolfe (ed.) *Estuarine variability*. Orlando, Academic Press, 241–60.

Thomas RH (1986) Future sea level rise and its early detection by satellite remote sensing. In: JG Titus (ed.) *Effects of changing stratospheric ozone and global climate. Volume 4: Sea level rise*. New York, UNEP and US EPA, 19–36.

Titus JG (1991) Greenhouse effect and coastal wetland policy: How Americans could abandon an area the size of Massachusetts at minimum cost. *Environmental Management* **15**, 39–58.

Titus JG & Narrayanan V (1996) The risk of sea level rise: Delphic Monte-Carlo analysis in which twenty researchers specify subjective probability distributions for model coefficients within their respective areas of expertise. *Climate Change*, **33**, 151–212.

Tooley MJ & Jelgersma S (1992) *Impacts of sea-level rise on European coastal lowlands.* Oxford, Blackwell.

Tooley MJ & Shennan I (eds.) (1987) *Sea-level changes.* Oxford, Blackwell.

Tsimplis MN & Woodworth PL (1994) The global distribution of the seasonal sea level cycle calculated from coastal tide gauge data. *Journal of Geophysical Research*, **99C**, 16031–9.

Turner RE & Rao Y (1990) Relationships between wetland fragmentation and recent hydrological changes in a deltaic coast. *Estuaries* **13**, 272–81.

UNEP (1990) *The state of the marine environment.* Reports and Studies No. 39, UNEP Regional Seas Reports and Studies.

UNESCO (1985) *Manual of sea-level measurement and interpretation.* Intergovernmental Oceanographic Commission, Manuals and Guides, 14.

Valentin H (1952) Die Küsten der Erde. *Petermanns Geographische Mitteilungen, Ergänzungsband* **246**.

Van Veen J (1954) Tide-gauges, subsidence-gauges and flood-stones in the Netherlands. *Geologies en Mijnbouw* **16**, 214–19.

Warrick RH, Le Provost C, Meir M, Oerlemans J & Woodworth PL (1996) Changes in sea level. In: JT Houghton, LG Meira Filho, BA Calander, N Harris, A Kattenberg & K Maskell (eds) *Climate change 1995: The science of climate change.* Cambridge, Cambridge University Press, 358–405.

Warrick RH & Oerlemans J (1990) Sea-level rise. In: JT Houghton, GJ Jenkins & JJ Ephraums (eds) *Climate change—the IPCC scientific assessment.* Cambridge, Cambridge University Press, 257–81.

Weggel RJ (1979) A method for estimating long-term erosion rates from a long-term rise in water level. *USACE-WES Coastal Engineering Technical Aid* 79-2.

Wigley TML & Raper SCB (1992) Implications for climate and sea levels of revised IPCC emissions scenarios. *Nature* **357**, 293–300.

Williams SJ, Dodd K, & Gohn KK (1991) *Coasts in crisis.* United States Geological Survey, Circular 1075.

Woodroffe CD & Grindrod J (1991) Mangrove biogeography: the role of Quaternary environmental and sea level change. *Journal of Biogeography* **18**, 479–92.

Woodroffe CD & McLean RF (1990) Microatolls and recent sealevel change on coral atolls. *Nature* **344**, 531–4.

Woodworth PL (1990) A search for accelerations in records of European mean sea-level. *International Journal of Climatology* **10**, 129–43.

Woodworth PL (1991) The Permanent Service for Mean Sea Level and the Global Sea Level Observing System. *Journal of Coastal Research* **7**, 699–710.

Wyrtki K (1979) The response of sea level surface topography to the 1976 El Niño. *Journal of Physical Oceanography* **9**, 1223–31.

Wyrtki K (1985) Sea-level fluctuations in the Pacific during the 1982–83 El Niño. *Geophysical Research Letters* **12**, 125–8.

Yohe G, Neumann J, Marshall P, & Ameden H (1996) The economic cost of greenhouse-induced sea-level rise for developed property in the United States. *Climatic Change* **37**, 243–70.

Zerbini S & 16 others (1996) Sea-level in the Mediterranean: a first step towards separating crustal movements and absolute sea-level variations. *Global and Planetary Change* **14**, 1–48.

Index